Technology in Modern German History

The Bloomsbury History of Modern Germany Series

Series Editors: Daniel Siemens (Newcastle University, UK), Jennifer V. Evans (Carleton University, Canada) and Matthew P. Fitzpatrick (Flinders University, Australia)

The Bloomsbury History of Modern Germany Series is an open-ended series of thematic books on various aspects of modern German history. This ambitious and unique series offers readers the latest views on aspects of the modern history of what has been and remains one of the most powerful and important countries in the world. In a series of books aimed at students, leading academics and experts from across the world portray, in a thematic manner, a broad variety of aspects of the German experience, over extended periods of time, from roughly the turn of the 19th century to the modern day. Themes covered include class, religion, war, race, empire, sexuality and gender.

Published:

Technology in Modern German History, Karsten Uhl (2022)

Forthcoming:

Sexuality in Modern German History, Katie Sutton

Dictatorships and Authoritarianism in Modern German History, André Keil

Imperialism in Modern German History, Eva Bischoff

Philosophy in Modern German History, Cat Moir

Revolutions in Modern German History, Andrew Bonnell

Technology in Modern German History

1800 to the Present

Karsten Uhl

BLOOMSBURY ACADEMIC
LONDON • NEW YORK • OXFORD • NEW DELHI • SYDNEY

BLOOMSBURY ACADEMIC
Bloomsbury Publishing Plc
50 Bedford Square, London, WC1B 3DP, UK
1385 Broadway, New York, NY 10018, USA
29 Earlsfort Terrace, Dublin 2, Ireland

BLOOMSBURY, BLOOMSBURY ACADEMIC and the Diana logo are trademarks of Bloomsbury Publishing Plc

First published in Great Britain 2022
This paperback edition published 2023

Cover image: The agricultural teacher Josef Annich explains two apprentices of the East German Barby agricultural college how to use Soviet milking machines, 27 November 1952. © German Federal Archives (Bundesarchiv), BArch, file 183-17369-0005.

A catalogue record for this book is available from the British Library.

Library of Congress Cataloging-in-Publication Data
Names: Uhl, Karsten, author.
Title: Technology in modern German history: 1800 to the present / Karsten Uhl.
Description: London; New York: Bloomsbury Academic, 2022. |
Series: The Bloomsbury history of modern Germany series |
Includes bibliographical references and index.
Identifiers: LCCN 2021036426 (print) | LCCN 2021036427 (ebook) |
ISBN 9781350053205 (hardback) | ISBN 9781350053212 (pdf) |
ISBN 9781350053229 (ebook)
Subjects: LCSH: Technology–Germany–History. | Technology–Social aspects–Germany–History. | Technology and state–Germany–History.
Classification: LCC T26.G3 U43 2022 (print) | LCC T26.G3 (ebook) |
DDC 609.43–dc23
LC record available at https://lccn.loc.gov/2021036426
LC ebook record available at https://lccn.loc.gov/2021036427

ISBN: HB: 978-1-3500-5320-5
PB: 978-1-3502-8994-9
ePDF: 978-1-3500-5321-2
eBook: 978-1-3500-5322-9

Typeset by Deanta Global Publishing Services, Chennai, India

Contents

Figures

Acknowledgements

I would like to thank my friends and colleagues who offered incisive readings of some of the chapters: Matthew C. Caldwell, Noyan Dinçkal, Robert Groß, Ana Honnacker, Nina Kleinöder, Michael Löffelsender, Detlev Mares, Stefan Poser, Nicolas Rohleder and Christian Zumbrägel. Your comments helped to improve my work, while as the author I do of course take sole responsibility for any mistakes or shortcomings. I am particularly grateful to the series editors Jennifer Evans, Matthew Fitzpatrick and Daniel Siemens who invited me to contribute this volume to the series. They read the manuscript very critically and thus gave important clues to further improvements of the argument.

Donna J. Drucker is the ingenious combination of a historian of science and technology and a native speaker of English teaching at a German university. She read most of the manuscript; I am grateful for her outstanding editing work considering both the content and the language.

Mikael Hård gave inspiration to me about what the history of technology can be since I got to know him in 2006. While our methodological approaches differ to some degree, he – as the book proposal's first reader – made an invaluable suggestion about the book's leitmotif. I am also grateful to the two anonymous referees of the proposal.

Most of all, I want to thank my wife Ronja Rückheim for tolerating the exhausting habits of the author trying to finish the manuscript.

Introduction

Technology shaped modern Germany. Non-Germans often associate post-war Germany with technology and with its products of mass consumption such as high-quality cars. Even pop music, most notably *Kraftwerk* (literally 'power station') with songs such as 'Autobahn', 'Radioactivity' or 'We Are the Robots', disseminates the stereotype of a close link between German culture and technology. *Technology in Modern German History* explores the origins of this association and the various forms of technology over more than 200 years of German history. It investigates the role technology played in transforming Germany's culture, society and politics from the early nineteenth to the early twenty-first centuries. At the core of this book lies an investigation into the mutual relationship between technology and culture. Its overarching question is twofold: How did technology shape modern Germany and how did modern Germany – and its specific cultural assumptions, social structures and political power – shape technology? In the course of this book, we will reconsider the master narratives of industrialization, urbanization and progress as well as areas that have long been neglected by historians such as the persistence of traditional and everyday technologies as well as rural technologies and technified bodies.

Scholars of German history still tend to neglect the importance of the history of technology for the development of the country as such. There have been broadly conceived studies that have proved how technology was crucial to the big questions of German history such as national identity, apprehensions of modernism and adaptations of Americanism. Aviation, for example, shaped the construction of a modernized German identity in the early twentieth century.[1] Furthermore, Taylorism and Fordism found many supporters in Weimar Germany who appropriated the American concept of industrial efficiency to specific economic, cultural and social conditions in Germany.[2] So too, as Jeffrey Herf has shown, 'reactionary modernists', most notably the Nazis, commandeered modern technology while at the same time rejecting the political values of modernity and enlightenment.[3] In addition to these major contributions, the present volume provides a broad introduction and synthesis upon which to build. It will show how technology permeated all facets of life in modern Germany. It is the first English-language book specifically on the role of technology in German history. This gap is surprising, as technology was crucial to the German economy, society and culture. The cooperation of two companies in the 1920s symbolizes different aspects of technology in modern Germany: Krupp and Kuka cooperated in manufacturing waste collection vehicles. While Krupp represented the German industry's past and present at that time, Kuka would become the most important player in industrial automation as a robot manufacturer half a century later.[4] With respect to the history of industry, Krupp represents the heavy industry of the nineteenth century that played a dominant role in German history well into the twentieth century. By contrast, Kuka was a newcomer that would become a major player after the end of the dominance of the old industries. Kuka also represents a later trend in global industrial development. In 2017, a Chinese corporation took it over, signifying the demise of Western industrial dominance. Moreover, the history of Kuka exemplifies that the history of technology covers the whole range of the history of both consumption – in this case, the final stage of consumption: waste disposal – and industrial production – in this case, its most modern version of automation by robots. It also demonstrates that the most sophisticated visions of progress – robotics – sometimes derive from everyday technologies such as the issue of waste disposal. Last but not least, the Kuka story reminds us that the urgent problems of the present time – especially environmental issues – were already very present in the early twentieth century when the established Krupp corporation and the champion of the future Kuka collaborated in solving the difficult issue of waste removal. While the Kuka company has been transformed over the past hundred years, one thing has not changed: Kuka's robot arms are still the same colour as the early refuse collection vehicles – orange. Today, the German chancellor Angela Merkel proudly presents Kuka's products to state guests, even if it is no longer a German-owned company but still a German-based high-tech player (Figure 0.1).

Historians of technology have opened the field to innovative approaches and shifted the focus of research within the last few decades. Since the 1980s, new research directions investigating the social construction of technology or the cultural appropriation of technology have gained prominence. Such new approaches have

Figure 0.1 German chancellor Angela Merkel shows Kuka robot arms used for Volkswagen manufacturing to the Mexican president Enrique Pena Nieto at the Hanover Fair, 23 April 2018. Bundesarchiv.

not gone unchallenged. According to the historian Joachim Radkau, the history of technology is still largely to be understood as a history of labour.[5] While I agree with a strong emphasis on labour, as the following chapters demonstrate, a mere focus on production tends to neglect the various appropriations of technologies and the agency of the technology users. The life cycles of technologies from the social and cultural structures of their design and construction over their usage to their disposal or recycling are of utmost importance to the history of technology. Ultimately it is only the decision of the historian that sees an emphasis placed on the technological artefact, the engineer or the users.[6] The present volume will shift between these different perspectives.

Sometimes, even the historical actors themselves switched between these perspectives. There is some evidence that German workers identified particularly strongly with the products of their labour and the notion of 'German quality work'. Ordinary workers were proud of working on technological icons of modernity, such as aeroplanes. For instance, the workers at the Focke-Wulf aeroplane company in Bremen celebrated the debut flights of 'their' aeroplanes. As historian Alf Lüdtke has shown, this was an expression of a real grassroots enthusiasm and pride in the labour of skilled workers.[7] Broadly speaking, the producers identified both with the artefact and with the efforts of the engineers and designers. Moreover, they imagined the usage of an aeroplane, although workers at that time had little chance of ever affording the high price of a plane ticket.

In another respect, the borders between producers and users are blurred; namely in the history of innovation. A closer look at the history of technology will not reveal many ingenious inventors. Instead, tinkerers most often improved technologies step by step. Many artefacts were the result of collective inventions.[8] More recently, contemporary feelings of ambivalence regarding technological progress have also been reflected in scholarship on failed innovations.[9] For example, the case of the *Transrapid*, a high-speed monorail train using magnetic levitation, clearly shows the importance of social, economic and cultural conditions for technological innovations. This refutes the still popular but naïve idea of the ingenious innovator whose unstoppable inventions succeeded exclusively because of their quality.

In general, the history of technology is not a mere history of innovations. David Edgerton has demonstrated the relevance of old and hybrid technologies for global history. Most importantly, one should not conflate novelty with significance.[10] And there are also specific aspects of old technologies in German history. According to Joachim Radkau, the very reluctance of Germans to adopt new technologies in the early nineteenth century and the rather slow adaptation of innovation were the recipe for success in the long run. They embraced well-established technologies and became a trendsetter for science-based industries in the late nineteenth century. Often missed, there have been 'old and growing industries' continuing until the present.[11]

Having said all this, I am not arguing for an entire shift of perspective from production to usage. For the overestimation of users' freedom bears new methodological problems, which will be discussed broadly in Chapter 7. For the moment, it is sufficient to point to users' agency in dealing with technologies on the one hand, and to the very limits of this agency on the other. In general, the history of technology is the history of the relationship between humans, the environment and technology. This approach is not to be mistaken for a mere history of ideas. While historical concepts of this relationship are important, the present volume seeks above all to analyse the changing relationship between humans and technology in social practices. Thus, technology in a broad sense is understood as the human production or usage of artefacts.[12] Historical actors claiming to introduce 'humane' technologies or fighting for a 'humanization' of technology characterize the global history of technology. As will be shown in the respective chapters, the humanization of industrial labour and of urban development was a particularly important topic in German history. Yet, the very notion of 'human', 'humane' or 'humanization' proved to be very flexible. Socialists, liberals and national socialists alike claimed to have a unique knowledge of the specific path towards a reconciliation of modern technology and human needs.

The most convincing approach to the history of technology is a history from below as the eminent work of Ruth Oldenziel and Mikael Hård has proved.[13] This approach focuses on common people. Of course, engineers, entrepreneurs and politicians deserve our attention, but workers, farmers, dwellers, housewives, leisure-time enjoyers, commuters and travellers all tell us a lot about the history of technology. This focus must not neglect the artefact. While technology does not determine the historical development of a country, it is by no means neutral.[14] As Langdon Winner has shown, artefacts have politics.[15] Once the political decision for a large technological system has been made, it is very difficult to counteract its momentum.[16] Of course, the social effects of technology are 'complex and contingent', but technology has social effects nonetheless.[17]

The issues of international transfers and transnational entanglements enjoy strong attention from historians of technology.[18] As David Arnold has demonstrated, it is, strictly speaking, the transnational circulation of technology and knowledge under consideration here, not a mere diffusion process.[19] These complex processes will be investigated throughout all chapters. Besides the obvious influence of British technology in the early and mid-nineteenth century, and the US impact on German technology development from the late nineteenth century onwards, many countries transferred knowledge and technology to Germany. They included wandering Belgian technicians in the early period of industrialization, Danish types of rural technological systems and forms of environmental protests. And we cannot forget that the Soviet impact on the development of socialist East Germany was crucial during the Cold War. This Soviet transfer of technology was not only limited to high tech but comprised also of technological artefacts of everyday use in rural settings such as milking machines (see Chapter 6, or the cover of this book). From the late nineteenth century onwards, Germany itself became a model for technology development to many European neighbours. And even the efficiency movement of the early twentieth century that has broadly been depicted as a matter of Americanization has many aspects of mutual transfers on both sides of the Atlantic. Another important consideration is, of course, the exploitation of the colonies and the ruthless global transfer of German ideas and artefacts, even when the transfer did not always reach its intended goal.

In this context, it is very important to investigate and distinguish between those technological developments that are part of transnational trends and those that constitute distinctly German developments. In the first place, there are three highly relevant and distinctly national characteristics in the development of technology in Germany. First, the crucial role of science-based industry for German industrial success from the late nineteenth century onwards resulted in a particular strong mentality of technocracy and belief in progress. As the philosopher Jürgen Habermas has argued, for some German technocrats the belief in the nearly unlimited means of technology and science was an *ersatz* ideology: they perceived innovative technologies as a means of constructing and ordering their society.[20] Second, throughout modern German history, the state and its bureaucracy played a very active role as the initiator of industrial technology. It is often overlooked that the state was also crucial for issues of urban planning or rural development. Third, the environmental movement had a particularly strong impact in Germany. It had roots in the nineteenth century, but it became most important in the late twentieth and early twenty-first centuries with regard to the contested issue of nuclear power.

This listing of peculiarities of German history might remind some readers of the *Sonderweg* debate that was prominent in the 1970s and 1980s.[21] Recently, the historian Jürgen Kocka took a look back on the debate. While it had some shortcomings in assuming that there had been a 'normal' (or liberal) path to modernity from which Germany had departed, some *Sonderweg* questions are still relevant. Most of all, the central point is still to understand why Germany, unlike France or Britain, was taken over by fascism.[22] Since Jeffry Herf's eminent study *Reactionary Modernism*, the debate has gradually modified with respect to the insight that there were manifold variants of alternative modernities. As a result, Germany is no longer perceived 'as a nation having trouble modernizing, but as a nation of troubling modernity'.[23]

The initial *Sonderweg* debate largely neglected issues of technology, and Heinrich August Winkler's eminent survey of modern German history gives little attention to German appropriations of Western technology. However, the political and social history of Germany that Winkler correctly describes as a 'long road West'[24] should be contrasted with the early Westernization of German technology. In general, there is some evidence that Germany's 'troubling modernity' was not least due to specific assessments of technology from the nineteenth century onwards. Most of all, as Herf has demonstrated, the reactionary embrace of technology was crucial for the formation of a 'German national identity' in modern times. Reactionary modernists sought to 'combine technology and the soul', but they were enthusiastic for only 'the technological aspects of modernity', as they rejected the values of the Enlightenment. During the Nazi era, reactionary modernism peaked. Some of its most important features were the celebration of 'creative capital' (opposed to 'parasitic', or 'Jewish', capital) as a 'source of utility, employment, and technological advance' as well as the 'aestheticization of the labor process'. Finally, the tradition of reactionary modernism culminated in the wartime rocket programme. For reactionary modernists, technology externalized the 'will to power'.[25] The historian Ulrich Wengenroth has supported this argument by pointing to the fact that important German actors, engineers and politicians alike, had chosen a self-imposed *Sonderweg* to modernity from the First World War onwards. According to Wengenroth, German industry became a victim of its late-nineteenth-century success because German industrialists and engineers preferred a rigid adherence to the recipes of the past. Until 1945, international competition gave little impetus to most branches of German industry for radical transformation.[26]

However, one should not mistake the disputed *Sonderweg* as a mere matter of a peculiar German mentality. Instead, issues of political structure were crucial. For instance, in contrast to Britain and France, there was no centralism in Germany. Thus, there was no harsh difference between the capital and cities with less political power. Rather, the levels of technological development in different German cities were relatively equal.[27] This does not mean, of course, that regional developments were not important. Actually, different regions of Germany were shaped by specific industries that developed in the early nineteenth century when the industrial revolution began (see Chapter 1).

This book starts in this very period around 1800 when important transformations in the history of technology occurred. Ulrich Wengenroth has argued that the occurrence of new technological ideas at that time had even more long-lasting effects than the classical technologies of early industrialization such as the steam engine or the spinning jenny.[28]

The present volume follows the general structure of the *Bloomsbury History of Modern Germany* series. It consists of two parts: section one, 'Tracing the history', critically lays out the 'narrative terrain' that is essential for an adequate understanding of modern Germany. It provides an overview of the key topics pertaining to technology in German history in the nineteenth and twentieth centuries. This part's subsections demonstrate in empirical detail how technology shaped modern Germany. At the same time, historiographical controversies will be introduced. Key topics in section one are the different stages of industrialization, the growth of networked cities and the triumph of a teleological narrative of technology as progress. Moreover, a critical revision of the history of high technology tells the story of high-tech euphoria shaping certain paths of history regardless of the respective technology proving to be successful or not. In particular, the first two chapters comprehensively discuss developments – industrialization and urbanization – that are essential to any understanding of modern German history. Necessarily, these chapters are more descriptive in nature and detailed.

Section two, 'New directions', discusses recent historiographical developments and approaches. This part is meant to counter the usual image of German technology as progressive, industrial, sophisticated and urban. It explores neglected areas such as rural technologies or the often-overlooked importance of everyday technologies: How did consumers or workers use new technologies? How did they appropriate and modify them? How were bodily experiences transformed in a highly technified environment? In general, it will consider the last decades of the twentieth century, and ask if they provided a significant new quality of technological change: to what degree and effects did computerization transform professional and private life? In culture and politics, reinforced by the German variety of environmentalism, the idea of progress was challenged, as the once-prevailing vision of progress gave way to new apprehensions of uncertainty.

As a whole, the second part reflects the findings of the first part: while Chapter 1 investigates the different stages of modern industry, Chapter 5 considers the impact technologies had on the bodies of technology users, both producers and consumers.

At the same time, the users' bodily practices and their appropriations of technology impacted the design and modification of artefacts. While Chapter 2 discovers the seminal importance of urban development for modern German history, Chapter 6 takes a close look at the often-neglected history of technological changes in rural areas that tell a fascinating story far beyond the cliché of rural backwardness. So too, Chapter 3 gives an account of German high technology, which contrasts with the everyday technologies explored in Chapter 7 that sometimes were much more relevant to the lives of ordinary people. Finally, Chapter 4 shows how visions of progress played an important role in technology decision-making, whereas Chapter 8 demonstrates that even in times of high optimism, protests against technologies were always present.

Paired in this way, the chapters illustrate that to understand technology in modern German history, an abiding sense of the role of industrialization, urbanization, high tech and visions of progress are still crucial, at the same time as often neglected, different paths and inconsistencies in the history of technology require closer scrutiny. At the end of this book, a short, annotated bibliography of the most critical works of research in English reflects the wide range of the themes and the methodological approaches covered by recent historiography.

Part I

Tracing the history

1

Industrialization and beyond

None of the great innovations of the industrial revolution had German origins, nor were the Germans particularly quick to adopt the new English technologies in the late eighteenth century. Only certain German regions were early adopters of industrial technology; Belgium, Switzerland and even some parts of France were more advanced than most of the fragmented states of Germany during this period. Proper industrialization first began in regions like Saxony and the Rhineland around 1815, nearly half a century after the industrial revolution began in England. Most parts of Germany, however, had their industrial take-off only in the later nineteenth century.[1] Nevertheless, by 1900 the newly unified German state had clearly caught up in the industrial race.

Prominent historical accounts explain this phenomenon of industrial backwardness in Germany as being something of an advantage; when new industries emerged at the end of the nineteenth century, Germany – like the United States – was free of the burdens the old industries had in England at that time.[2] In this context, historians had a lively discussion in the 1980s debating whether there was a unique German path, or *Sonderweg* to modernity, one that would explain the failure of the first German democracy and the rise of National Socialism.[3] The long-lasting supremacy of Prussian elites, a strong, but undemocratic, bureaucracy and the enormous power of heavy industry magnates were crucial elements of the *Sonderweg* argument. Even today, the most famous of these companies, Krupp, still symbolizes not only 'German industrialization, but German history per se', as former German finance minister Peer Steinbrück declared at a G-7 meeting in 2007. According to the Social Democratic politician Steinbrück, Krupp represented both the best and the darkest sides of German industrial history. On the one hand, Krupp was an industrial pioneer with a developed sense of social responsibility for his workers; but on the other, the company played a notorious role during the Nazi era: it was crucial for armament and employed thousands of forced labourers.[4] This is one illustration of how political, social and economic history continues to be intertwined in public debates on Germany's industrial past.

This chapter investigates the initial situation in the late eighteenth century as the English economy took off and inspired European industries to study its example. There were enormous economic and technological differences between various European regions. For German historians in particular, it makes little sense to reconstruct a national overall picture, not least because the regional economic divergences were exacerbated by the political fragmentation that endured until the foundation of the German Empire in 1871. In many regards, Prussia is correctly viewed as a technological innovator, but the picture would be incomplete if other German regions were neglected. Some technological and economic developments can be explained only by studying the south-western German states. In the course of this, the overall role of the state in introducing and developing industrial technology will be considered: it is quite controversial whether the German states, especially Prussia, dominated the German course of industrialization or if private enterprises were more important. Furthermore, the very issue of German particularities points to a central topic of the history of technology: the relationship between man and machine. In this chapter, special emphasis will be placed on human factors as a constant throughout the development of German industry; from its proto-industrial beginnings to the age of automation. There is good evidence for this line of continuity, starting with a notable persistence of traditional craftsmanship in early factories during the emergence of science-based industries in the late nineteenth century. This enduring debate about the humanization of work has persisted all the way up to contemporary discussions about skilled labour in automation.

The very term 'industrial revolution' must not be misunderstood; it necessarily describes a long process, but also represents a substantial 'break with the past'.[5] The manifold British inventions in the decades before 1815 were indeed 'transformative'. Unlike so many technological inventions of the European past, the effects of the industrial revolution neither burned out nor faded away.[6] However, conditions particular to Britain initially enabled its industrial launch, from both European and global perspectives. Globally, the 'great divergence' (Kenneth Pomeranz) between Europe and Asia, which occurred between 1750 and 1850, can be explained by 'technical innovation and geographical good fortune': crucial factors were steam and coal. Contrary to the Chinese situation, British coal seams happened to be located rather close to industrial centres. Moreover, the technical challenges differed fundamentally. Chinese mines mainly faced the problem of ventilation, while British and European mines were in need of constant water drainage. Therefore, there were decisive incentives for the invention and advancement of coal-fired steam engines in Britain that within a few decades became important power sources in factories, and these provided the basis for modern railway transport.[7] Moreover, the British Empire profited from the plentiful resources of its North American colonies. Coal, steam and raw materials in the New World enabled Britain, and later Europe, to avoid the staggering labour-intensive path of East Asian industrial development.[8]

From a European perspective, British geographical circumstances and technological advancement were also intertwined. Cheap coal in northern and western England gave a strong impetus to new technology.[9] Continental Europe, on the contrary, turned to coal significantly later during the nineteenth century, as peat-burning had satiated the fuel demand of important centres like the growing Dutch cities. In a hypothetical theory, the historian Robert Allen concluded that 'the industrial revolution might have been a Dutch-German breakthrough rather than a British achievement' if German coal use had been developed 300 years earlier than it actually was.[10] But this did not happen and therefore the industrial revolution – technology and knowledge – had to be imported from England.

Although the industrial revolution marks a significant break from the past, traditional structures by no means vanished quickly. Even in 1820s England, new technologies in the textile industry were still compatible with old structures of work organization. Modern factories emerged, but decentralized workshops and putting-out systems did not disappear, managing to integrate with more modern technologies. In contrast to independent artisans, those employed in cottage industries had their raw materials delivered by a putter-out, or merchant, who also took care of sales. In this way, cottage workers became de facto wage-earners, and thus a capitalist relationship of dependence arose even before industrial factories did.[11] In some cases the dependence was especially severe: the putter-out provided tools that determined quality standards and dictated the time of delivery.[12]

In light of the later sweating system and its use of intermediaries, certain aspects of the cottage industry can be seen as predecessors of the work discipline that would be systematically developed in factories. The putting-out system came into existence as early as the fourteenth century, and was widespread in the sixteenth and seventeenth centuries, particularly in textile production. These old networks for distributing textile goods finally disappeared only in the twentieth century.[13] As the historian Maxine Berg put it, 'it was competition and capitalist pressures, not new technology itself,' which resulted in the development of the factory system.[14] The economic historian Jan de Vries too provided good evidence that important changes occurred in the century before the industrial revolution began. This early modern 'industrious revolution' combined the supply and the demand side: households intensified production for the market and simultaneously had an increased demand for consumer goods. According to de Vries, the industrious revolution 'preceded and prepared the way' for the industrial revolution.[15]

Even in England – let alone in Europe or globally – there were 'many "paths" to the industrial revolution'.[16] Despite this heterogeneous picture, which does not allow simple techno-centric explanations, technology is still central for understanding the industrial revolution, and especially its diffusion. In particular, the impressive nature of new technology like steam engines or spinning machines led many foreign visitors to Britain to try and copy or even import it, despite the British technology export ban that was imposed to defend British economic uniqueness.[17] In this regard, technology was essential for industrial and social development. Technology not only had a direct influence in a social and economic context, but its indirect effect on culture was equally significant.

Smuggling new technology to Europe did not automatically result in economic success on the continent. Britain's high level of performance must be attributed to 'more than machines': the successful adaptation of English technology also required the skill and efficiency of English workers and managers.[18] Therefore, a transfer of technology, knowledge and people was required.

The early industrialization in Germany: More than technology transfer

In the decades around 1800, the transfer of technology and know-how resembled a one-way street from Britain to the European continent. The picture is incomplete, however, if one does not look back to some earlier developments, which were essential for England's industrial take-off. The French historian Fernand Braudel suggested that the industrial revolution had its prelude when German precious-metal mining (including gold, silver and copper) had reached a high technological level, resulting in a boom in the late fifteenth century. This trend was halted only after 1535, when silver mining began to expand in the New World and to compete with Europe. During this early time, German mining technology had been quickly adopted in England.[19] In this sense, early modern transfers in mechanization and technological progress were more important for the later British industrial revolution than usually stated.[20]

With regard to the so-called industrious revolution there is some evidence that the concept is adequate to Britain and the Netherlands, but of somewhat limited use when looking at Germany. There were serious constraints on the rising 'desire to work and consume in the market' by guilds and other traditional institutions well into the nineteenth century.[21] Despite this, German society was by no means monolithically old-fashioned and anti-progressive. In regions like Saxony, the guilds themselves partially promoted industrialization by establishing weaving schools where apprentices and journeymen learned about technological improvements.[22] Whereas this is an example of a hesitant modernization approach intended to preserve the old order, there was also wholehearted support of new technology from diverse sectors of society; bureaucrats, scientists, entrepreneurs and craftsmen alike were intrigued by the new technologies from abroad, and disseminated knowledge of it. One of the most important of these German travellers who reported from England in the late eighteenth century was the eminent scientist Georg Christoph Lichtenberg who was highly interested in steam engines and the new factories. In his letters home, he described details of industrial production, obviously impressed by the strict division of labour.[23]

In a letter from 1770, Lichtenberg reported how an English Lord, 'a great admirer of the Germans', pointed out alleged differences between the two nations regarding their relationship to technology. According to this Englishman, the German inventors had deep theoretical insight, whereas the English were mere tinkerers. The English Lord could not comprehend how his compatriots managed to construct machines, 'which they nevertheless often explain and understand quite incorrectly'.[24] Retrospectively, this narrative gives an accurate account of how the industrial revolution became a British success story of gradual improvements. British tinkerers were less interested in theoretical insight than in economic output. Five years later, Lichtenberg visited the 'famous manufactory' of Boulton and Watt near Birmingham and was fascinated by their most important product, 'a new kind of fire- or steam-engine'.[25] At that time, it was obligatory for tourists of technology to visit the factory – until Boulton became afraid of industrial espionage and closed the plant to visitors at the end of the 1780s.[26]

Even while a famous scientist like Lichtenberg could write about the impressive things he had seen in England, the decisive agent of technological transfer was most often the common artisan. Watchmakers and other skilled mechanics visited England, learned from English engineers like Watt and transferred new technology to Germany when they came home. In particular German salt mines benefitted from this transfer of knowledge, and became modernized in the late eighteenth century.[27] Moreover, some

German craftsmen spent time in English factories and returned to Germany as skilled experts who knew how to use, repair or even install the new technology.[28] The personal transfer of technology and know-how was of utmost importance in the early phase of industrialization, but individual technical knowledge and tacit knowledge remained essential even in later periods of science-based technology.[29]

During the decades around 1800, technology transfer from Britain to Germany developed in three stages: in the first stage, Germans were wholly dependent upon machine imports and the accompanying immigration of experts. After the British export ban was imposed at the end of the eighteenth century, a second stage of industrial espionage meant that German entrepreneur-technicians took study trips to Britain, with some even trying to sneak into factories. Finally, during a more advanced state of technological development, knowledge transfer took place by publications, design drawings and the like.[30] Johann Beckmann, the German pioneer of technical sciences and therefore nicknamed 'father of technology', promoted this latter type of transfer. In his books and journals, most notably his main opus on the history of inventions first published in 1780, Beckmann informed German experts about various industrial and technological advances in England like canal construction, coking and copper production. Of course, Beckmann also described the steam engine production at Boulton and Watt.[31]

The acquisition of technological know-how was a slow and tedious process. In the early nineteenth century, German industry hinged upon the imported technical knowledge of British and Belgian artisans and experts.[32] The first mechanical spinning mills in Germany were modelled on British examples and built by British experts. The early success of the Wuppertal cotton weavers can be largely attributed to the Prussian state, which supported the weavers by providing a semi-mechanical loom in 1821 that was easily copied.[33] As late as 1835, the first German railway line between Nuremberg and Fürth was completely imported from Britain: rails, wagons, the locomotive, even the train driver, stoker and coal! Within the next decades, German engineers in Berlin and Chemnitz managed to build replicas.[34]

Well-known accounts of technological transfer in the industrial revolution era have rightly stressed stories of early adaptations: for instance, the Belgian mechanic Jean Wasseige built the first Newcomen-type steam engine in Germany at a lead pit near Düsseldorf in 1751, nearly forty years after the first Newcomen engine had been invented in England. Wandering Belgian artisans played a major role in introducing new technology because Belgium was more technologically advanced and many Belgian craftsmen had become the first continental experts in British innovations.[35] The first Watt-style steam engine, which was more efficient than its Newcomen predecessor, did not have such a long delay on its way to Germany. It was transferred to Hettstedt near the Harz Mountains and used for drainage in a copper mine in 1785 only a few years after the first British-made Watt engine was constructed.[36] Three years later, a Watt-style engine was built in Germany for the first time, in a lead-ore mine near the Silesian town of Tarnowitz, for the purpose of mine drainage. The engine's parts, however, had been manufactured in England.[37] Obviously, the technological gap was anything but closed. Indeed, the transfer was possible only because steam engines were not covered by the British export ban: their proliferation abroad was not seen as a threat to British economic dominance, because at this stage of industrialization, the steam engine had a rather marginal impact on economic activity other than mining,[38] although it would become essential for further industrial development in the long run. German engineers managed to build a home-grown steam engine only in 1791 – and this was fashioned after the outmoded Newcomen model. This engine provides a good example of the difficulties that new inventions sometimes faced: its completion was delayed, and the mine that it was designated for rejected it. Another mine took the steam engine only after eight years of storage.[39] What appears to be a failure in the early stage of German industrialization nevertheless turned out to become the starting point for modern mining in the Ruhr area, enabling German industrialism to take off in earnest. This very steam engine allowed deep coal mining from 1808 onwards.[40]

Apart from this example, it is crucial to consider technological life cycles, thereby overcoming the limitations of approaches, which only focus on technological innovations. In several cases, Newcomen engines were used for a long period of time – independent of Watt's invention. Moreover, British technology often made a detour through other parts of Europe before reaching Germany. Such is the case of the first Prussian steam engine installed in a lignite mine at Altenwedding in 1779 for water drainage, thereby replacing horses. It was a Newcomen engine probably of Slovakian origin where mining had made great advances since the early eighteenth century. This engine was a fine example of the persistence of technology-in-use: the Newcomen engine was already outdated in 1779, but as late as 1800, there were more Newcomen engines than Watt machines in use, even in England. The old Newcomen engine in use at the Altenwedding mine was not replaced with a new Watt engine until 1828.[41]

These initial efforts in the late eighteenth century laid the foundations for an industrial take-off in the German states. The first continental mechanical cotton mill in the Rhineland town of Ratingen, which opened in 1784, is a perfect example. The German mill owner even called this factory 'Cromford' after the location of the English plant it was modelled upon;[42] its namesake was founded by the eminent English inventor Richard Arkwright in 1771, and can be understood as the first modern factory, integrating a comprehensive division of labour: line production and the concept of partial automation.[43] Germans were making headway in the iron industry, too. After a few short-term trials, all of which failed, the first fully functional German coke furnace was put into service in 1796 in Upper Silesia.[44] Besides technological and economical imitation, Germans also learned from the English experience how best to avoid mistakes, or so they thought. The issue of how to avoid the worst social consequences of industrialization was widely discussed in Germany. As such, England was not only a positive role model but also offered lessons regarding the undesirable developments of the industrial pioneer. There were nonetheless notable similarities, for instance, concerning the issue of child labour. Although Prussia lagged behind the British in industrial development, some of the most important Prussian factories employed children to such an extent that they made up nearly a third of the total workforce. In both states, legislation restricting the practice came about in the 1830s: the British Factory Act of 1833 outlawed child labour, while similar laws in Prussia did so in 1839.[45]

In general, Britain was both Germany's role model and rival,[46] but certain conditions in Germany were clearly different. Unlike in Britain, where the capitalist entities were the major innovators, the Germans states – most notably Prussia – promoted industrialization through their own state bureaucracies, and organized the transfer of knowledge from England.[47] Thus, German industrialization was not initially the result of capitalist principles but was initiated by statesmen who conceived of themselves as representatives of liberal progress; thereby, the ideas of liberalism and mercantilism were merged. German contemporaries were well aware of their differences to Britain: many important thinkers and reformers admired English industrial success, but at the same time, they were not convinced that simply copying the English model would work. Germany – or more precisely the German lands that were to become the German Empire in 1871 – had to find its own path to industrial growth. Consequently, what took place around 1800 was more than a mere transfer of technology and economic concepts, but was rather a comprehensive cultural exchange. Leading figures, such as the eminent Prussian reformer Karl August von Hardenberg, had been familiar with industrial developments at an early stage. Hardenberg took three study trips to England between 1773 and 1781.[48] Knowledgeable officials like him and Alexander von Humboldt stimulated Prussian industrialization from the late eighteenth century onwards, employing small-scale strategies of technological innovation. Germany's leading role in global industry during the late nineteenth century was the result of an evolution that began a century earlier.[49] After a military defeat, the government in Saxony had promoted industrialization to re-establish the economy after the Seven Years War (1756–63) through the establishment of a state economic authority that paid premiums for implementing new products, processes and technologies. Forty years later, after having lost a war against France in 1806, the Prussian reform movement similarly accelerated the development of industrial technology.[50]

Despite such seeming patterns, the history of German industrialization does not follow a linear or homogeneous path. There were significant regional differences in its development, and its progress was varied. This has not always been clear to some historians who have depicted the German states as 'the first underdeveloped country' in relation to England. According to these accounts, Germany was 'aware of its backwardness and deliberately modelled itself on' England.[51] In this regard, it is necessary to distinguish between the history of ideas and the economic history of technology. While there was an emerging discourse on progress, which made Germans envious of England's prospering industry, this does not mean that German industry was necessarily uncompetitive. Alongside this, it is problematic to assume that German industry was universally backward, because there were regional differences. The political scientist Gary Herrigel pointed out that there were two main industrial orders in Germany, and only one matched up to the common picture of the so-called German mode of production. This includes formerly rural landscapes, which rapidly transformed themselves into industrial centres dominated by large corporations, notably in the Ruhr area.[52] The other pattern of industrial development was equally important for German industrial history, but is often overlooked: regions of the 'decentralized industrial order' where small- and medium-sized enterprises maintained their competitiveness without discarding traditional structures of handicraft and specialized production. Neither German nor foreign contemporary competitors conceived of these enterprises as being 'backward'. Areas of this type of rural industrial order were mainly based in south-western Germany, in Saxony and in the countryside of western German around Wuppertal and Siegen. The early rural industrialization was characterized by a 'slow transition' towards the factory system and a surprising persistence of the traditional 'decentralized, production-fragmenting, subcontracting networks' throughout the nineteenth century and well into the twentieth century. This type of 'alternative industrialization' met the challenges of European industrialization and managed to integrate some of the new technologies while preserving the basic organizational structures.[53]

Neglect of these regional differences led historians to wrongly identify *one* special pattern of German industrialization with *the* particular 'German' path of industrialization. Accordingly, when describing the alleged characteristics of German industrialization scholars often erroneously referred to the economist Alexander Gerschenkron's classical definition of 'economic backwardness', which suggested that ostensibly 'backward' countries like Germany had concentrated on importing the 'most modern and expensive technology' and preferred 'large-scale plant'.[54] Gerschenkron and his followers conceived this pattern as a completely rational way to compensate for initial backwardness. This line of analysis, however, neglected that there was a variety of German paths to industrial modernity. Gerschenkron's depiction was applicable only to regions of industrial urbanization such as the growing Ruhr cities, which necessarily depended on the import of skill and knowledge; namely British machines, engineers and workers.[55]

Until recently, most historians were also convinced that Germany's industrial success was to be attributed to a particular German style of banking, the universal bank. The economist Caroline Fohlin has recently shown, however, that universal banks grew 'during the latest stages of industrialization, not as a precursor to it'. In a certain sense, technological innovations gave rise to the universal banks, and not vice versa, as had commonly been believed. Moreover, much of the industrial growth before 1850 'derived from small and medium-sized producers in metalworking, textiles, and other light industries', not from the large heavy-industry enterprises German industrial growth has usually been attributed to.[56]

An important distinction also needs to be made between the mere transfer of technology as opposed to its proliferation. This can be exemplified by the case of the textile industry in Saxony. The first steam engine in a Saxon spinning mill was introduced in 1820, though widespread adaptation was slow: by 1848 only eighteen cotton mills used steam engines, and of these the majority still used waterwheels to generate supplemental power. Thus, the problem in Saxony was not the transfer of technology but its slow spread. Moreover, new loom technology did not necessarily lead to the establishment of factories; instead, it was simply adopted in small workshops by groups of hand weavers. Again, new technology was easily combined with old structures. This version of a slow path of industrialization proved to be quite successful.[57] In a way, early adoption of technological innovation played a crucial role in this process: The early diffusion of the first spinning machines in Saxony, the semi-mechanical 'Spinning Jennies', slowed down the evolution of fully mechanized processes because decentralized production with older technologies persisted for a longer period of time.[58] Furthermore, Saxon state business policy fostered the coexistence of cottage industry alongside factories as late as the mid-nineteenth century by promoting both factory construction and technical improvements in those cottage industries.[59] However, the 1850s were a decade of technology proliferation and by 1861, most Saxon textile workshops used steam engines.[60] These changes occurred in the textile industry all over Germany: hand spinning was largely replaced by water power and steam-driven spinning machines.[61] In other German regions, the 'persistence of small firms' also went hand in hand with technological innovation: small firms in the areas of Remscheid and Solingen were among the first users of small steam and gasoline engines in the latter part of the nineteenth century and also of small electrical motors in the 1890s; which is remarkable considering that most German factories only switched to electrical motors after 1905.[62]

Nevertheless, these two towns of the Lower Rhine area mechanized their industry quite late. Both had been traditional centres of cutlery production since the Middle Ages and had been very prosperous since the late eighteenth century. They stayed competitive without changing industrial structure: both were shaped by 'small scale production, small shops and hardly any sign of mechanization until the introduction of the steam-engine in the 1850s'. However, the artisans were extremely specialized and the division of labour was high even if no proper transition to factory organization had taken place.[63]

In general, German industrialization was 'characterized by a long survival of traditional institutions' which slowed down economic and social change.[64] Cottage industries and factories coexisted and competed for a long time during the nineteenth century.[65] As has been shown above, however, this slow transformation cannot be equated with a general technological and economic backwardness. In a certain manner, agents of technological transfer paid particular attention to the issue of human factors of production. Thus, the slow technological change in the early nineteenth century paved the way for rapid transformation based on human capital in the late nineteenth century. Apart from the rural artisans who managed to partially maintain their organization of work for decades, the Prussian reformers also took part in this process of slow transformation. It was not the Prussian king but ministers and officials who allied with scientists and technologists initiating a step-by-step strategy of industrialization in the late eighteenth century. This alliance lasted for several generations, pushing Prussian industrialization and institutionalized scientific-technological education and research. Finally, this innovation strategy led to the science-based industries Germany was so very fond of around 1900. In a sense, this had been the birth of 'technoscience' before the concept existed, making up one of the crucial peculiarities of Prussia's path to industrialization. Germany's world-renowned engineering sciences of the late nineteenth century had Prussian predecessors called functional or 'useful' sciences.[66] In the early 1820s Prussia established schools to train technicians for private industry: provincial vocational schools on a lower level and the Technical Institute in Berlin for higher technical education for factory managers. Most German states followed this example.[67]

The states of Baden and Württemberg in the German southwest took another road to industrial modernity. It ultimately converged in the same 'German' pattern of industrialization by focussing on the human factor of production, that is, the technology user, the worker. Unlike the more elite technical education established in Prussia, this southwest-German region provided a broad educational system originating in handicraft training. Some scholars interpreted the long persistence of small firms and handicraft as evidence for a long survival of anti-modern institutions in German history. In contrast, the historian Hal Hansen has shown that 'the evolution of Germany's corporately organized craft sector' was not at all 'an anachronistic throwback' but one of a number of viable paths to liberal industrial modernity. Industrial associations replaced the guilds in mid-nineteenth century. At first glance, this appeared to be a continuation of the old practices. In fact, contrary to the guilds, these new industrial associations accepted the market concept and prepared its artisans to be competitive within it. In the rural Southwest, 'human capital accumulation' by means of institutions for training skilled workers was industry's crucial factor for success.[68] Large investments in new technology to grow businesses was not an option for the small- and mid-sized enterprises in any case. In the long run, regional economic disadvantages enabled the establishment of institutions, which step-by-step fostered the 'collection and dissemination of information, technology, and better business practices'. Initially in the state of Baden, a wide range of schools emerged to continue the momentum of progress – from institutions of higher learning, such as the polytechnic at Karlsruhe (the predecessor of the later technical university), all the way down to lower-level trade schools. This integration of training, teaching, research and the beginning of cooperation between polytechnic schools and local business formed a basis for the particular German pattern of science-based industry, which became a model for its international competitors after 1900. This approach, which originated in the southwest, soon became a role model for neighbouring states like Bavaria, Hessen and Saxony, before it was adopted in Prussia at the end of the nineteenth century.[69] In this sense, a particular 'German' path to a very successful form of industrialization emerged from various regional developments, which had begun in the early nineteenth century.

Given its long pre-history, Germany's rapid industrial rise in the late nineteenth century was not at all surprising: there was a comprehensive system of technical education, the artisans had technical know-how and comprised a large reserve army of industrial labour. Beyond this, the German states also had good fortune with natural resources. The regional rivalry between the German states before the unification of 1871 also produced strong competitive pressures, which turned out to be an important factor of success.[70] Another factor, usually associated with German industrial success in the twentieth century, also emerged in this early stage: namely an emphasis on quality work. While Britain had established factories, traditional manufacturing prevailed in Germany. This was, however, not backwardness, but rather part of an international division of labour that saw the German states import British semi-manufactures to be further processed by Germans 'on the basis of inherited skills and low wages' and exporting these products to Eastern Europe.[71] Thus, in the mid-nineteenth century, German industry profited from cheap labour, but its success is equally attributed to 'high-quality labour'. Effective technical education in special institutions and on-the-job training meant that an intensified interest in human capital was key to German industrial success.[72] As will be demonstrated later, however, the marriage of mass production and quality labour took a complex and twisted path; manifold problems arose along the way to progress, as the importance of quality work versus mass production became a contested issue.

Coal mining and railways: Paving the way for the new industries

In general, the modern state was essential for the development of the market and the process of capital accumulation, thereby setting the framework for industrialization.[73] With regard to monetary integration, the German Customs Union, started by Prussia and some of its allies in 1834 and joined by most German states by 1867, was of utmost importance.[74] The German states regulated currency conversion only in 1851. There had still been fifty-six different standards of currency conversion in Germany in the 1840s. Full monetary union came as late as 1873. Measurements and weights were also standardized in this period.[75] To some extent, the strong impact feudalism had in Germany in the early nineteenth century prevented the British private enterprise path to industrialization. This meant that the German states played an important role by easing feudal restrictions, standardizing weights and currencies and establishing freedom of trade as a basis for industrialization.[76]

However, at least in some cases, private business initiatives were more important than political reforms: in Solingen for instance, 'powerful putters-out had already dismantled most of the guild restrictions and established a new order with themselves at the top' before the trade reforms that came to French-occupied Rhineland in 1809.[77] As such, the cottage industry, organized on the putting-out model, was more important for early capitalist development in Germany than the first factories.[78]

There is good reason, therefore, to question the centrality of the states' role in modernizing and industrializing. On the one hand, private initiatives were essential, on the other hand, quarrels within the state bureaucracy make it difficult to identify one singular objective of *the* state; rather, various bureaucrats followed their own agendas.[79] Nevertheless, with respect to two key sectors of mid-nineteenth-century industrial development, the German states deserve closer attention: coal mining and railways. While both fields were not determined by governmental action, their interconnected expansion cannot be explained without special regard to state policy. This was often ambivalent: most Prussian mining bureaucrats in the late eighteenth century shared an attitude which the historian Brose described as a 'backward looking progressivism'.[80] The 'early leaders of Prussian mining and metallurgy' around 1800 were not modern in a narrow sense, but rather on a mission 'to restore Germany's mines to an *earlier* level of technological competence'.[81] However, there were crucial impulses from practitioners: the Prussian minister and mining reformer Friedrich Anton von Heynitz sent a mining official to England to study modern ironworks technology twice in the 1780s. This official, Friedrich August Alexander Eversmann, convinced his superiors that pig iron production by coke furnaces was essential for Prussia.[82] Thus, coal became more important, even if coke replaced charcoal in German iron production only during the nineteenth century while the first blast furnace had operated in Britain at the beginning of the eighteenth century.[83]

Such developments were decisive for German industry, even if the implementation of new technologies took time. In the 1820s and 1830s, the German economy was caught between the old and the new order. Thereafter, driven by the interwoven sectors of mining, railways and heavy industry, the German economy faced revolutionary change in the middle of the century. Important technological innovations transformed the mining and steel industries and gave birth to rail transport: deep mining, coke furnaces, puddle and carbon steel and last but not least steam engines.[84]

Although cultural memory has tended to overvalue the steam engines, its long-term effects both for mining and for industrialization more broadly were enormous. The steam engine's success was a historical watershed for the concept of energy because it transformed heat into movement. The significance for transport – railways and ships – and for machines of different sorts was eminent.[85]

As noted earlier, the first steam engine operated in Ruhr mining during the first decade of the nineteenth century. A modest twenty-six steam engines were in use in the Ruhr area by 1828, increasing to ninety-five in 1843. They were essential for large production sites, especially for pumping water, but also for coal production, shaft sinking and securing fresh air.[86] Although there was a long mining tradition, there was almost no coal trade in the Ruhr well into the early nineteenth century. Farmers produced coal before 1800, but only for self-supply or as a side job. Furthermore, the river Ruhr became navigable only as late as 1780.[87] Only the aforementioned technological innovations allowed deep mining. Earlier, mining had meant mainly going after metal ores; afterwards the age of coal began. To be sure, the Ruhr area was an industrial latecomer; the age of coal started here about 1840, half a century later than in Britain. Eventually, however, the Ruhr profited from its advantageous geography: large, easily exploitable coal reserves of excellent quality, good transport links by the now navigable waterways and early railroad lines by the mid-nineteenth century, as well as nearby markets for sales, which were also experiencing rapid growth. Above all, the sheer size of the Ruhr coalfield was a big advantage.[88]

In this sense, German mining was 'backward' to a certain degree, but as a latecomer, the Ruhr profited as the mines could rely on established technologies. Altogether, this yielded dynamic growth during the nineteenth century and peaked around 1900, with the highest output of all mining regions worldwide. While only 1,356 miners worked in the Ruhr in 1792, the number grew to 15,212 sixty years later. By 1912, 371,059 workers in the Ruhr were employed as miners. The increase in production was even more impressive, starting with 177,000 tons of coal per year in 1792, then 1.9 million tons in 1852 and finally 103 million tons in 1912.[89] Apart from the Ruhr, the other large coal reserves of Germany – in Silesia, the Saar region and the Aachen district – provided similarly excellent conditions to further industrial growth in the second half of the nineteenth century. The Silesian and Saar regions had even dominated in the early nineteenth century, but, although they benefited from a sustained period of expansion, their increase was not as dynamic as the Ruhr's. By comparison, France and other European countries also lacked the benefit of a dominant coal industry and developed considerably slower.[90]

Coal mining is crucial for understanding political history in the region. In many ways, Germany's rise to the position of a great European power was largely due to the age of coal.[91] The exploitation of coal and to a lesser degree lignite was the foundation for the German industrial upswing. Coal served as an energy resource for German industry and as a basic raw material for the iron and steel industry in the mid-nineteenth century.[92] Although the steam engine was essential for modern deep mining, it should be emphasized that mining technology changed only gradually well into the twentieth century. Old technologies persisted, and much of the work was done the traditional way. Due to the differing geology at various sites, different local mining techniques and labour practices, there had never been one singular state of the art in mining technology; rather, a variety of mining technologies coexisted at different sites. In general, mechanization was a peripheral phenomenon in mining before 1918, while manual labour prevailed.[93] Labour productivity had almost doubled during the four decades after 1850, but then stagnated until 1914.[94] Neither steam nor electricity was used by faceworkers before the First World War: they still worked with hoes, club hammers, hand drills, saws, picks and shovels.[95] In contrast to factory workers, miners remained broadly in control of the labour process in the first stages of mechanization; machines became a determinant factor in the work process only from the 1960s onwards.[96] Moreover, sometimes innovations were introduced, but were not accepted by the miners. In Prussia, a mechanized process had supplanted the practice of descending by hand rope by 1859, but many miners resisted this change. Up to half of the miners still preferred ropes well into the 1880s.[97]

Concerning power, compressed air, first introduced to mining in the 1880s, was the second most important type of energy that superseded steam in mining. On this basis, pneumatic drills were subsequently developed and became commonplace. From the 1890s onwards, a third type of energy – electricity – was adapted for mining. In spite of these developments, steam engines were still used in Ruhr mining as late as the 1950s. Only then did electricity become the most important energy source to power mining in the Ruhr.[98] As mentioned earlier, compared to the overwhelming impact of coal, lignite had less effect on Germany's rapid industrial growth, but the figures are still impressive. In contrast to coal, lignite was extracted in open strip mines. Saxony-Anhalt was the most important region for lignite production in Germany. The introduction of steam-powered bucket-wheel excavators mechanized surface mining in the late nineteenth century. Primarily, lignite was formed into briquettes for use as domestic fuel, but it also gained industrial relevance by fuelling power plants after 1900. Lignite extraction in Germany increased from 4.4 million tons in 1860 to 87 million tons in 1913.[99]

Coal was also essential for the interrelated industries of railroad, iron and steel. With the initial boom in German coal mining during the 1850s, demand for railroad transport sharply increased.[100] The historian Dieter Ziegler stresses that the industrial revolution, especially in Germany, cannot be explained without reference to the railroad.[101] In some ways, the railway industry's prominence distinguishes German industrialization from that of Britain: in Germany, the railways facilitated the upswing of coal, engineering and timber industries, also pushing the iron industry and wagon construction to higher levels of productivity.[102] The interconnectedness of coal and railroads had self-reinforcing effects: coal-fuelled trains, and the railways both reduced transport prices and opened new markets for more coal. Ruhr coal replaced its British competitors in the north German market due to targeted price reductions in transport. These new markets in turn became an incentive for further modernizing Ruhr mining. A decrease in freight costs from the 1850s onwards accelerated the process of transformation.

As mentioned earlier, most locomotives and railway tracks were imported from Britain in the early days of the German railroads. Only one of the first fifty-one locomotives that operated in Germany before 1841 was produced domestically; forty-eight were imported from Britain, and two came from Belgium. Locomotive construction turned out to be a real technological challenge for German engineers who made fruitless attempts to build locomotives at the end of the 1830s. Their early failures, however, paved the way for later successes in the following decades. By the 1850s, locomotives were being built in all of the major German states, while British imports had become rare.[103] Thus, railway construction gave a strong impetus to the engineering industry, for both locomotive and steam engine production.[104]

In contrast to Britain, the German states had to be more active in railroad development. The British practice of giving private companies the right to expropriate landowners was not common in Germany. Instead, the establishment of Germans railroads occurred in a more diverse manner, and varied from state to state. Mixed systems of private and state companies were the most prominent form, as easy profits were not to be expected in most cases. The initial heavy investments necessary to start a railroad often yielded meagre earnings, and these could take years to materialize. Therefore, it was common for municipalities and regional authorities to be major stakeholders in the railways. In the initial period of the 1830s, railway construction began as a mainly private enterprise, but as those private investments became rarer, governments became more engaged in the business, realizing the new technology's economic potential. Thus, by international comparison, the German states played a preeminent role in railway establishment.[105] This is true even of Prussia, which had initially experimented with private railroads, but had effectively created a state monopoly by the 1870s, after problems with private railroads had become obvious.

In general, government involvement in German railway politics served as an antidote to certain problems ancillary to the industrial revolution. The states of Prussia, Saxony, Bavaria, Oldenburg and Hessen wanted to use rail systems to strengthen the rural regions in relation to the growing industrial centres. They improved conditions for rural areas through state support for light industry, agriculture and the food industry. At the same time, this had the positive effect of securing food supplies for the cities, thereby pacifying fears of social unrest from the growing industrial labour movement.[106]

After 1850, the rapidly increasing demands of the railways stimulated the modernization of another latecomer industry of the Ruhr area, which would grow enormously in the following decades: iron and steel. It was only then that coke smelting and refining supplanted charcoal technology in the German iron industry.[107] The resulting crucible steel was of utmost importance for the industrial era due to its manifold uses. Crucible steel had first been developed in mid-eighteenth-century Britain, and in its beginnings was still produced by pre-industrial means: it required no more manpower than traditional glassworks. Moreover, it was done in simple clay pots using traditional furnaces. What was different now was crucible steel's applicability, which perfectly fit the evolving industrial environment. It was the first malleable iron that could be shaped in liquid condition. Furthermore, it guaranteed an extraordinary level of purity, which made it useful for new applications coming later in the nineteenth century. Crucible steel's success story ended only in the twentieth century when it was replaced by electrical steel.[108]

The traditional iron industry based on water power and charcoal persisted in many German regions in the first half of the nineteenth century, becoming only slightly more modernized from the 1830s onwards.[109] By the first decades of the nineteenth century, all of the early German crucible steel factories failed, with one notable exception: Krupp, established in the heart of the Ruhr area in 1811. Its success can be attributed to specific physical characteristics of Krupp steel, which were developed by accident. Krupp's crucible steel cured in a peculiar way. After quenching, the steel grew much harder. This made it useful for purposes which English crucible steel was unsuited to. Otherwise, Krupp's steel had some clear limits. It could not, for example, be used for knives. The technological qualities of Krupp steel explain why the firm, unlike its German competitors, survived the early years against all odds. Good fortune also played a crucial role, and Krupp's survival in its initial stages depended on finding new industrial applications for this new grade of steel. In a certain sense, Krupp's peculiar path to becoming a modern factory was exemplary for the process of industrialization. It was not the work of a genius. Rather, its company founder Friedrich Krupp initially misjudged many aspects of the process and found the appropriate technical, economic and organizational solutions only by trial and error.[110] However, this entrepreneurial mentality – and good fortune – was decisive for its success.

In addition, neighbouring industries were indispensable. Machine building created new demand for steel tools. *Gutehoffnungshütte* started steam engine construction in 1819, offering newcomers to the iron and steel industry like Krupp a growing market for their products. In the following decades, the rapidly increasing demand for military and industrial steel products resulted in Krupp's expansion from a 'successful artisan workshop' of 130 workers to a 'gigantic enterprise' of 12,000 workers within a period of only twenty years.[111] The founder's son, Alfred Krupp, restructured the factory after 1848, leaving custom-built production behind while moving forward to series production, mostly machine construction and trains, but also cannons. Krupp steel became part of the German national myth by virtue of its military products. In particular, the cannons it produced were widely believed to be an essential part of Prussia's victories during what was later called the German Wars of Unification (1864-71).[112]

Beyond the cultural myths about Krupp steel, the mere economic figures were impressive. Krupp and its Ruhr competitors enhanced pig iron production from 11,500 tons in 1850 to 7.6 million tons in 1912, almost drawing level with Britain (8.75 million tons), which had been far ahead in 1850 (2.2 million tons). The modern iron and steel industry established itself during this period. By 1871, sixty-four furnaces operated in the Ruhr area, almost all of them working on a coke basis. This accounted for a third of the Prussian pig iron production. However, for further processing, puddled iron still prevailed by 1871.[113] Afterwards, the modern 'age of steel' began with mass production of steel and a leading role for the steel industry both economically and technologically.[114] This transformation had effects on the global industry as a whole: Britain lost its role as the leader in industrialization, while Germany and the United States surpassed it.

This outcome was by no means technologically determined. Germany's success on the global steel markets was less due to a supposed technological supremacy; it was based on strong state and banking support and, above all, the power of cartels. Then again, in contrast to traditional puddled iron, the new steel technology of the Bessemer process or the open-hearth process privileged cartels because of the sheer size of the systems and the problems to adjust the systems in times of decreasing sales. According to the historian Ulrich Wengenroth, an inferiority complex survived within the German steel industry even after it overtook the British competitors. To a certain degree, the avoidance of domestic competition and the resulting cartels were the key to success. Furthermore, German economic success before the First World War was mainly due to low prices: German raw material resources privileged the basic Bessemer process, which produced steel of a very basic quality, but was much cheaper to make than the British high-quality steel made in the open-hearth process.[115]

As has been shown, industrial take-off in Germany started in the mid-nineteenth century. The large investments in heavy industry and railways during the 1850s and 1860s made a remarkable economic upswing possible, not least in conjunction with the politics of liberalization.[116] Although government action facilitated the process of industrialization, its importance must not be overestimated because German industrialization was well underway before the political unification of 1871.[117] The industries of coal, railway and steel stimulated three sectors, which pushed Germany to its new status as an economic and technological world power in the early twentieth century based on chemistry, machine-building and electrical engineering. As mentioned earlier, coal and railways mutually stimulated this economic rise, and steel was essential for the prospering machine-building industry. For its part, coal was the foundation for many chemical developments. Railway lines had to be equipped with telegraphy for safety reasons, thereby supporting the electro-technical industry and especially the later world-renowned corporations of Siemens and AEG. Moreover, all of these new industries depended on skilled labour and applied sciences.

Scholars widely agree that there were three principal periods of industrialization in Germany during the nineteenth century. A first and rather shaky stage of early industrialization between 1815 and the revolutionary years of 1848 to 1849 was followed by the industrial take-off. In this second phase, German heavy industry achieved a breakthrough. It was only shortly interrupted by the depression of 1857 to 1859, but by and large economic growth continued until the 1870s. A third period of high industrialization began after German unification and the great depression of 1873 to 1879.[118] This German industrial take-off had some peculiar features. The overseas emigration of two million Germans between 1840 and 1880, for example, caused a serious lack of manpower in many new industries. Skilled workers were particularly rare in machine building and steel industry. Technical education and labour training, which persisted in many German regions during early industrialization, became crucial factors of economic success. From that time onwards, the 'accumulation of human capital' was a defining feature of German industry and German economic success.[119]

There was also a particular basis for this upturn. The historian Joachim Radkau describes the alleged German technological backwardness as a myth. According to him, the German industrial take-off in the latter half of the nineteenth century was possible only due to an existing culture of small-scale machine building.[120] Compared to the French and Belgian engineering industries, however, which became independent of British innovations in the 1820s, German machine building was a relative latecomer, reaching this level of independence only by mid-century.[121] Consequently, the early German engineering industry depended on the skills of well-paid British foremen or craftsmen. In this period, German machine building was largely based upon producing copies of British models. These initial difficulties actually had some lasting positive side effects. The British experts were especially valuable as teachers who trained a generation of German skilled workers, many of whom later became entrepreneurs.[122] Moreover, there was another long-term benefit, as has been demonstrated via the example of German machine-tool makers. In many cases, 'good imitators become good innovators'.[123] Significant periods of 'learning by imitating' paved the way for catching up, closing the technological gap and finally becoming innovators in their own right.[124] In the following decades, the German engineering industry established itself in two ways: the first type evolved in regions without craft tradition, characterized by production in larger facilities using imported technology. The second type tied in long-lasting artisan traditions and started from the workshop level.[125] As early as 1856, Saxon producers built more than 75 per cent of the steam engines used in domestic textile industry.[126]

In general, factories started to play an important role in the German economy only after 1850, but then the numbers of workers employed by them increased rapidly. While in 1835, 400,000 workers were employed in factories or other kinds of large companies, the figure had risen to 600,000 by 1850; to 1.8 million in 1873, and to 5.7 million in 1900. Thus, at the turn of the last century, 22 per cent of all employees worked in factories.[127]

Science-based industries and rationalization

The 1870s and 1880s marked the beginning of a new stage of industrialization sometimes dubbed the 'second industrial revolution'. Some scholars claim that this stage, dominated by the new chemical, engineering and electrical industries and the new industrial world powers of Germany and the United States, had more far-reaching effects than the first industrial revolution. In particular, the historian Cornelius Torp points to the crucial impact that 'the first wave of economic globalization before 1914' had on Germany. Beyond ongoing technological progress, new economic institutions

– corporations, cartels and global structures – were established and, finally, knowledge became an essential factor in production.[128] In terms of specific industrial applications, scientific knowledge had a rather minor impact during the first stage of industrialization. Nevertheless, science offered the kind of utopian imagination that has shaped ideas of the industrial revolution since the late eighteenth century.[129] Science did, however, have a discernible impact on industrial production in the late nineteenth century. By that time, scientific knowledge was at the heart of the new industries. Germany, which had been on the periphery of the first industrial revolution, had become an important player in modern industry and technology. The German experience during the second industrial revolution had two dimensions. The first was the establishment of a mode of production which became a global trend in the twentieth century: science-based industry. The second dimension, while closely related to the first, was the starting point for a peculiarly German mode of production, which prevailed well into the twenty-first century: namely diversified quality production.

In a nutshell, two seemingly unspectacular incidents are symbolic of the transformation of German industry during the second industrial revolution: open letters from a German engineering professor in 1876 and British legislative action of 1887. While they come from different spheres, both dealt with the same issues of quality and competition. During the Philadelphia World Exhibition of 1876, an American newspaper mocked the German exhibits, mostly arts and crafts, as being 'cheap and nasty'. A German visitor, the famous professor of engineering Franz Reuleaux, popularized this verdict in Germany by affirming it (translated as *billig und schlecht*, 'cheap and shoddy') in a speech later published under the title *Letters from Philadelphia*. Reuleaux admitted that sadly, the quality of German industrial products actually had been poor because German industry wrongly emphasized price competition instead of focussing on quality. This caused quite a stir in Germany, but Reuleaux had hit the nail on the head. German industrial production of that time still primarily consisted of cheap imitations of British products. Only increasing labour costs resulted in a reversal and subsequently led to a new German competitiveness in high-quality production. Companies also had to revise their strategies. With increased wages, competition based on price was no longer sustainable. Thus, quality production became the new basis of their competitive advantage.[130]

Of course, these changes are not to be solely attributed to Reuleaux's influence, but were the result of other temporal factors, most notably a labour shortage. Four years earlier, the electronics enterprise Siemens & Halske had already reacted to this problem, which had been preceded by foremen strikes. Siemens had established the so-called American Hall at the Berlin plant that signified a transition to mass production, in which the United States had become the trendsetter. In a situation of labour shortages, this new mode of production offered a solution. As telegraphy pioneer Werner von Siemens wrote in a letter to one of his brothers, by 'the implementation of American working methods' and American-style special machines it would be possible to 'produce quality products even with poor workers'.[131] To be sure, mass production was still very rare in German industry. Only the most modern plants in modern industries like Siemens or the sewing machine and armament manufacturer Ludwig Loewe – also Berlin-based – Americanized their factories at the time. Siemens' 'American Hall' was not even representative of the Siemens plant, as the foremen still dominated production in most factory halls.[132] This was a clear feature of Germany's path to mass production. Skilled labour persisted in most industries, and American-style mass production was not simply transferred unaltered to Germany but rather hybridized. Nevertheless, Siemens' different path from Reuleaux's demonstrated the entrepreneurial point of view. Being both inventor and entrepreneur, Siemens decided to prefer neither quality nor price competition; he wanted high quality in large quantities to be produced as cheaply as possible. At the same time, he also sought to restrict his dependence on foremen. This would enable management to assume greater control over its labour force. Early adopters like Siemens did not represent the prevailing practices of German industry at large, however, and highly mechanized production and handicraft coexisted side-by-side in German factories well into the twentieth century.[133]

Professor Reuleaux, teaching mechanical engineering, and Siemens – 'an inventor as well as a scientific researcher in his own right'[134] – did not merely represent the quality approach in general. They both promoted models for how science-based industries could thrive. Both had based their efforts on prior educational reforms, particularly the Prussian government's establishment of engineering schools in the early nineteenth century, paving the way for the notable success of new industries around 1900.[135] More broadly, German scientists had long enjoyed a high international reputation. Britons had admired German science since the 1830s, in particular the Prussian state support for science, which had become an important role model for Great Britain.[136] In fact, German technical universities (*Technische Hochschulen*) developed into the global standard for technical education and research. Nonetheless, it was by no means a one-sided transfer of German innovations, but was rather an example of transnational entanglement, with American experimental laboratories an early international example to the first German technical universities.

German technical colleges had high standards of research and teaching, but above all the sheer size of the German higher technical education system was impressive. In the first decade of the twentieth century, 30,000 engineers graduated from German institutions, 9,000 more than in the United States.[137] Many students came from abroad, and in the case of the electrical engineering courses, half were foreign-born, primarily hailing from eastern and south-eastern Europe. Driven by chauvinistic fears of knowledge transfer, Germany introduced restrictions on the number of foreign students, which limited their numbers after 1905.[138]

By establishing new institutions, the German states paved the way for modernizing the German economy. The technical universities created a 'new middle class of engineers, technical bureaucrats, and industrial researchers'.[139] In addition, a new type of lower-level engineering college had emerged since the 1870s, which 'supplied industry successfully with engineers trained for the shop floor'.[140] This demonstrates once more that it was not the 'advantages of backwardness' which helped Germany take a leading role in some industries of the early twentieth century, but rather its institutional innovation. Research-oriented universities and science-based industry were among the keys to its success.[141] Siemens and the electro-technical industry were at the heart of the process. This sector was as essential for German industry as a whole as telegraphy was vital for the further growth of railroads.[142]

Basic research was the key to the technological revolution caused by the electrotechnical industry in the late nineteenth century. The transformation of mechanical into electrical energy facilitated by the research of Werner Siemens meant a breakthrough for telegraphy and initiated a growth cycle in the entire German economy.[143] Since that time, close contacts arose between electrical industry and electrical engineering science.[144] By 1887, Siemens himself strongly supported the establishment of the Imperial Institute for Physics and Technology, because such an institute offered the opportunity to 'carry science into technology'.[145] Science-based industry became established, but its relationship to scientific institutions was mutually beneficial. Strictly speaking, according to the historian Wolfgang König, 'industry-based science' would be the more appropriate term, at least with regard to electrical engineering, as the technical universities 'profited more from the knowledge being produced in industry than vice versa'.[146]

The three-part relationship of science, technology and industry became particularly manifest in the chemical industry. According to the historian Abelshauser, the emergence of the chemical industry marks the beginning of the 'knowledge society, in which science and research-based innovation have become the key determinants of economic growth and social development'.[147] Chemistry seemed to offer the solution for the German problem of limited resources. A German chemical manufacturer stated around 1900 that in contrast to the United States or Russia, Germany could not rely on sheer endless amounts of raw materials but had instead to count on 'the intelligence and industriousness of its population'.[148] This assertion, while based on clichés, nevertheless illustrates the understanding undergirding the actions and beliefs of those who came to place a high value upon the human factor of production and the technical knowledge of employees. Many German industrial and political leaders saw the chemical industry as the 'industry of unlimited possibilities', alluding to the German dictum of the United States as the 'land of unlimited possibilities'.[149] Although human capital was appreciated, the employee's specific economic value depended on his or her respective position. While the chemists and technicians were essential factors of production, most employees in coal-tar industry were still unskilled workers of little individual value to the companies.[150]

Germans were particularly successful in the research-based production of synthetic dyes, producing more than 75 per cent of the global output by 1913.[151] In retrospect, however, this requires a more nuanced assessment. At the beginning of the 1860s, even BASF, which would later become a world-famous company, imitated coal tar dye producing processes from abroad and took over patents. Only systematic research yielded profits far above those 'offered by mere experience and tacit knowledge'.[152] The historian Ursula Klein has demonstrated that early modern developments prepared the ground for the upswing. The science-based industries could rely on eighteenth-century traditions of chemistry as an 'early form of technoscience'. Even if the contemporary chemists did not add significantly to the transformations of the industrial revolution around 1800, chemists nonetheless contributed to marginal production spheres at the time.[153] Scientific innovations did not totally replace the pre-existing industries of the late nineteenth century. Instead, the old and the new coexisted in German economy. On the one hand, there were 'the most sophisticated technologies and giant corporations'

Figure 1.1 Female workers at the Osram battery factory, Berlin, 1930. Photograph illegally shot by the former worker Eugen Heilig entitled 'piecework in the dirt'. Galerie Arbeiterfotografie, Cologne.

in branches like chemical or electrical industry, while on the other hand, handicrafts continued to be economically relevant.[154]

In this context of traditional German industries, British legislation intervened in 1887. The British 'Merchandise Marks Act' labelled German imports 'Made in Germany' to warn consumers against buying products of minor quality. However, the label soon became a means to advertise German products.[155] The act especially targeted the Solingen blade industry to protect its Sheffield rivals from the German low-price competition. However, due to the presence of skilled labour, the Solingen producers reached a high level of quality, which made them competitive during the 1890s. By the early twentieth century, Solingen blades had become global market leaders. This was especially due to the blades' success in African markets, and in a certain sense, the persistence of handicraft production in the German hinterlands was partly based on colonial global commerce.[156]

So what happened to the cottage industry in this period of science-based industries? While one might assume that they vanished under the pressure of technological progress, they did not. Even as factories prospered around the turn of the last century, numerous small firms survived throughout Germany. For example, electrification further decentralized Saxony's textile industry around 1900, because many small businesses were able to operate with only one machine. Large factories often outsourced some parts of their production by employing cheap, mostly female labour working at home. The cottage industry survived because it changed. Just as big businesses modernized rapidly, so too small-scale industries also adjusted to the new circumstances.[157]

Furthermore, there was a significant persistence of skilled labour in this period of high industrialization. The establishment of factories did not revolutionize the labour process immediately. Artisans persisted in Germany's engineering and even automotive industries well into the 1920s. Regarding the tasks of smiths and cartwrights, there was hardly any difference between the early automobile factory and the traditional craftsman's workshop. Foremen still dominated plant production, and old structures persisted.[158]

Nor did technology seamlessly transfer between nations. The car manufacturer Daimler had already adopted the most modern German and American machine tools from 1903 onwards. However, Daimler did not apply these machines the 'American way, which meant their purpose was not to increase efficiency and output. Instead, the American machines were used to enhance high-quality, manual production. Mechanized production was suspended regularly as craftsmen controlled their work pieces. New technologies were put to use in a traditional manner without significant changes in the labour process. In a certain sense, the artisan workshop persisted throughout the early stages of automobile manufacturing. Only wartime production motivated a gradual transition to mass production.[159]

A closer look into the plants of different industries reveals another German particularity. Machines were grouped by type, and the work pieces were moved. This nodal pattern survived during the first stages of industrialization, and was not replaced by linear flow production for quite some time. According to David Landes, a 'stronghold of skilled craftsmen' making up an 'aristocracy of the labour force' remained a typical feature of German factories. Even one of the most modern Berlin machine-tool works – Loewe – switched over to flow production only as late as 1926.[160] The particularly high degree of skilled labour in German industry did not even decrease during the first stage of intensified rationalization in the 1920s. On the contrary, skilled workers were held in high esteem as guarantors of German quality work.[161]

When examining the German path to industrialization, what at first sight appears to have been features of a failed or incomplete modernization was often an alternative path to modernity. From the macro-economic perspective, it can be said that Germany developed from 'free competition to the corporate market economy' after the 1870s depression. This must not be mistaken for Germany failing to establish liberalism, as *Sonderweg* historians have claimed. Rather than preserving a 'pre-industrial value system', Germany established a new system and became 'the first post-liberal nation'.[162] After the First World War, German industry focused on 'diversified quality products' because the American path of Fordism was not open to Germany: there was not enough capital for high-tech innovations and a lack of consumer demand, preventing the establishment of an American-style Fordist economy of mass production and mass consumption. Thus, German industrial traditions paved the way to a successful alternative: established technologies, skilled labour and customized production formed the foundation for a 'German' system of 'diversified quality production', which was to prevail during the twentieth century.[163]

Nonetheless, the early efforts of entrepreneurs like Siemens or Loewe to introduce mass production were by no means isolated phenomena; further attempts to improve productivity and efficiency followed in the early twentieth century. Soon after the turn of the century, flow production was introduced in some German factories, initially limited to certain industries like packaging or food; women nearly always performed this new type of work.[164] The Hanover food company Bahlsen even introduced a conveyer belt in 1905. Its owner, Hermann Bahlsen, was clearly inspired by the slaughterhouses of Chicago, but beyond this American influence, British role models like the famous confectioners Rowntree and Cadbury also had a significant impact. These English companies particularly influenced Bahlsen's paternalistic, social personnel management, which aimed to create a joyous atmosphere of work to enhance workers' motivation, while also providing a bulwark against the encroachment of labour unions.[165]

It should be emphasized that even the debate on efficiency, which began in the United States in the late nineteenth century, became relevant in Germany only during the interwar period – although it had been adopted by some German engineers and factory owners prior to the First World War. Frederick W. Taylor's writings on scientific management were essential because they provided a practical means for enhancing efficiency in industry: time-and-motion studies allowed operations to be fragmented in simple motions. In that way, the goal became for the operating process to be perfected by finding the 'one best way'.[166] Taylor soon found German followers, among them Ernst Abbe, research director and later co-owner of the optical firm Carl Zeiss, who had suggested a high degree of division of labour and a transition to series production as early as the 1860s. Before the First World War, the Zeiss Company was among the first German adaptors of Taylor's scientific management.[167]

In some respects, though, Abbe provided an 'alternative' version of scientific management, stressing the significance of the human factor by treating workers as 'responsible individuals' and enhancing their motivation to work. This modification was partly due to technological constraints caused by the nature of the optical industry, which simply was not a large enough sector of the economy for an extreme version of Taylorism or a maximum level of division of labour.[168] The electrical company Bosch was another adopter of Taylorism, reorganizing its production and conducting time-and-motion studies in 1907.[169] However, it is often forgotten, that at around the same time, Bosch also improved its working environment through technological innovations like air-conditioning.[170] The concurrence of these developments demonstrates that in many German industrial settings, efficiency and the humanization of work went hand in hand. They were conceived as different means to the same end: increasing productivity.

It is widely known that the First World War gave the German government – as well as the nations it was at war with – cause to promote industrial efficiency. For this purpose, the War Raw Materials Office was established, directed by the chairman of the electrical corporation AEG, Walter Rathenau, who later became the Foreign Minister of Weimar Germany.[171] However, like all subsequent German proponents of Americanization, the pioneer Rathenau was already convinced that Taylorism had to be adapted to German conditions. Rathenau wanted to modify modern technologies to fit the prevailing practices in Germany; namely, he wanted to maintain certain German cultural peculiarities.[172] Less well known than the German government's support for Taylorism is the close connection between the rationalization and the humanization of work, which evolved as complementary concepts during the war. In this context, gender played a prominent role: the female workers who replaced the soldiers raised awareness about working conditions. The new gender dynamic expressed itself in two key ways: first, several workplace designers and experts of work were considerate of female labour. They believed women workers needed a more humane workspace than men did. Thus, the German government – as its American, British and French adversaries – introduced female welfare supervisors who inspected factory shops and interested themselves in the well-being of female workers. Second, this engagement with the problems of the work environment fostered considerations that benefitted the labour force as a whole, both men and women. Industrial experts considered atmosphere and environment essential for job satisfaction that, in turn, would motivate workers to active cooperation with management.[173]

After the war, ideas about industrial efficiency gained even more traction, primarily due to the fact that a new model had entered the shop floor: Henry Ford's system of mass production. This innovative system of assembly-line production proved much more practical than scientific management and was highly successful. As the historian Daniel Rodgers has put it: 'Fordism invaded Europe as a progressive idea: future-oriented, flexible, and melioristic.'[174] This was especially true for Germany, a nation desperately in need of economic recovery after losing the war. Americanization, represented by Taylorism and Fordism, seemed to be the adequate remedy. It is remarkable and essential for their impact that both were compatible with very different political directions. Liberal and conservative businessmen, unionists, social democrats and even reactionary modernists could embrace the new wave of production. Taylorism and Fordism provided a technological means that could be directed to achieve very different ends.[175]

However, the idea of overcoming Taylorism unified both supporters and opponents of Americanization. German engineers and managers agreed that American models could not be adopted wholesale in Germany; scientific management's 'brutality' might be adequate for the United States, but it could not be combined with German social values.[176] In his book published in 1922, economist Fritz Söllheim postulated upon the humanization of industry, suggesting that the implementation of the Taylor system in Germany had its limits: 'After Taylor has taught us how to think economically, we must learn to economize humanely.' He pleaded for 'more occupational satisfaction and vocational happiness' and overall for a new type of 'human economy' (*Menschenökonomie*).[177]

In contrast to the United States, the catchword in the German debate was not 'efficiency' but 'rationalization'. This term managed 'at one and the same time to incorporate, transcend, and Germanize various versions of Americanism'.[178] The particular German way to rationalization, which was to some degree a European way, was characterized by more flexibility than the American example. The difference was not due to an ostensible technological backwardness, but depended on economic factors. Both lower sales potential and a lower degree of concentration in manufacturing forced the companies to install flexible systems of flow production, with strict assembly-line production remaining an exception.[179] These findings again point to the aforementioned German mode of 'diverse quality production'. The German promoters of rationalization pursued the convergence of humanization and efficiency, which started during the war. Apart from enabling economic recovery, they also promised to 'solve "the human problem" in modernized factories'.[180] In general, the human factor of production became more important after the First World War, especially under the new keyword 'social rationalization'.[181]

A new German trend of applied psychology, 'industrial psychotechnics', also offered an alternative approach to scientific management. This way of thinking devoted more attention to the full person along with psychological needs. This innovation primarily impacted public debates, however, and was scarcely implemented.[182] Nonetheless, industrial psychologists' demand for 'a more humane form of scientific management' had some long-lasting effects.[183] Two institutions, established in 1921 and 1924 respectively, were the most important disseminators of scientific management and the idea of efficiency in German industry: the National Efficiency Board (*Reichskuratorium für Wirtschaftlichkeit*, or RKW) and the National Committee for Work Time Determination (*Reichsausschuß für Arbeitszeitermittlung*, or REFA).[184]

In general, 'there was more talk about rationalization in the Weimar era than actual measures implemented,' although the first steps were taken.[185] Only 80,000 workers – less than 1 per cent of all employees in firms with a workforce higher than fifty – were working at assembly lines or flow production in German industry around 1930.[186] By 1931, apart from electrical corporations, only the metal processing industry had made use of flow production on a larger scale. Except for car and bike production, assembly lines were a rarity. Even in this sector, only 11.6 per cent of all departments in these plants had established assembly-line production.[187]

The German automobile industry was not a significant sector of the economy until the 1930s. Assembly-line production was introduced to the first German automobile plants in 1924, but in 1929, the labour force still consisted largely of skilled workers. To manufacture a single car, American Ford needed a day's work of nine men, while an English plant needed thirty-five to forty-four workers; a German factory employed eighty to ninety men for the same task.[188] For instance, at Opel's Rüsselheim factory, the principle of flow production was far less advanced than at Ford: Opel's final assembly line, established in 1924, was only forty-five meters long and it moved slow and jerky – just once every thirty minutes. The speed of the line was not automatically set but was dependent upon the first line of workers. These relatively inefficient manufacturing practices continued until 1929, when General Motors bought the majority of Opel shares.[189] By contrast, the rationalization of work in the mining industry progressed rapidly, and Ruhr mining successfully modernized itself during the 1920s. By the end of that decade, purely manual labour had all but vanished from coal mining in Germany; in 1929, fully 91 per cent of German coal was mined using pneumatic picks, compared to 80 per cent in France and Belgium and only 28 per cent in Great Britain at that time.[190]

Fordism did not begin to become entrenched in German industrial practices until after the Second World War; although it had beginnings during the Third Reich, this belated modernization was not a failure of technology transfer. In contrast to the first stage of Americanization between 1870 and 1914, when primarily large German firms acquired US machinery, small firms also participated in a 'second wave of Americanization' during the 1920s.[191] However, there was a significant productivity gap between German and American industry. Regarding less dynamic sectors such as the textile and primary metal industries, this was due to differences in production techniques. In this sense these German industries may have been relatively 'backward'. However, in newer, dynamic industries like chemicals, electrical engineering and machine building, the same advanced technologies were being used as in America. Here, the productivity gap arose from the inefficient use of American techniques, as new technologies had to be adapted to German conditions before they could be effective. American assembly-line production techniques, for example, could not be replicated in German factories without special modifications, as those facilities were still reliant on non-standardized batch production. The particular strength of German industry – diversified quality production – hampered the successful adaptation of Fordism. Nonetheless, it could be argued that this early, though inefficient, adoption of new production technology during the 1920s and 1930s actually contributed to Germany's post-1945 industrial growth.[192]

The Nazi rearmament boom had ambivalent effects on the German post-war economy. Some historians stress that 'West Germany's economic miracle was prepared by large-scale investment' in the military build-up prior to and during the Second World War. German industry experienced modernization in terms of both quantity and quality, becoming more efficient as a result. However, the Nazi wartime efforts were enormous; more capital per capita was spent relative to the United States. Additionally, American factories tended to be tooled single-purpose machines that were better suited for mass production. The Germans only made limited use of these more specialized machines, while still largely investing in general-purpose machines. All in all, German manufacturers slowly learned to unite their skills of 'flexible specialization' with mass production goals during the war.[193]

Furthermore, the manufacture of aircraft had an interesting side effect on West Germany's auto industry after the war. Messerschmitt, the major supplier of fighter aircraft for the Luftwaffe, had established a 'supplier network' resulting in a deep 'inter-firm division of labour', which became crucial for the success of car-making in the 1950s.[194] Seen from this perspective, the Nazi armament boom provided an infrastructural basis for much of West Germany's eventual industrial success, but it also caused some problems and slowed down economic structural change. After the Second World War, the country's entire economy became largely dependent upon its strength in manufacturing. Although further industrial rationalization would follow, traditional modes of labour-intensive production remained prevalent in the 1950s.[195]

Contrary to popular belief, German industry was by no means devastated at the end of the Second World War. Despite the fact that the war destroyed 15 per cent of East Germany's industrial facilities and 22 per cent of the factories in West Germany, gross capital stock was actually 20 per cent higher at war's end than in 1936. This was due to the enormous armament and autarkist investments of the Nazi administration before and during the war. In addition to this fact, millions of Germans had abandoned the eastern territories lost in the war and had migrated within the borders of what would become the newly formed two German states. Prior to the construction of the Berlin Wall in 1961, two million and 700,000 more fled the GDR for the West. These political developments culminated in a surplus of available, experienced workers. As such, West German industry, with a comparatively large stock of skilled labour at its disposal, had a relatively favourable starting position for its post-war economic recovery.[196]

Hybrid Fordism, automation and humanization after 1945

Although there were important continuities in German industrial history after the Second World War, one issue changed: German industry had to regain its international appeal. From that time forward, American production technology and organization of work became the sole model upon which other capitalist countries based their industrial development.[197] West German industry drew on the interwar experiences of Americanization, but was pressed into further economic change by the Allies after the war. The occupying powers broke up the old industrial cartels, while the Americans brought German engineers, managers and trade unionists on study trips to visit US factories through a programme sponsored by the European Recovery Program, commonly known as the Marshall Plan.[198]

Once again – and to a much larger extent than in the 1920s – German industrial experts travelled to the United States to study efficiency and automation. Although mass production had been implemented in some German plants prior to 1945, those efforts were isolated to a few production sections, while traditional craftsmanship was still an essential component of even the most modern industrial facilities.[199] Furthermore, the transition to mass production happened slowly, as many West German companies were unable to invest in new technology in the immediate post-war years. Accordingly, West German industrial production in the early 1950s more closely resembled the pre-war German model than the American-style mass production. This changed during the economic recovery, as Americanized production techniques became more prevalent by 1960, albeit not in all industries and not in all companies.[200]

The best-known icon of the West German economic miracle is the Volkswagen Company, symbolized by its most popular product: the Beetle. Originally formed under the Nazi regime, the company had only a modest output of vehicles before the end of the war, mainly consisting of a few car prototypes. Nevertheless, American influence had an impact on Volkswagen's very beginnings: Ford's River Rouge facility was the role model for the first Volkswagen plant and American experts helped launch the factory.[201] After the war, Volkswagen successfully adapted American modes of mass production in West Germany, establishing its own version of Fordism in the mid-1950s. On one hand, the Volkswagen mode of production resembled Ford's original product – its Model T, built from 1908 to 1927 – by producing just one standard model, the Volkswagen Beetle. The company did not market a new car model until the 1960s. On the other hand, Volkswagen also relied on elements of the German tradition of diversified quality production, namely in-house industrial relations, characterized by close cooperation between management and the trade unions. This, along with a global service network, ensured high product quality.[202]

Despite its successes, Volkswagen's path to mass production was not a simple one. In the first years after the war, Volkswagen started making cars in batch production, with no intentions to switch to mass production. Volkswagen had initially acquired American technologies for the purpose of quality control: for instance, automatic machines for motor balancing were more reliable than semi-mechanical operations. These investments in modern Fordist technologies were expensive, however, and made economic sense only when put to use in mass production schemes. As such, competitive pressures and rising sales compelled Volkswagen to shift its manufacturing processes from batch production to mass production in 1954. In these early stages, Volkswagen made further strides towards automation by implementing production lines that linked different processing stages, and made widespread use of single-purpose machines to increase its productivity.[203] However, most of the technological changes were limited to Volkswagen's bestseller: although the Beetle had been mass-produced since the mid-1950s, the manufacture of the new Volkswagen van used older technologies, because its production numbers were too small for new investments in automation to pay off. In all, the United States still served as the role model for technological development in post-war Germany. The same characteristics of the first stage of industrialization in the nineteenth century were again essential for the successful transfer of technology in the twentieth: study trips, machine acquisitions and the occasional hiring of foreign experts – with the difference that British expertise had now been replaced by American know-how.[204]

There were 'two kinds of Fordism' in post-war Germany. In the West German automobile industry, mass production was combined with co-determination, diversified quality production and a high degree of inter-firm division of labour. Meanwhile its East German counterpart often had a high degree of vertical integration in the process management of its factories, which, in a sense, was a more authentic representation of traditional Fordist organization.[205] Certainly, this East German variant of Fordism was woefully inadequate; from the 1970s onward, its factories and workers had to make do with outdated technology, resulting in low levels of productivity.[206] Moreover, over 90 per cent of workers in most East German factories were actually skilled labourers, a significantly higher number than in West Germany.[207]

While Fordist mass production and price competitiveness principles changed West German industrial practices in the 1960s and 1970s, this did not signify a break with its tradition of diversified quality production.[208] Mass production and automation were heavily dependent upon the use of highly customized machine tools. After the oil crisis of 1973, a deliberate re-orientation towards diversified quality production helped German industry outperform many of its international competitors. In particular, the flexibility to shift between mass production and diversified quality production became a successful strategy, particularly when compared with countries like France, where diversified quality production was even more widespread than in Germany. The German companies' emphasis on quality rather than on customization proved to be a crucial advantage.[209]

However, there were winners and losers in the post-war period of industrial growth and the evolution of mass consumption. While women and migrant workers made up the majority of the remaining unskilled labour force, the number of male

German unskilled workers decreased from 34 per cent of the total male industrial workforce in 1925 to merely 20 per cent by 1970. Moreover, half of the male skilled blue-collar workers became technical specialists during the 1950s and 1960s. Thus, Fordist technologies of production led to a remarkable trend towards a polarization of skills.[210] From a macroeconomic viewpoint, capital and consumer goods, as well as the automobile industry performed particularly well in the modern economy, while the pioneers of German industrialization, namely mining, iron and steel, lost ground.[211]

While two thirds of the Ruhr employees worked in coal and steel in 1950, only 25 per cent of the region's workers were employed in these industries by the year 2000. The steel industry's production technology radically changed after 1960. Nevertheless, manual labour did not completely disappear due to automation: modern machinery could accomplish much of the back-breaking work of the past, but certain tasks in the steelmaking process could not be automated. These jobs still required low-skilled workers to perform manual labour. In spite of the revolutionizing effects of automation, proletarian workers continued to be a necessary part of the industry.[212]

Coal shows a similar picture: after the rationalization of the 1920s, very little technical change had taken place. However, during the 1950s, a second wave of mechanization occurred in Ruhr mining, as older steam engines, which had still been common in the post-war years, were replaced by electrically powered machines. Due to a boom in coal production because of the Korean War, as well as an energy crisis in 1950–1, the German government fully supported the modernization of mining. In 1960, 40 per cent of coal production became mechanized by the use of hydraulic shield supports; a decade later, that figure had grown to 92 per cent. Ultimately, coal mining became fully mechanized by 1980. Automation, process optimization and computer technology yielded further increases in productivity after 1990, although the long post-war boom had run its course. Notwithstanding these advances, the history of German coal mining ended in 2018 when the last two mines ceased operations.[213]

Another nineteenth-century champion, the textile industry, underwent a technological transition after decades of stagnation. The textile industry, which had been the starting point for both the British and European industrial revolutions, remained an important sector of the German economy well into the twentieth century, employing around 1.2 million workers in the mid-1920s. However, the efficiency movement of the interwar period had a very limited impact on the textile sector. After the end of the Second World War, excess capacities prevented large-scale acquisitions of new production technology in German textiles, as there were no apparent financial incentives for such high investments. As such, few significant technological changes occurred until the 1960s and 1970s, when automation revolutionized the sector. As the first wave of robots and microelectronics took over the production floor, many jobs were immediately cut; a trendsetter in the use of automation, textile factories were among the first to be able to produce goods and function virtually without workers. By the end of the 1980s, only 200,000 people worked in the German textile industry. Apart from modern production techniques, the introduction of synthetic fibres provided the basis for further technological transformation. The textile industry, historically a labour-intensive enterprise up until the 1960s, had become a modern, capital-intensive, high-tech business.[214]

The chemicals sector, which had been the very model of German science-based industry at the turn of the twentieth century, maintained a good position in global markets after the Second World War, mainly due to the persistence of its competitive strengths: customized quality production, solid research and advanced automation by 'sophisticated process technology that guaranteed low production costs'.[215] Regarding its use of raw materials, changes occurred more gradually. Although the shift to petrochemicals began in the 1950s, the industrial use of coal-based chemistry remained prevalent in Germany until the early 1960s. After that time, innovations in production technology gained traction due to the fact that 'epochal innovations in chemistry', which had shaped the early part of the twentieth century, were no longer to be expected. Therefore, when BASF, the leading chemical firm, 'transferred the principle of innovation to the process of mass production', the fully integrated production complex ensured global competitiveness beyond product innovations.[216]

Thus, the German adaptation of Fordist mass production principles went far beyond the concept of assembly-line manufacturing. Automation at the highest level possible had always been the ultimate goal of mass production; the vision of 'factories without men' was essential to the Fordist period.[217] This vision came true, albeit partially: when not in entire factories, then at least in some departments, for certain products. The food industry was among the early adopters of automation before the advent of the digital age. For instance, the Hanover chocolate factory Sprengel introduced a fully automatic production line for chocolate bars controlled by a punch-card system in 1967 (see Figure 1.2). At that time, the promise of automation reflected the high

Figure 1.2 Automatic control centre at the Sprengel chocolate factory, Hanover, 1967. Original title: 'The brain of the new factory'. Rheinisch-Westfälisches Wirtschaftsarchiv, file 208-F744

political aspirations of the day. Economic growth and social progress were the expected by-products of the new, modern facility when the prime minister of the state of Lower Saxony opened the new plant.[218]

In this political setting, reinforced by the reformist spirit of the new federal government elected in 1969, the 'humanization of working life' became a buzzword. On one hand, the 'humanization of working life' referred to an American personnel management approach of 'human relations'. On the other hand, there remained a strong line of continuity from the German interwar period, again resulting in a hybrid model of American concepts, modified to fit German cultural and social conditions.[219]

The research programme for the 'humanization of working life', established by the Social Democratic and Liberal coalition government in 1974, signified a corporatist compromise between the state, private companies and trade unions. The programme allowed the different participants to follow their own different agendas, ranging from traditional union demands for occupational safety and co-determination, to management claims that rationalization or automation by their very nature would bring about improved working conditions. In this era, even unionists and Social Democratic politicians had hopes that automation would end the monotony of assembly-line work and create more highly skilled jobs.[220] One of the government posters advertising the 'humanization of working life' declared that humans were 'in the limelight', which was signified by the silhouette of a gigantic human head in front of factory machinery. Moreover, automation technology – punch cards, widely used for numerical control machines at that time – was depicted, toy-like, at the human head (see Figure 1.3).

Figure 1.3 Poster promoting 'the humanization of working life: humans in the limelight', *c.* 1976. Archive of Social Democracy / Friedrich Ebert Foundation.

Despite the fact that the 'humanization of working life' was generally seen in a positive light, some managers opposed the research programme. Kurt Pentzlin, a member of the food corporation Bahlsen's executive board as well as an expert on efficiency concepts, denounced the rejection of assembly-line production and experiments with teamwork as machine breaking, or Luddism.[221]

As is generally known, advances in computer technology enabled a quantum leap in automation. Although German physicist Konrad Zuse invented one of the first computers in 1941, the German computer industry never became competitive with the United States after the war: due to restrictions imposed by the Allied Control Council, German microelectronic research was severely limited. Furthermore, aviation research and nuclear physics, both important fields of computer application, were banned in Germany. Capital was limited, as was demand for the machines. Legal restrictions were lifted after the ratification of the Bonn–Paris convention in 1955, which put an end to the Allied occupation of West Germany. Thereafter, a special West German path towards the information society was established: in contrast to the United States, there was almost no private market for computers in West Germany at the time. Therefore, a public agency, the German Research Council (*Deutsche Forschungsgemeinschaft*, or DFG), championed the cause of information technology. By 1960, fully half of West German universities had functioning data centres complete with mainframe computers.[222]

However, the home-grown computer industry got started on a rather small scale. Anticipating the end of legal restrictions, Siemens had already decided to enter the computer market in 1954. A few other companies joined the competition within the next years, but by the end of the 1960s, only two German computer manufacturers survived: AEG and Siemens. As pioneers of the electrical industry in the late nineteenth century, both companies had long been significant to the modernization of German industry. The beginnings of the IT field in Germany were initially dependent upon the transfer of American knowledge, enabled by cooperation with American corporations like Westinghouse and General Electric.[223] West German computer development began in earnest shortly thereafter: first, government-funded programmes enabled AEG and Siemens to develop their own computers to compete with the IBM mainframes, which were ubiquitous in German public administration. Eventually, Siemens increased its market share and established itself as a relevant industry player during the 1970s.[224] Second, midrange computing played an important role in the diffusion of computers in West German business. Midrange computing was a technological derivative of booking machines and resembled electric typewriters, but they were developed into powerful computing machines in the 1970s. Small- and medium-sized companies were the typical clients for these computers, which were cheaper to purchase and enabled a gradual transition to the computer age. The adaptability of midrange computers to the specific needs of these companies proved to be more important than the higher performance of the larger, more expensive mainframes.[225]

In socialist East Germany, the advent of computing differed only slightly from its development in West Germany but took longer to become established. A publicly owned company, Carl Zeiss built the GDR's first computer in 1954. Despite beginning more or less at the same time as in the West, many years passed before mass production of the first data processing equipment began in 1968. During this initial stage, the nascent GDR computing industry conducted frequent exchanges with international computing experts; not only with other specialists from socialist East European countries but also with West German computer scientists. Both German states emulated American role models, in particular corporations such as IBM and Control Data Corporation. In the late 1960s, both Soviet and East German IT experts agreed that socialist countries had to manufacture IBM-compatible mainframes to achieve any significant global market share.[226] Political change in the GDR leadership temporarily put an end to the national microelectronics programme in 1972. Under the administration of Erich Honecker, the new General Secretary of the state party, priorities shifted from high-tech research to mass consumption and expansion of the welfare state. Yet, only five years later, it became obvious that computer technology was an economic necessity. Accordingly, the GDR resurrected its computer research and production, but by then it had fallen far behind international standards. Thereafter, any aspirations that the GDR could become a leading exporter of IBM-compatible mainframes to the Global South remained an unattainable illusion. The historian Dolores Augustine has even declared that the staggering costs to prop up its computer industry 'contributed to the downfall of the GDR'.[227] Microelectronics was an essential issue for the GDR and the Eastern Bloc as a whole, as they desperately attempted to keep up with the economics of the West.[228] Remarkably, nearly a quarter million employees worked in the East German IT business in 1989. However, this impressive number should be taken with a grain of salt: many employees of East German tech companies possessed no high-tech expertise. About 15 per cent of the workers at the Sömmerda office machine plant in the early 1980s did simple manual work, like coil winding.[229]

Despite its humble beginnings, the origins of German computing had roots that preceded the First World War: German public authorities and even some private companies had been using punch cards for administrative purposes since 1910. Their use was modest in scale compared with the United States, explained by the longer tradition of European public administration that operated efficiently even without punch cards.[230] After a first post-war period of usage in West German public administration, military, research and banks, large industrial companies increasingly invested in mainframes from the mid-1960s onwards. A West German labour shortage beginning in the late 1950s provided a further incentive to foster automation and computerization. At the end of the 1970s, all large West German enterprises with more than 500 employees used computer mainframes. Meanwhile, employment dynamics were also changing. As in most Western societies, West Germany began to face the problem of mass unemployment. Moreover, in the late 1970s, other industries began to be affected by digitization. The German watch industry was struggling due to the new East Asian competition of digital clocks; manufacturers of cash registers and office machines were being undercut by ever smaller and cheaper computers. Additionally, the printing industry was disrupted by the triumph of computer typesetting. In 1978, the printers' strikes were the first large-scale public protests against the social effects of digitization in West Germany. The broad success of microprocessors and personal computers during the 1980s provoked comprehensive rationalization and automation in many different industries. The digital age arrived in Germany, which reached a new stage of development through the spread of the internet and the proliferation of mobile communication devices since the mid-1990s.[231] The fears that computers and automation would destroy millions of jobs did not come true. Finally, digital technological change was generally accepted.[232]

Numerical control (NC) signified an essential stage in automation, which began before digitization, but was later refined by microcomputers. It had huge significance for the West German economy, because it was crucial for core industries. Soon after the introduction of the first NC milling machines in the early 1950s, the diffusion of NC gained momentum, although it primarily confined to machine construction, as well as the automobile, electronic and aviation industries. The second generation of computer numerical control (CNC) caught on at the end of the 1970s and featured integrated computers. Thus, skilled workers sometimes programmed the CNC machines directly on the shop floor. In a certain sense, the machines became tools in the hand of skilled workers, rather than a means to make those workers obsolete. In many cases, management followed the path of 'weak automation' and actively sought to design ways that would involve their workers in machine programming. This was rooted in the German tradition of integrating the 'humanization of work' in its early rationalization efforts. There were even companies that took a personnel management approach to CNC work, a concept derived from the interwar period in Germany: skilled workers inspected the quality of their own work.[233] Furthermore, the tradition of vocational training seemed to be compatible with the age of automation. It is telling that the pioneering vocational training centres for CNC operation and programming were located in the German southwest,[234] where the institution of modern vocational training and technical education had flourished in the nineteenth century.

Broadly speaking, computers had taken root in West German industry by the early 1980s, but it took some time for the most advanced automation technologies to catch on. In 1983, more than 80 per cent of companies in the metal industry used visual display units while two-thirds of the companies had NC or CNC machines. At the same time, only 12 per cent of these companies had robots. The use of industrial robots in West German industry roughly doubled every two years, beginning with a modest 133 in 1974 up to 6,600 in 1984, which still was far less than earlier optimistic expectations.[235] Volkswagen drew public attention in 1983 when it announced the completion of a fully automated final assembly line operating with robots. Initially, this factory hall was celebrated as a technological marvel, but other problems soon arose: in hindsight, Volkswagen's own managers were critical of this innovation, believing that they had chosen a path of automation that had carried Taylorism too far. As the aspirations of automation without men had failed, workers re-entered the factory hall in larger numbers. The 'human factor' became an omnipresent catchword in West German automation debates from the 1980s onwards. Obviously, technology could not completely replace the need for car workers – at least not for complicated tasks, and not by that time.[236] West German automobile management discovered the 'humanization of work', a previous demand of the unions, during the technological changes of the 1980s and Taylorist management practices suddenly had a limited appeal due to increasing automation, which required integrated jobs and workers' responsibility. Instead, teamwork, reduced hierarchies and qualification programmes became the order of the day.[237]

As mentioned earlier, due to the interruption of its microelectronics programme, the socialist East German industry lost ground developing state-of-the-art technology. The lack thereof meant that productivity suffered. Computers were introduced to the GDR's industry relatively late and only in moderate numbers, when compared to West Germany. Computers first appeared in GDR factories during the mid-1970s, numbering just a few thousand. Next came the first industrial robots, which were mainly used for welding: by early 1979, about fifty robots were present in East German plants. All of them were imported, mostly from socialist countries, but twenty-two came from the capitalist West. In the next year, GDR robot production started with an output of nearly 2,198 units, clearly short of the stated goal of 7,000 units. In the 1980s, GDR politicians believed the so-called key technologies – microelectronics, fibre optics, computer-aided design and manufacturing and robotics – were all means to solve its economic problems. However, the politicians harboured these high expectations without realizing the qualitative dimension of digitization.[238] Eventually the whole programme turned out to be a very costly failure.

By contrast, West German production technology was well advanced. Nonetheless, the core West German industries – automobile, machine-tool and electrical industries – struggled during the 1980s and lost ground to international competition. In a certain sense, the 1980s were a lost decade for the leading German corporations in these key sectors, because they were unable to keep pace with global developments in automation, information and communication technologies, which led to a significant productivity gap.[239] Although there were important differences between American and West German industry at the end of the Cold War, West German industry was Americanized to a much larger extent than in Britain or France.[240] The economic historian Werner Plumpe has even taken the controversial view that any German industrial particularities had vanished by the 1970s, because a globalized economy tended to converge different models of corporations towards a common free-market model. Plumpe does concede, however, that German small- and medium-sized companies have maintained their traditional model of corporate governance, which is significant for diversified quality production.[241]

From a macroeconomic perspective, there is good evidence that Plumpe's theory is right but that he errs with regard to production technology; German industry still maintained some traditional peculiarities. First and foremost, Germany has not yet become a post-industrial society. Even though the industrial labour force has decreased since 1973, there has been no major decline in German industry. Admittedly, the growth sectors had a high affinity with new technologies, and did not generate many more jobs. Nonetheless, the growing service sector was partially intertwined with production industry.[242] As such, in comparison with other Western countries, 'Germany has remained a dominant industrial economy.'[243] The German model of organized private enterprise and diversified quality production did not vanish after German reunification in the 1990s; rather, the German model became more flexible and integrated more parts of the Anglo-Saxon model of liberal market economies.[244] This flexibility constitutes the specific strength of German industry to combine mass production and customized production, while also providing producer services, which have been strongly in demand since the end of the cold war.[245] The self-image of quality production has even become something like a second nature to German industry. As has been shown earlier, the British Merchandise Marks Act of 1887 had labelled German products of mostly minor quality as 'made in Germany'. A hundred and twenty-one years later, the Association of German Machine Builders complained about Chinese low-price competition and pirated knock-offs of its own products. These German industrialists ignore the fact that their nineteenth-century predecessors too had 'relied on counterfeiting strategies to catch up to their British and American competitors'.[246]

Some highly innovative sectors were nonetheless lost. In the 1990s, the pioneering firms of telegraphy – Siemens and AEG – left the communication market: AEG ceased to exist in 1996, and Siemens lost the mobile phone, semiconductor and computer markets, despite managing to hold on to significant market shares in other sectors and maintaining its position as a global player nonetheless.[247] With regard to its implementation of the newest production technologies, German industry is again among the global leaders. In 2017, more than 21,000 industrial robots were sold in Germany, making the country the fifth-largest robotics market in the world, behind only China, Japan, the Republic of Korea and the United States. Regarding robot density, German industry is even third after Korea and Singapore, having 322 units per 10,000 employees in its manufacturing industries. For all countries, the highest level of automation is in the automobile industry. In 2017, Germany had 1,162 robots per 10,000 automobile workers.[248] The imagined next stage of automation, however, towards the vision of 'factories without workers' is still far away. The adaptation of the latest digital technology, namely cyber-physical systems, has been rather slow and gradual: in 2016, less than half of German manufacturing enterprises had seriously dealt with the possibility of using digital manufacturing technologies. Another 37 per cent of the companies digitized their production, but only to a limited degree. However, digital technologies are of vital importance in certain industries, for instance, in machine and vehicle construction, but also in chemistry and pharmacy.[249] Thus, there is much talk about the 'third industrial revolution' and 'cyber-physical systems', but this specific technological revolution has yet to reach the shop floor in more than a very limited degree.

2

Urban technologies

The iconic technologies of the industrial revolution paved the way for the rapid urbanization of the nineteenth and early twentieth centuries. The majority of early factories had still relied on natural power resources, above all water power, and had thus usually been situated in rural areas close to rivers. By contrast, steam engines, after their breakthrough in the mid-nineteenth century, provided the means of power for the establishment of urban factories. Furthermore, steam engines drove locomotives, resulting in a fast-rising railway network, which in turn supplied the growing industrial centres with food, coal and raw materials. Within the emerging cities of the nineteenth century, technological networks organized supply and disposal, which were no longer handled individually. Instead, these processes now took place collectively in a mechanized manner better suited to the challenges of the industrial age. The most prominent examples of these efforts were the central supply of energy and water as well as sewage disposal. These processes of urban transformation into networked cities are not identical to the processes of industrialization, yet both developments were deeply intertwined. To a large degree, however, industrialization had caused the very problems of overcrowding, hygiene and transport that were later answered by urban networks of technological infrastructure. This chapter discusses the resulting ambivalence of how urban technology networks solved some problems of the ever-growing cities, while also creating new ones.[1] In general, an unspoken preference for a 'technological fix' dominated the debates on urban development: throughout the nineteenth and twentieth centuries, politicians and engineers believed in the ability to solve technologically caused problems with the establishment of new technologies.

In the early nineteenth century, the German towns occupied by Napoleon's armies still resembled medieval towns.[2] The German railway after 1840 proved to be essential for urbanization: railway construction intensified the process of industrialization, and the heavy industries of the Ruhr and the Saar districts as well as Upper Silesia profited because raw materials were needed virtually everywhere. Although urbanization on a large scale began only after 1850, the towns in those districts grew rapidly from 1840 onwards. Early urbanization also occurred in the centres of the textile industry (Elberfeld, Barmen, Krefeld, Mönchen-Gladbach and Plauen); the old centres of the iron industry (Hagen, Iserlohn, Lüdenscheid and Solingen) and some traditional commercial towns such as Cologne and Nuremberg. Hence, the first industrial entrepreneurs laid the material foundations for the rise of modern cities. On the one hand, the railway boom was crucial for urban growth; on the other hand, urbanization pushed railway development because the surrounding agrarian areas alone could no longer supply the food demands of urban centres. Fast-growing railway networks ensured that agrarian products could be delivered to the large cities from faraway rural regions.[3]

Later than Britain, Germany witnessed exceptional urban growth rates in the late nineteenth century. When the German states were unified in 1871, there were only eight large cities of more than 100,000 inhabitants in the whole empire. By 1910, this number increased to forty-eight large cities that made up 21.3 per cent of the German population. In this thirty-nine-year period, German urbanization rates were significantly higher than in other European countries, and Germany was especially hard hit by the social consequences of this wave of urbanization. Middle-class contemporaries associated both the so-called social question, caused by the poor living conditions of the ever-increasing working class, and the resulting menace of a social uprising, with the process of urbanization.[4]

Germany's path to modern urban technologies broadly resembles the history of its industrial beginnings. First, German cities followed English ones, then American examples in the late nineteenth century. From the turn of the twentieth century onwards, German cities themselves became role models for international urban development. Again, similar to the case of science-based industry explored in the last chapter, scientification was key to the German path of urbanization, or so both German and international observers thought at that time. While impulses from English reformers were important for the urban development of British and continental cities in the nineteenth century, Germany gave birth to the academic discipline of urban planning at the turn of the twentieth century.[5] In particular, the German cities appropriated the British model of publicly owned suppliers, renowned as 'municipal socialism', which in its German reincarnation was, however, more comprehensive and became an international role model itself.[6] In its fully developed form, municipal socialism pushed certain technologies and hindered others. However, rationally planned urbanization is not to be mistaken for the whole of German urban history. The rapid growth of Ruhr villages such as Bochum and Essen in the mid-nineteenth century was often anarchic, missing any hint of urban planning. Factories, traffic facilities, supply lines and spoil heaps as well as whole residential districts were mixed in a haphazard way. These chaotic areas provoked an outspoken cultural pessimism and hostility against cities, which lasted well into the twentieth century and had long-lasting effects on political culture.[7]

In general, there was not one singular model of urban growth for German cities. Instead, the multiple paths to modern urbanity depended on local traditions, political decisions and different cultural appropriations of modern technologies. The stories of traditional commercial centres like Hamburg or Frankfurt clearly differed from those of Ruhr cities that rapidly increased from small villages to urban agglomerations within only a few decades. Nevertheless, some developments, such as the electrification of tramways around 1900, characterized German cities in general. Other technological developments distinguished cities by the time of implementation; while some cities were early adopters, other cities only followed some decades later. Additionally, each technology was adapted to specific local conditions. Water supply and sewer systems, for instance, which were broadly discussed from the mid-nineteenth century onwards, had been already been introduced in Hamburg by the 1840s. By contrast, other towns waited until the end of the century until they installed these technologies. Moreover, municipalities, engineers, hygienic experts and businessmen all made local decisions, which resulted in different technological solutions in different regions.

This chapter will explore how German cities grew and changed their appearance, not least due to the arrival of new technologies. The city extension brought with it new centres, which developed around the train stations built on the former outskirts of the old towns.[8] Furthermore, detached houses soon became the minority in the shadow of fast-built tenement blocks, such as the infamous *Mietskasernen* or 'rental barracks' of Berlin.[9] Yet, the story of urbanization also has a part that is not noticed at first glance. Urban technologies made up a 'second city' beneath the city, among them tramway tracks and sewage systems as well as water and gas pipelines.[10] Accordingly, the history of urban technologies must emphasize energy, transport, communication, hygiene, waste and housing. The first period of urban growth was characterized by steam and gas, but earlier stalwarts such as horses also played a crucial role. In the late nineteenth century, the emergence of electricity marked a watershed for urban technologies because it provided the means for the further transformation of cities. Only from the mid-twentieth century onwards, and thus relatively late compared to other countries, did mass motorization take place in Germany. Around the same time, telecommunications reached the masses and transformed everyday life. Both of these technologies had their early successes in cities. At this point it became obvious that the urban technology policy of the past had caused some challenges for contemporary traffic and housing politics.

Challenges of early urbanization

The issue of transport was of utmost importance for urban history. On one hand, efficient urban transport kept the city economically, socially and culturally alive. On the other hand, good traffic connections were vital for each city in economic competition with other cities. Although railways are often thought of as the dominant transport innovation of the nineteenth century, before them, road construction gave an impetus to the old technology of horse and horse-drawn transport. Before Napoleon's army occupied parts of western Germany, there had been hardly any paved roads, with only a few exceptions in the southwest. After the French occupation ended, however, the German states continued the French efforts in road construction from the 1820s onwards. They borrowed the French term *chaussée* and connected Berlin with the Prussian provincial towns in the east. When the German Customs Union was established in 1834, the road infrastructure in northern Germany had already significantly improved.[11] At that time, France remained the leader in road construction, introducing the road roller in 1835 and the steamroller in 1861. As a consequence, asphalt roads became widespread only in the late 1870s, a few decades later than in France.[12] By no means did the railway boom put an end to enhanced road construction. On the contrary, Prussia increased its road infrastructure significantly between 1837 and 1895: from 12,888 kilometres to 83,000 kilometres.[13] As transport rates were ever-growing due to railway and canal development, feeder roads also gained importance.

The driving forces behind railway development were the interests of local factory owners and merchants who sought to improve their city's transport connection. Thus, urban location policy was far more important for the expansion of railways than the vague visions of public figures such as the famous economist Friedrich List, who conceived of the construction of a German railway network as a means to promote national unity.[14] In fact, the first German railways were established instead to connect cities in pairs, while no one was concerned with the installation of a comprehensive network at that time. Nevertheless, sometimes railway hubs paved the way for small villages to grow and to become towns themselves.[15]

The towns of Germany underwent a rapid transition after a long period of continuity. While in the early nineteenth century, most European cities did not differ fundamentally from their appearance in the Middle Ages, this changed rapidly by mid-century: to cope with the rapidly increasing numbers of passers-by and vehicles, new roads were built and existing roads were both paved and broadened, while new technological devices, namely gas lamps, illuminated many roads.[16] Moreover, the mere size of cities, with Berlin as the most prominent example, multiplied. At the beginning of the nineteenth century, any place in Berlin was within walking distance: usually, people lived near work, but even the edge of town was never far away, as the city had a diameter of four kilometres from north to south and three kilometres from east to west. This changed from the 1880s onwards, when new factories were established at the urban fringe and commuting became ever more common. The widespread demand for day and weekend trips to the outskirts gave incentives to the development of public transport.[17]

Yet surfaces had to be improved first. By the mid-nineteenth century, dirt roads were still common in German cities. Bad weather turned these roads into swamps because the existing gutters were insufficient for drainage. Not only did this make urban transport difficult, but the poor urban infrastructure caused severe hygiene problems (discussed further in this chapter).[18] At the same time, the preconditions for improvements were now available, with railway transport providing the cheap shipment of asphalt from rural mines to the cities.[19] After the construction of a central sewage system, Berlin's roads were paved beginning in 1876.

The steam engine, namely the locomotive, did not initially replace the horse, but rather contributed to the expansion of horse transport through shipment of the resources for road building. As the historians Clay McShane and Joel Tarr point out, both the steam engine and the horse – 'a flexible, evolving technology' – were 'crucial to the evolution of the modern city'.[20] Horse-drawn omnibuses operated in Berlin starting in 1846, a quarter of a century later than in London. Thereafter, public transport developed unevenly. In 1864, 393 horse-drawn omnibuses operated in Berlin, while in 1876, the number had decreased to 373, although urban growth had continued rapidly. The private omnibus business essentially depended on economics: and the boom of the early 1860s was followed by severe economic problems due to high competition a few years later.[21]

The horse-drawn omnibus was in many ways a 'transit technology', a technology that resulted in the transit to modern public transport. While it was virtually just a modified stagecoach, the horse-bus already had predetermined routes and fixed schedules.[22] However, after the late 1860s a competing means of transport was much more successful: private enterprises introduced horse-drawn cars pulled over rails. The first of these tramlines were established in Berlin, Hamburg and Stuttgart. By 1880, horse-drawn trams, which had been introduced in New Orleans in 1835 and made their way to Europe only in 1853 (Paris), operated in more than thirty German cities. They fulfilled more requirements of modern urban life than the horse-buses: they were both faster – with a maximum speed of eight kph – and cheaper.[23] However, prices were still too high for workers in this early period of horse-drawn trams. Despite this problem, horse-drawn trams already initiated a certain process of suburbanization: some middle-class dwellers moved to the outskirts at that time, because the urban fringe had become newly accessible. Thus, between 1865 and 1890, hence before tramway electrification started, horse-drawn trams were the prevalent means of public transport in German cities. The tramways connected the main thoroughfares, namely train stations, popular destinations and exclusive residential areas with the city centre. Even after electric tramways had replaced horsecars at the turn of the twentieth century, horse-drawn buses persisted in Berlin until the First World War. As late as 1914, most Berlin passengers preferred the horse-drawn variant to the new motorbuses.[24]

Often neglected, this period of urban modernization meant at the same time an increase of horses in the city: in 1881, there was one horse per thirty-seven citizens of Berlin (the horse figures were even higher in New York and Paris, with 1:26 and 1:31, respectively). Around 1890, horse-drawn trams operated in sixty-two German cities with passengers totalling 353 million per year. Unfortunately, the high amount of horse manure threatened to worsen urban hygiene, which had just been enhanced (discussed later).[25]

By and large, McShane's and Tarr's assertion that the horse-drawn tram had been 'a path-breaking technology' is correct insofar as urban space and even the city's outskirts were now accessible by this new means of public transport. The horse trams introduced a new kind of semi-public space, for which a new regime of considerate 'transport behaviour' was required. Announcements in the cars' interiors asked passengers for polite behaviour. As a matter of fact, however, the horse tram was still relatively slow, and access was mostly limited to the middle class due to the relatively high fares.[26] Strictly speaking, it was not yet a means of modern mass transport. Nonetheless, the transformation process towards modern transportation started with the horse-drawn trams. Even before electrification, technological changes had improved the trams. On one hand, harnesses and vehicles as well as road surfaces were enhanced; on the other hand, the horse itself had to a certain extent 'become a machine' with enhanced features due to 'better breeding and feeding'.[27] The novelty of the horse tram caused some protests in German cities. In the early years before 1880, coalitions between citizens, the city council and police delayed the development of tramlines in the inner cities of Dresden and Munich. Nevertheless, the trams were soon widely accepted.[28]

The emerging urban transport was both a reaction to growing numbers of city dwellers and a further reason for the growing attractiveness of cities. The case of Berlin was most spectacular. The city increased from 170,000 inhabitants in 1800 to 420,000 in 1850 and even more rapidly to 1.9 million in 1900. By then, Berlin had become the third-largest city in Europe.[29] Both its density and size grew as former suburbs and villages became integrated into the city of Berlin. On a larger scale, mass housing began in Berlin and other German cities in the latter half of the nineteenth century. Already by the 1820s, however, the very first tenement construction projects were realized at the outskirts of cities such as Berlin or Elberfeld (which today belongs to the city of Wuppertal). In Berlin too at that time, five multistorey buildings were built that housed up to 2,500 people in 400 rooms, which were sometimes even furnished with a weaving loom.[30]

However, these were exceptions, and detached houses still dominated German cities. This changed drastically during the fifty-year period following a construction boom from the mid-nineteenth century onwards. In 1910, Bremen was the last refuge of an urban housing style similar to Dutch, Belgian or British cities, where detached houses were prevalent. By then, tenements were omnipresent in all other German cities.[31] House building had changed since the mid-nineteenth century. According to the trend of urbanization, tenements proliferated in the growing cities. Notwithstanding the aforementioned exceptions, the third upper floor was literally unknown by the 1840s. Yet many urban dwellers lived on the third floor at the end of the century. The new tenements aroused criticism right from the beginning, with the high population density sparking both hygienic and moral concerns.[32] As early as the 1860s, reformers criticized the conditions of the Berlin working-class districts and clamoured for housing with more light, better air and heating.[33] In this context, contemporaries attributed the pejorative term *Mietskaserne* ('rental barracks') to those tenements.[34]

These buildings enjoyed a bad reputation well into the twentieth century, but how adequate was this label? First of all, the main tenements facing the streets were usually decent, while only the tenements in the back buildings housed the poor in rather dark and unhealthy rooms.[35] In the initial years, tenement building was largely unrestricted and in large German cities such as Berlin, Hamburg, Danzig, Breslau, Dresden and Munich, tenement blocks with up to six floors were constructed. By the 1870s, most cities had adopted regulations restricting house construction to four floors. As early as 1862, Berlin's development plan organized the construction of tenements in the new city districts. Like many European metropolises, Berlin took Georges-Eugène Haussmann's reform of Paris as an example.[36] Yet by 1910, statistical evidence proved that Berlin had by far the highest average number of residents per building in the Western world. However, this rate did not signify an especially high population density. Rather, the Berlin 'rental barracks' were simply very large buildings. The urban historian Brian Ladd asserts that these tenements were even of a 'fairly good quality', because they were 'solidly built and featured large rooms and high ceilings'.[37] (This feature explains the surprising comeback of these tenements after the German reunification in 1990, when the refurbished tenements were in high demand.[38])

The officials responsible for Berlin's urban development were well aware of planning theory and policy. Berlin urban planner James Hobrecht, who later became city mayor, was a strong proponent of tenements, and opposed the British model of detached houses, as he was convinced that this would establish social separation and result in districts inhabited and controlled by the 'dangerous classes'. In contrast to single-family houses, tenements had been a means to foster social exchange between the classes. Thus, Hobrecht envisioned the tenement barracks as a technology of social inclusion, or a melting pot. However, Hobrecht held a minority position among German officials and faced strong opposition by the many critics of the rental barracks.[39] All in all, however, it was the poor technological amenities of tenements in German cities that justified their bad reputation. Usually, they did not even have a separate bathroom.

Life in the growing cities was certainly not healthy, especially in working-class dwellings. Further industrial expansion and increasing use of coal for heating resulted in 'a general deterioration of the urban environment' in cities such as Hamburg between the late 1850s and 1890.[40] In 1871, the German urban mortality rate was significantly higher than in rural districts. This changed, however, by 1901 when the rate decreased in both areas, but more rapidly in cities thanks to the urban progress in hygiene. The mortality rate decreased in German cities from forty per thousand in 1870 to twenty-four per thousand in 1900.[41] However, while urban infrastructure improved, this was not the case everywhere and for everyone equally. Access to a hygienic standard of living was dependent upon one's class. Furthermore, large city dwellers benefitted from the higher pace of modernization in those places compared to medium-sized towns. The urban technologies of gas, water supply and sewage disposal had a broad impact only incrementally, finally breaking through in Germany at the end of the nineteenth century. Large cities established the new amenities faster, in particular in the middle-class quarters of towns.[42]

Railways and gas lighting are probably the best examples of those urban technologies that derived from industry. Like the railways, gas lighting originated in industry and depended on coal. Coal gas, which illuminated gas lamps that were initially used for factory illumination, was a by-product of coke and its use for industrial lighting was an obvious option, taken from English plants beginning in 1805.[43] Public gas supply copied the concept of in-house water supply, which existed in London from the early eighteenth century onwards. When the gas industry took off in the early nineteenth century, many English middle-class houses had already established an in-house central water supply with pipes in many rooms. This was still unknown to Germans, however, and an English treatise on the operating principles of gas supply that used the analogy of water supply common to English readers had an explanation of this new technology added to the German translation of 1815.[44]

Compared to its British counterpart at the time, the German gas industry was backward. Consequently, it was an English company, the Imperial Continental Gas-Association, that established the first gas works in German cities, beginning with Hanover and Berlin in the mid-1820s. At that time, the most prevalent usage of gas was for street lamps. Soon thereafter, a few cities such as Aachen, Cologne, Dresden and Leipzig followed those gas pioneers. However, the actual founding boom of gas works took place as late as the 1850s when the number of German gas works multiplied to about 200. The reliance on English expertise continued, however, and well into the second half of the nineteenth century the German gas industry was still dependent upon English technicians. Yet being a latecomer had its advantages. German municipal administrations learned their lessons from Britain's example, where too much competition between gas companies slowed down overall development. Thus, each German municipality gave a licence to only one gas firm. This policy resulted in a reasonably successful catch-up process. By the early 1860s, any German town with more than 20,000 inhabitants offered a gas supply to its residents.[45] Despite this, Germany's total gas consumption was only half of London's, while 40 per cent of coal used for German gas production was still imported from England.[46]

In this period, so-called municipal socialism began to establish itself in Germany. From the 1860s forward, German municipalities opened new gas works or took over private businesses. By 1862, municipalities owned a quarter of all gas works. This number increased to two-thirds in 1908, three-quarters in 1920 and four-fifths in 1930, while the remainder comprised a few gas works owned by private companies.[47] Due to municipalization, the municipal services hired engineers and thereby gained expertise on these grounds.[48] Usually, the process of municipal takeover took the following course: most private gas works began operation with a municipal licence for twenty-five or thirty years in mid-century. These gas works were then municipalized after the licence period ended, when the municipality modernized the gas works and pipelines and cut prices. Roughly the same thing happened to waterworks at the same time.[49]

While gas works proliferated during the second half of the nineteenth century, the appeal of gas finally declined at the turn of the twentieth century, not least due to the new – clean – competitor: electricity. While gas lighting enjoyed a reputation for cleanliness in the early nineteenth century, by the end of that century it was classified as unhygienic and dirty.[50] Until the end of the nineteenth century, gas lighting was a rather simple technology comparable to a Bunsen burner: coal gas downgraded the air quality and dirtied the homes. Gas lightings' distinctive disadvantage was high oxygen consumption, making it rather inadequate for indoor lighting. Moreover, noisy and smelly city gas works caused environmental problems with their emissions.[51] In addition, even at the turn of the twentieth century, most workers could still not afford gas supply. Thus, the old – and by no means cleaner – technologies of petroleum lamps as well as coal or wood for heating and cooking persisted.[52]

In many ways, the history of water supply had close ties to the history of gas. As mentioned earlier, public gas supply borrowed its main principle from earlier forms of in-house water pipelines. Three decades later, the balance had changed. Now, central water suppliers benefitted from the advances in pipeline construction that the prospering gas industry had made in the meantime. Accordingly, water and gas experts cooperated closely and established the 'German Technical and Scientific Association of Gas and Water' in 1859.[53] However, central water supply was roughly two or three decades behind its gas counterpart. Most German cities built waterworks in the 1870s and 1880s, while only Hamburg, Frankfurt and Leipzig had established central water supply before 1870. It took until 1907 before any large German city maintained waterworks in the same way as 93 per cent of the medium-sized towns with 20,000 to 100,000 residents had. However, only 57 per cent of small towns had an established central water supply. At that time, almost every second German still lived in villages. As such, as late as the turn of the twentieth century, wells were still the prevalent means of water supply. City dwellers, however, mostly used a central water supply, but there were strong regional differences. While most western industrial regions had central systems, the East Prussian rural areas lagged behind and were still dependent upon traditional wells.[54]

In the early nineteenth century, this sound infrastructure of wells was decisive for the rather slow establishment of central waterworks. In Berlin, for instance, wells met the requirements of urban water supply until the mid-nineteenth century. From 1800 to the 1850s, the number of wells increased from 5,500 to 9,900. However, the per capita ratio declined in the rapidly growing metropolis: while there was one well per thirty citizens in 1800, by 1856 the ratio decreased to 1:45.[55] Before central water supply systems were established, urban dwellers in cities such as Hamburg had taken their drinking water either from wells, rivers and canals or bought it from water carriers. From the 1830s onwards, several private water companies piped unfiltered river water into the houses of their Hamburg middle-class customers.[56] Similar companies operated in several German cities before central systems were established; the first German steam engine–powered water supplier operated in Magdeburg from 1819 onwards.[57]

Central systems did not initially enhance the water quality compared to well water. Hamburg, which had been the first German city to install a central water supply in the late 1840s, fed unfiltered river water into the system. In this period of transition to central supply, hygienic considerations were marginal. Although filter systems were available, they were usually rejected due to high costs,[58] despite the fact that London's positive experience of water filtration, which was originally invented in the Scottish town of Paisley in 1804, was well known in Germany. The English-owned Berlin Water Company, which had established the Berlin water supply in 1852, did use sand filtration and the English engineer William Lindley, responsible for the Hamburg water supply and sewage system, also strongly recommended this filtration technology. The demand was rejected, however, on the grounds of cost, and the municipal authorities cancelled the consulting agreement with Lindley. The tragic outcome of this decision is well known: in 1892, the last European cholera epidemic killed thousands of Hamburg's citizens while far fewer people died in the neighbouring town of Altona, which had been using sand filtration since 1859. Although the Hamburg City Assembly had finally decided to start construction of the filtration plant in 1890, it was far from completion when the epidemic broke out.[59]

As the Hamburg reactions to sanitary needs show, lack of scientific insight cannot explain the belated development of water filtration. In this case, the classical narrative of steady scientific and technological progress is insufficient. The experts of urban development as well as the City Assembly members were wholly aware of the fact that the filtration plant was a necessity. The challenge they faced, however, was enormous, as an ever-increasing wave of urban growth had led to a cost explosion for the projected facility. Once initial funding for a filtration plant had been ensured, this amount was already insufficient, and much higher rates of water supply had already become an urgent need.[60] In some ways, there is a certain parallel between this situation in late-nineteenth-century urban hygiene and today's challenge of global warming: there was consensus that change was essential, but due to high costs, vested interests slowed down technological change.

At first glance, central water supply seems to have been just another example for technology transfer from Britain to Germany in the nineteenth century. In many ways, this assumption is correct. Given the lack of German technological know-how, British engineers and companies were often hired to install the first water supply systems.[61] In general, the British concept of sanitary reform seems to have been the role model for its subsequent development in German urban areas. Soon after the London cholera epidemic of 1831, which caused 50,000 deaths, the British public health movement had established itself. In particular, the British Poor Law Commission undersecretary Edwin Chadwick criticized the sanitary conditions of the urban working class. His sanitary report inspired the British Public Health Act of 1848. Chadwick's concept of sanitary reform placed the social question of the poor living conditions of the working class in a technocratic context. Chadwick and his followers conceived of urban engineering as a means of solving most social problems via water supply and disposal.[62] It was therefore not accidental that a Chadwick disciple – William Lindley – built the first German water supply system in Hamburg. Afterwards, Lindley advised several German cities on issues of water supply and sewage systems.[63]

However, a closer look reveals local reasons for installing the new technologies that somewhat differ from the English example. In most German cities, it was not primarily fear of epidemics or the poor living conditions of the working class that animated urban administrators to engage in water supply and sewage systems. Rather, Hamburg's engagement with central water supply was due to the desire for fire protection. In the aftermath of a major fire in 1842, which destroyed the houses of 20,000 people or 10 per cent of the city's inhabitants, Hamburg was among the very first European cities to follow the British example of sanitary infrastructure. Lindley, who was coincidently living in Hamburg at that time for the construction of a railroad line, was put in charge of reconstructing and modernizing urban infrastructure. He consulted with the chief engineer of the New River Waterworks in London and argued for the establishment of a 'state-controlled, central water-supply system' in Hamburg.[64]

By contrast, Berlin's main motive for the establishment of a central water supply was city cleansing. Its urban gutters were infamous for dirt and odour. Thus, cleaning the gutters was the main driver for the installation of waterworks, rather than the supply of drinking water, for which the wells still worked sufficiently. The city sought a central water supply as a means for flushing the gutters into the river and getting rid of the sewage. Municipal authorities also wanted Berlin to become a modern metropolis, and a central water supply was an essential signifier of this status, with waterworks representing technological progress at that time. Plans for this new technology, which was finally installed in 1852, had been made since the 1830s. While the issue of drinking water supply gained influence within the following years, flushing the gutters and fighting the odour remained the key argument for the very establishment of waterworks. Crucially, the new water system did not actually solve this problem, because piped water laid the ground for water toilets, which enormously increased the amount of sewage flushed into the gutters.[65]

As such, it was not primarily a fear of epidemics that animated German urban administrators to engage in water supply and sewage systems. At that time, medical experts had only a limited impact on concrete decisions about urban hygiene. Instead, it was a new sensibility towards annoying odours in the growing cities of the early nineteenth century that paved the way for the hygiene reformers. Middle-class pressure groups of mostly homeowners and businessmen made a huge impact on the construction of water and sewage systems. Only the city of Danzig was an exception. There it was clearly the very high mortality rate that motivated urban hygiene initiatives. In all cases, however, engineers became important players only after the general decision was approved, and when experts for concrete construction were needed. Thus, while urban hygiene in England was the product of administrative reform, the situation in Germany differed, where the urban middle class was the driving force for water supply and sewage disposal. Medical experts' impact on urban hygiene must not be overestimated. In fact, medical evidence was not needed, because in popular opinion dirt and illnesses were already linked. Medical theories were listened to only if they corresponded to existing middle-class experiences and values. Only in the late nineteenth century, after germ theory had been established, did medical experts gain ground in the practical debates of urban development.[66]

Although central water supply had become increasingly accessible by that time, severe hygienic problems remained unsolved. As late as 1880, less than 5 per cent of the dwellings in most German cities included a private bathroom. By the end of the century, public bathhouses had become common in almost every city.[67] Part of the problem was that the private water companies, before being municipalized after an initial period of twenty-five years, had rather high prices, which restricted general accessibility. The Berlin private waterworks company, for example, privileged wealthy districts for profit reasons because the water price depended on the average rent. Correspondingly, working-class districts were not connected initially. Only after municipalization from the 1880s onwards, and the establishment of a fixed water price dependent on the amount of water consumption, were these areas serviced.[68] Unsurprisingly given their concentration on well-heeled parts of the city, waterworks usually made good profits. During the first years after establishment, the enterprises suffered expected losses from their initial costs, but after a few years with an increasing customer base, the waterworks paid off very well. The water systems then became widespread, and this proliferation of the new technological system of water supply transformed the urban scenery: pump stations and water towers were among the new technologies that accompanied its establishment.[69] The water quality significantly improved after the establishment of water deferrization, which provided the means of iron removal for central systems based on ground water wells after 1890.[70] At that time, Hamburg was the only major German city that exclusively depended on river water for central supply. Additionally, about 30 per cent of the middle-sized Prussian cities with more than 15,000 inhabitants still depended on rivers for drinking water in 1903.[71] On the eve of the First World War, water supply and disposal were established in every major German city. Water consumption reached a level between 110 and 229 litres per capita daily. This number is remarkably high given that the average German consumed 122 litres of water daily in 2012.[72]

At base, the politics of urban hygiene were a matter of class politics. On one hand, the new technologies provided sanitary ways of living even to the urban poor. On the other hand, those very technologies were instrumental to middle-class policies of social engineering. Thus, historical accounts emphasizing notions of technological progress overlook a crucial aspect of sanitary history. The technocratic middle-class reformers of the nineteenth century made their approach explicit: by offering the working class access to water supply and sewage systems, they sought to implement not only hygienic behaviour but also middle-class values in general. Thus, the struggle for hygiene allowed the municipal authorities a 'benevolent siege' of the working class.[73]

From a technical point of view, water supply and disposal are intertwined. Central water supply was a necessary prerequisite for sewer systems, which were technically not compatible with the traditional well-based water supply because the sewer systems needed high quantities of water and lowered the water table. However, the municipal authorities usually prioritized water supply. Thus, most German cities had already established water pipelines when sewer systems were introduced several years afterwards. After central water supply was widely established, the high quantity of wastewater clearly called for an accompanying sewer system. From the 1870s onwards, bathrooms and toilets became more common in tenements, although rather slowly. As a result, the establishment of sewage systems started to become an urgent need. Most German cities installed sewage systems around the turn of the twentieth century. Some cities had established sewage systems before 1870, but Frankfurt, for instance, had to modernize its first system after only two decades of usage. By 1907, all large German cities had developed comprehensive waterborne sewage systems of the combined sewer type, which carried both waste and storm water. However, small- and medium-sized cities lagged behind. While more than 1,800 Prussian municipalities had established central water supply by 1900, less than 200 of them operated sewer systems.[74]

Beyond sheer size, another factor essential to sewerage development was whether a city could rely on an established bureaucratic structure to build and maintain them. Most of the large German cities in the mid-nineteenth century – namely Hamburg, Munich, Dresden, Leipzig and Cologne – were characterized by a diversified economic structure. These cities employed established urban administration, which facilitated the establishment of urban hygienic infrastructures in the second half of the nineteenth century. By contrast, Gelsenkirchen was one of the new industrial Ruhr cities that did not have these traditional structures. Such cities suffered a delay in the establishment of hygienic infrastructures as late as the eve of the First World War.[75]

By the mid-nineteenth century, before sewer systems became widespread, the urban practice of street drainage still resembled medieval practices. This traditional infrastructure was 'hopelessly overburdened', according to Ladd. Open gutters disposed storm water from the streets, often polluted by excrements and household waste, into 'a few underground sewers', which were usually in a poor state. In other cases, the streets were drained through open ditches. In both cases, the wastewater poured either into the groundwater or into rivers and canals, where additional excrement was thrown indirectly. This traditional and fragmentary infrastructure was partly dependent upon the unofficial profession of women emptying latrine buckets into the local rivers.[76] Crucially, however, by the 1870s when the first German cities established modern sewer systems, the final destination of sewage disposal had not changed. Most cities still piped waste water unfiltered into local rivers. Only the technological means had changed, with combined sewer systems replacing buckets. The ongoing trend of rapid urbanization further impacted the construction of sewage systems as middle-class residents moved to the new suburbs, which did not even provide the imperfect inner-city system of the old canals. Accordingly, the stench on suburban streets was even worse than in the inner cities. Due to the strong complaints of citizens, such as those in Munich, the municipal authorities installed sewage systems in the suburbs even before a central water system had been established. Nevertheless, the odour worsened, and the problem of sewage disposal remained unsolved.[77]

Once again, technological know-how was transferred from England. In the mid-nineteenth century, most German cities relied on the expertise of British specialists like Lindley or James Gordon, who advised the city of Munich.[78] After 1860, several German cities sent commissioners on trips to England to study drainage systems.[79] The lack of technological know-how was a severe problem for sewage planning until the 1870s and state-of-the-art technology was usually adopted with a time lag. As late as around 1900, German cities applied English innovations in sewage technology, which had been established there by the 1870s.[80] There was an absence of German sewage experts throughout the nineteenth century, as neither university- nor apprenticeship-based education programmes focusing on sewage were yet available. In England and the United States, specialists for public health or sanitary engineering were graduating from colleges by the end of the nineteenth century. By contrast, in Germany civil engineers were the only available experts for urban technologies. It was virtually on the eve of the First World War that German civil engineers learned about public health and closed the gap with their specialized American colleagues, with German cities approaching common American standards.[81]

The issue of sewage disposal was handled ambivalently. During the 1870s, the Prussian administration opposed sewage systems without filtration and thus delayed the establishment of sewage systems in cities like Frankfurt. From the early 1870s onwards, the city of Berlin, which did not have the option of a large local river for draining the sewer, adopted the concept of percolating fields or sewage farms, which had first been applied in England thirty years before. Consequently, the municipality bought agricultural lands for sewage disposal. In this way, as Chadwick had suggested, the sewage was recycled as fertilizer. The tainted soil, however, transferred germs to the farm vegetables. Moreover, in the process of further industrialization, the composition of the sewage changed, and it later contained increasingly harmful substances.[82] In the case of Berlin, the aftermath was realized only 100 years later when the construction of the large settlement Marzahn on the outskirts of socialist East Berlin began: 'the overturned soil revealed the contamination that had been hidden from residents.'[83]

In the short term, however, the Berlin sewage system was successful and became a model for other cities. By the early 1880s, 1.5 million residents were connected to the system.[84] Cities such as Danzig, Breslau, Dortmund, Königsberg, Münster and Darmstadt followed the Berlin example of sewage farms, while most German cities were reluctant to follow suit due to the high costs.[85] Frankfurt was one of the cities that discussed the establishment of sewage farms but rejected the idea for cost reasons. Instead, Frankfurt opted for an alternative technology and established the first sewage treatment plant in Germany in 1887.[86] Critics nonetheless complained about the high costs of this technology too. As late as 1912, the eminent chemist and industrialist Carl

Duisberg denounced sewage treatment as a 'waste of national capital'. In his opinion, it was sufficient to discharge industrial sewage into the rivers where it would get adequate, simple and cheap treatment.[87]

Each city claimed to be adapting to their local conditions. Consequently, there were some exceptions to the dominance of the modern combined sewer. In some cities, sewage systems such as the Heidelberg bin system persisted, which separated wastewater from faeces. This system had the advantage of providing fertilizer for agriculture. The cities of Leipzig and Dresden in Saxony were the only large German cities that had not chosen a comprehensive sewage system by that time. Instead, both cities commissioned a company for fertilizer export to dispose of the sewer waste.[88]

In hindsight, the urban technologies of the nineteenth century discussed so far in this chapter formed a large technological system together. Soil sealing by road pavement set the stage for city cleansing. Combined with central water supply and sewage disposal system, this large technological system replaced natural hydrology.[89] Once established, modern cities were dependent upon these technological environments. It was, however, a twisted road to modern urbanity. At least three aspects were ambivalent. First, not everyone was convinced about the new technologies. For example, the construction of the Berlin sewage system in 1873 had serious opponents who advertised in several newspapers warnings against these 'pestilence pipes'.[90] Second, although science gained more influence over concrete hygienic measures at the end of the nineteenth century, technological change was far more complicated than a somewhat misleading understanding of technology as applied science suggests. In fact, the most important impact German science had on urban hygiene was based on a fundamental error: the eminent hygienic expert Max Pettenkoffer directed Germans' attention to the contaminated soils, which he thought were the cause of epidemics. This proved wrong only in the 1880s when bacteriologists demonstrated that contaminated water caused typhus and cholera. Although Pettenkoffer erred, he gave the right impetus to improve urban hygiene.[91] And third, sewage systems were not exactly a clean solution. The hygienic problems, which the cities thought they had solved by water supply and sewage systems, returned via the river water in the early twentieth century.[92] It took nearly the whole twentieth century until sewage treatment worked sufficiently on a wide scale.

Nonetheless, the technological progress of the late nineteenth century impressed contemporaries. To them, it seemed that industrialization and modern technology, which had produced the problems of the modern city, also held the key to the solution. Urban technologies such as sewage systems, waterworks and supply systems paved the way for functioning urban areas. Experts euphorically praised technological progress, convinced that there were technological fixes for nearly any social problem.[93] In particular, the belief in a technological fix to water pollution problem replaced any preventative measures in the medium term.[94]

This ideology of technological progress was especially vivid in colonial politics. The very urban technologies that Germany had itself adopted from Britain just a few decades before played a crucial role among the 'tools of empire'.[95] The colony of Qingdao is a good example. After colonizing the region in 1898, the German colonial administration of the Chinese town of Qingdao was obsessed with the issue of hygiene and the transfer of German infrastructural amenities to a Chinese town. The new-built city of Qingdao was 'designed so as to demonstrate the German empire's technological superiority, ultimately turning its infrastructure into an export good for the Chinese market'. Thus, a technological network was built that consisted of water supply and sewage disposal as well as gas and electricity supply, accomplished by roads and railroads.[96] However, the city was clearly separated along lines of race and class: the European town was well equipped, whereas the town of the Chinese middle class followed up with a less sophisticated, but still remarkable, level of technological infrastructure. By contrast, the village of the Chinese workers was only connected to central water supply several years after the wealthier parts of town. Furthermore, there were severe natural restrictions against the colonists' vision of a modern city: sewage treatment was hard to realize due to the region's geography.[97]

The age of electricity

Before the first inner-city streets were illuminated by electric light around 1880, telegraphy had already made commercial use of electricity from the mid-nineteenth century onwards. Before that, the semaphore telegraph had been established connecting Berlin with the western German towns of Cologne and Koblenz. This technology had originally been developed in France in the late eighteenth century. Inspired by British improvements to the French system, the Prussian king ordered the establishment of a semaphore system in 1832. The semaphore telegraph accelerated pre-railway communication enormously: instead of several days, messages could be received within a few hours because visual messages could be sent through mechanical devices on semaphore towers. However, even under perfect weather conditions, one signal from Berlin to Koblenz took seven and a half minutes. Thus, in 1848 a telegram of thirty words took one and a half hours.[98]

Nevertheless, the existing infrastructure of the Prussian semaphore system laid the foundations for the change to electrical telegraphy. The Prussian government sought to install an electrical telegraph line in reaction to the 1848 revolution in winter 1848–9. Although the plan initially failed under time pressures, a new protagonist entered the stage: namely Werner Siemens, a Prussian officer, inventor and co-founder of the Siemens & Halske telegraph company in 1847.[99] After Siemens became responsible for the project, the transition to electrical telegraphy was successful in 1852. Simultaneously, the network expanded. Despite this success, it is worth pointing out that Britain's network preceded this by five years.[100]

Indirectly, telegraphy was essential for urbanization because further railroad development, one of the foundations for urban growth, was dependent upon the establishment of communication networks. It was a safety issue: telegraphy provided the means for safe two-way traffic on what were mostly single tracks.[101] In the next three decades, the system expanded and connected virtually every German city to any other. Furthermore, after ocean telegraphy broke through in the 1860s, a 'system of global communication based on electric speed' emerged. The submarine telegraphy of the late nineteenth century was, however, 'by no means a Victorian Internet'. Despite its rapid expansion, telegraphy was still very exclusive and connected only the commercial centres of the world, bypassing the less prosperous rural areas.[102] Nonetheless, usage increased in the latter half of the nineteenth century. While only one telegram was sent per thousand citizens annually in 1850, it was one telegram per every second German in 1890. Even so, old communication technologies were still prevalent. In 1850, two and a half letters were sent per citizen, but the number rose to an impressive thirty-three in 1890.[103]

The Siemens company, however, was there to stay. Siemens soon became a multinational corporation and had a crucial influence on Germany's history of electrification. The electric generator was a simultaneous invention: among others, Werner Siemens formulated the concept of the dynamo in 1867, allowing steam engines to be used to generate strong electric power, which was initially practical only for very bright arc lamps, but not for home illumination. By 1868, electric light was being used for bridge construction work in Munich. Around 1880, grand boulevards such as the famous *Unter den Linden* in Berlin were electrically illuminated. Yet only Thomas Edison's system consisting of electric bulbs, generators and distribution systems, which he presented at the Paris World Exhibition in 1881, showcased the possibilities of what an integrated system could do. His New York electric central station, which was established in the following year, turned electricity into a promising competitor to gas lighting.[104] Edison's importance was based less on a single invention than on his improvements to existing parts and the combination of these technologies into a comprehensive system. Last but not least, he was simply a brilliant propagandist of the new technology.[105] Edison's patent licensing agreements soon spread the new technology around the globe.

In 1884, the first German public power plant opened in Berlin. It was constructed by the Emil Rathenau–founded German Edison Society for Applied Electricity, which would soon become the world-famous *Allgemeine Elektrizitäts-Gesellschaft* (AEG) in 1887. This first power plant was only a small lighting plant providing power for a café and its surrounding block. The first central station opened in 1885. A rather low voltage of 100 volts also limited the supply area to the plant's neighbourhood. Nonetheless, Germany soon became a trendsetter in electricity. The AEG's Berlin electric utility developed a specific technical style based on 'unprecedentedly large generators' powered by enormous steam engines. Already in 1890, the Berlin utility was a model of efficiency for the Edison Illuminating Company of New York. In this initial period of electricity, the customers were mostly high-class shops, restaurants and theatres in the city centre because electricity was much costlier than gas lighting; however, it was safe and signified a modern lifestyle.[106] In general, the replacement of gas lighting by electric light in municipally owned facilities for safety or practical reasons was a common feature of German urban electrification in the 1880s and 1890s. In this period, theatres and granaries too changed to electricity. There was, however, no strong private demand for the new technology yet.[107]

By and large, early electrification was all about luxury illumination. By 1888, the whole city of Berlin had only two electric elevators, a few fans and several sewing machines. At the same time, AEG's bulb production increased. In its first year of production in 1885, the AEG produced 60,000 bulbs; the annual production numbers quintupled within only two years.[108] Apart from the metropolis of Berlin, the wish to establish a modern image also motivated some medium-sized cities to embrace electrification. The city of Darmstadt was an electrical pioneer. Its technical university established the first German chair of electrical engineering in 1882. Soon thereafter, the city established a power station.[109] The early electrification of Darmstadt did not, however, yield a more rapid diffusion of electricity compared to other German cities. In Darmstadt, like other places, the electricity supply was mainly used for luxury illumination.[110] With regard to organization, though, Berlin was a model for most other German cities. The cooperation between the AEG and the Berlin city council resulting in the foundation of the electricity utility *Berliner Elektrizitätswerke* (BEW) was conceived as the road to success.[111] Initially, only private companies dominated the German electricity market, while after the mid-1890s, public utilities were established. On the eve of the First World War, two-thirds of electric utilities were at least partly municipally owned. Similar developments took place in Britain at the same time.[112]

In the early days, direct current was unrivalled. Then, from the late 1880s onwards, two new systems competed with direct current, namely one-phase alternating current and three-phase alternating current. Both newcomers enjoyed a competitive advantage in comparison to direct current in that the power stations could be settled at the outskirts, whereas direct current power stations had to be situated in the city centres where the electricity was consumed. These inner-city power stations caused environmental problems, including an increase in noise, traffic, smoke and vibrations which would not have been tolerable in the long run. Even in this early period, some resistance against inner-city power stations arose as they caused smoke, carbon black and noise. For example, Stuttgart residents complained that the new power station had transformed their middle-class town into an industrial city. Both alternating current and three-phase current, however, had the disadvantage of being non-storable; they had to cover the demand peaks, which caused the need for larger engines. Moreover, transformer stations and long-distance transmission increased the price compared to direct current.[113]

By 1891, the war of the currents seemed to have ended. The Frankfurt Electrical Exhibition of 1891 signified the victory of alternating current over direct current. In contrast to direct current, which was applicable only within an area of a few hundred metres, both alternating and three-phase current had a decisive advantage: as demonstrated at the exhibition, they could be transferred over very large distances. A power station at Lauffen provided three-phase current over a distance of more than 170 kilometres by an overhead power line to Frankfurt. This large hydroelectric power station and the technology of power lines yielded important changes. These power stations were outside of urban areas and near the natural power source (mostly rivers or mining areas). Large power stations also did not supply energy for merely one city, but for whole regions. This centralization of energy, however, also meant the danger of blackouts for whole regions.[114]

As long-distance transmission of electricity was made possible by alternating and three-phase current, the rural hydroelectric or lignite-fired power stations could be used for urban power supply. The first German run-of-river power station of Rheinfelden went online in 1898. It gave an example of the establishment of similar power stations within the next years. Which system was chosen – alternating or three-phase current – depended on the respective needs of a city: some industrial cities preferred high-voltage three-phase current for the needs of their industrial bulk purchasers. By contrast, several trading cities such as Frankfurt chose low-voltage alternating current to prevent the settlement of new factories in the city.[115] Despite its obvious disadvantages, direct current did not disappear immediately. The persistence of direct current is an example of the phenomenon the historian Thomas P. Hughes described by the term 'momentum': 'a conservative force reacting against abrupt changes in the line of development'.[116] Direct current persisted in several cities, including Darmstadt, because it worked well for the main local applications of inner-city lighting and tramways. High costs prevented the change to alternating current. Only the establishment of a new main railway station at the outskirts and the rise of new industrial areas at the periphery made the establishment of a new three-phase alternating current power station necessary. Nevertheless, the old direct current power station supplied the Darmstadt inner city well into the 1930s.[117]

In general, urban electrification was a complicated process, far beyond a simple diffusion of a pre-existing technology. Similar to the earlier establishments of water supply and disposal, the municipalities were dependent upon the expertise of advisors. These experts, mostly engineers or professors of engineering, played an important role in widening the scope of electricity beyond mere illumination at the end of the nineteenth century. By contrast, urban administrations had initially ignored any other applications. Furthermore, each city developed specific technical styles in relation to local conditions with consultants influencing their decisions.[118]

During the last years of the nineteenth century, Germany witnessed a boom in electrification. This was largely due to the establishment of electric tramways. While in the first half of the 1890s, only thirty-five power stations were opened annually, by 1898 the annual figure quadrupled to more than 150. The electricity crisis of 1901–2 was a direct consequence of this supply boom. However, the crisis resulted in the consolidation of the two main players in Germany's electric industry: Siemens and AEG.[119] By 1910, any German town with more than 2,000 residents had established a power station. At that time, usage patterns had shifted: electricity was no longer restricted to luxury illumination. Instead, it had become an important infrastructural feature for towns to recruit industry. Statistics give evidence of this shift. While light current consumption increased in Germany by a factor of 37 between 1891 and 1913, power current rose by the remarkable factor of 676.[120] Originally, the utilities' interest in new customers beyond nightly illumination was raised by the need to sell energy in off-peak hours. From 1888 to the end of the First World War, the Berlin electricity utility BEW offered a special tariff for small engines. This policy was aimed at craftsmen to increase daytime consumption.[121]

The main characteristic of the early history of electric supply is ongoing growth. On the eve of the First World War, most large power stations were situated in the Rhineland, the eastern German lignite fields or southern Germany. These areas continued to be vital for German energy supply until the early twenty-first century. Only the turn towards renewable energies began to change the topography for new areas like the offshore wind farms of the North Sea.[122] However, the electricity supply of the twentieth century was shaped by a concept that had evolved right from the start in the late nineteenth century: scale. The 'idea of increasing the size and scale of electricity systems thus was an inherent part of the electrical engineering philosophy'.[123] Indeed, engineers and corporations conceived of international cooperation as a means to improve efficiency. Among the first international projects was an 800-kilometre-long overhead line between the Austrian alpine hydropower plants of Voralberg and western Germany.[124] As such, the long history of European network plans with manifold protagonists started in the first half of the twentieth century. Liberals from several European countries initiated the first network plans in the 1920s, but the Nazis also continued this planning in the 1930s and during the war. Thereafter, the post-war politics of European electricity integration could partly rely on those European experts who had already managed similar plans in Nazi-occupied countries during the war. The Marshall Plan provided further know-how by recruiting European electricity experts to study trips in the United States in 1949. In the end, national energy sovereignty was combined with 'far-reaching forms of technical cooperation' beyond borders.[125]

In fact, in the early twentieth century, international transfers in different directions had already shaped the development of Europe's electrical supply. Once German cities became role models soon after they had appropriated Edison's technology, German technical universities were among the foremost places in the world to study electrical engineering. In the early twentieth century, scientists and engineers who had studied at German universities made up the core of General Electrics' laboratory.[126] On the eve of the First World War, the Berlin electric utility was one of the largest of its kind worldwide. Thus, urban engineers came from faraway places such as Melbourne, Australia, to study it. According to Hughes, Berlin had developed 'the technical style of a large-scale, capital-intensive, science-based enterprise'. In particular, the close relationship between the Berlin municipality and the corporation was remarkable. This paved the way for municipalization in 1915. Germany was considered a successful example of public–private partnerships; in contrast to Britain, German electric utilities were not hindered by the primacy of political considerations.[127] Instead, German municipalities preferred economic calculation. In the early twentieth century, municipal gas and electricity works were important sources of revenue for most German cities. The same was true for waterworks, tramways and slaughterhouses.[128] The improvement of urban technologies served both hygienic and social ends; additionally, it simply paid for the projects of municipal authorities.

However, apart from these impressive developments, the United States was still the role model for German electrification and private energy consumption. In 1910, when even small German towns had their own power station, only 10 per cent of households were supplied with electricity.[129] Nevertheless, cities' appearances had already been transformed in the late nineteenth century by arc lamps and illuminated advertisements, such as the famous Berlin Manoli tobacco revolving wheel in 1898. Furthermore, transformer stations, electricity pylons, electric clocks, tramways and subways shaped the appearance of modern cities.[130] Nonetheless, for many Germans, the United States signalled the future. In particular, the early introduction of electric tramways in the United States during the 1880s made a strong case for German supporters of the electrification of horsecars.[131]

Without bulk purchasers such as the tramways, there would have been no rapid establishment of power stations in German towns around 1900.[132] Thus, the same companies that owned power stations and produced both generators and electric equipment invested in tramway development: Siemens and AEG. Siemens experimented with electric tramways as early as the 1880s. However, both the tram of Berlin-Lichterfelde in 1881 and the tramline between the neighbouring cities of Frankfurt and Offenbach in 1883 did not stand the test. The Berlin tram, which was powered by a third rail, was electrically unsafe, whereas the Frankfurt tram, which had a supply system of two overhead lines, was prone to failure. Leaving aside these failed experiments, American cities were the real pioneers of electric tramways with twenty-one systems established in 1888. In the same year, the AEG started to purchase German horsecar companies and projected the transition to electric tramways. Three years later, the city of Halle introduced the first effective German tramway line, operated by AEG. The company had meanwhile signed a patent licensing agreement with US tram pioneer Frank Sprague. At the end of 1891, two more German cities had established electric trams, but then the number increased rapidly to a total of 104 towns at the turn of the century. This proliferation of tramways was a general trend in Europe: by 1910, most European cities had established electric trams.[133]

Initially, there was some resistance to the new technology. In the 1890s, the magistrates of Berlin and Hanover opposed trams' overhead wires in representative districts of the inner cities for 'aesthetic concerns'. In those cities, the short-lived technology of accumulator cars was unsuccessfully tried but replaced after a few years. However, the 'clear public demand for electric tramways' overruled those initial complaints.[134] For most contemporaries, the electric tram symbolized that a town had finally become a major city.[135] Between 1890 and 1920, the electric tram became the dominant means of urban transport in Germany, although the alternative of horse-drawn trams persisted in many towns. Bit by bit, municipal authorities took over electric trams, but private firms still prevailed in urban transport. During this period, the demand for leisure traffic strongly increased.[136] Moreover, the trams contributed to a general trend of further electrification. Munich, for instance, introduced electric streetlights in 1894, although electric illumination was far costlier than gas lamps. Yet the increased brightness of electric lights was considered crucial in times of fast and busy tram traffic.[137]

Initially, electric tram fares were quite high. According to a trade union survey of 1901, half of the workers in Frankfurt never took the tramways due to the high costs. Thus, even long walks to work persisted for many in the initial times of modern transport.[138] Yet the municipal authorities reacted and offered discounts for workers from 1903 onwards; other cities soon followed this example. Originally, the concept of special fares for workers derived from England and was first adopted by the railways in Prussia. Initially, this policy was unsuccessful because most German cities lacked suburban connections. Only the establishment of tramways gave impact to the building of light-rail lines to the suburbs.[139] The local decision to introduce worker's fares to tramways was clearly related to the issue of urban class structure. While an industrial city like Mannheim subsidized special fares for workers from the outset, the middle-class city of Darmstadt introduced worker's fares only on the eve of the

First World War. Even then, the Darmstadt tramways largely ignored workers' needs; the tram operator was reluctant to connect working-class quarters and industrial districts.[140]

In general, the new means of transport paved the way for mass commuting, which had become a need for a growing number of workers. Many of those had migrated to the growing industrial cities from rural parts of Germany but lived at the outskirts due to the housing shortage in many city centres. Hence, around the turn of the century, significant parts of the urban population and new arrivals settled on the outskirts, while the city centres gradually transformed into business districts. German contemporaries were aware of this transformation and dubbed it by the anglicism *Citybildung*. Commuting became a mass phenomenon at the end of the nineteenth century, when railway prices fell, and timetables took industrial labourers' working hours into account. For the first time, railways became an option for commuting workers. Still around 1875, many miners of the Saar area commuted only once a week. Daily commuting became an option only around 1900, when commuter trains were established in many German regions. On the other hand, transport often determined working hours. For instance, a Hamburg spinning mill started work when the morning train arrived.[141] In several cases, cooperation between politics and business eased the efforts of commuting. For instance, due to the relocation of an artillery workshop from Berlin to the neighbouring town of Spandau (which was later incorporated), the railway company offered special trains for which the municipal authorities provided subsidies. These special trains corresponded with the working hours of the labourers.[142]

Modern transport thus transformed urban life. On the eve of the First World War, electric tramways had become an essential part of urban everyday life, for both work and leisure. While workers increasingly used the tram on their way to the factories, more and more middle-class city dwellers chose to live in the suburbs. Some exclusive residential areas were established as a consequence of the new tramway connections to the city. Suburban areas were now more easily accessible, but at the same time the once-low land prices rapidly increased. As such, new technologies caused urban segregation in some cases. Both the middle and the working classes made use of the tram for weekend trips to the countryside. For many middle-class women, the tramways opened new possibilities of independent transport.[143] During the First World War, the tram even offered new job opportunities for women; the first female tram driver started to work in Berlin in 1915 (see Figure 2.1). Within a few years the city had changed its appearance. Apart from overhead lines, tram tracks, stations and railway viaducts, public transport had additional side effects such as kiosks, public clocks, traffic signs and policemen. In 1924, automatic traffic lights were installed in Berlin, without precedent in Europe.[144]

The electric trams both accelerated urban life and disciplined the townsfolk, who had to obey new traffic rules. Urban life became more dangerous and trams caused serious accidents with many deaths. The antidote to this issue was tram drivers' excessive warning-bell ringing, which made urban life noisier.[145] The tramway required urban discipline not only from passers-by but also from passengers. They had to board and disembark rapidly and had to cease the horse tram tradition of jumping on or off the wagon.[146]

Figure 2.1 The first female tram driver of Berlin, 1915. Bundesarchiv.

The urban transport revolution of the early twentieth century opened new possibilities of mobility, but at the same time modern transport created new everyday dependencies upon technologies. Thus, urban dwellers were 'ever more vulnerable to technical malfunction'.[147] In 1920, the young Austrian journalist Joseph Roth, who would later become a famous novelist, worked for the Berlin daily newspaper *Neue Berliner Zeitung*. He wrote a feature article about his morning tramway ride from his Berlin flat to the editorial office. By that time, the tram had become so common to the everyday lives of Berlin residents that Roth did not even once explicitly mention the 'tramway'. Instead, he laconically listed the numbers indicating the lines he used. Roth's article also pointed to the undesired effects of the standardized commuting routines. Commuters had become totally dependent on trams being on schedule; otherwise, they spent half an hour waiting for the right line, like Roth did on that day.[148]

By that time, public transport had been diversified. Light rail had already been established in Berlin parallel to the rise of (horse-drawn and electric) tramways from the 1880s.[149] During the next decade, local and regional light rail trains proliferated throughout Germany.[150] In 1902, the first Berlin elevated railway and subway line was opened. Inhabitants of Berlin's middle-class quarters had protested successfully against elevated railways and urged the operator to alter their plans. Despite higher costs and delayed construction, some sections of the line went underground. Protests against public transport disappeared within the next decades, however, as this new metropolitan infrastructure was accepted and became an indispensable part of urban life. In fact, the beginnings of mass consumption and public transport in the city were intertwined. In 1901, the department store Wertheim intervened in the planning of subway lines and succeeded in having a central stop right near the store; obviously, this annoyed the competing retailers. Nevertheless, in 1923, a new Wertheim store was again directly linked to the subway because Wertheim offered cost sharing. Even more spectacular was the opening of the grand department store Karstadt in 1929, which was directly linked to the subway station by escalators.[151]

In 1905, public transport became more comprehensive when motorized omnibuses began to operate.[152] The first motorbus began to operate in the western German district of Siegen as early as 1895. However, due to technical problems the bus line operated for only a short time. The national postal authority established several bus lines before the First World War, but buses did not play a major role in public transport in most German cities until 1919. Only then did the federal state of Baden establish bus networks, which after a decade consisted of a fleet of 300 buses. Also during the 1920s, vehicle constructors began to improve travel comfort by lowering the chassis, introducing pneumatic tyres and enhancing the suspension.[153] By that time, lorries were also established as a means of urban transport. While lorries were no competition to rail transport yet, they were important for short distances in the inner cities. While in 1913 only 8,000 lorries were registered in Germany, the number rose to 155,000 by 1933; however, that number was still significantly lower than in Britain.[154]

Urban traffic had become very busy by the turn of the century. During the first decade of the twentieth century, Potsdamer Platz in Berlin was 'Europe's busiest intersection with up to 18,000 vehicles daily'. This square was surrounded by two train stations and frequented by thirty-five tramlines. At that time, old and new transport technologies still competed (see Figure 2.2).[155] The very busy road Leipziger Straße, which connects to that square, is an example of transit diversity in the mid-1920s. During one peak hour, 599 vehicles and 273 bikes were counted. The observed group of vehicles consisted of 265 cars, 177 trams, 71 omnibuses, 36 lorries and 17 motorcycles as well as still 33 horse carriages.[156]

Urban technologies, most of all transport, also brought the collateral damage of increased urban noise. This started with the clattering of horses' hooves, followed by the ringing of tram bells, electric shocks of power lines and finally the drone of motor vehicles.[157] At the end of the Weimar Republic, urban noises became political: the Nazis made excessive use of the new technologies of motor vehicles and amplified sounds for their election campaigns. In 1932, the Nazi party rented loudspeaker vans by Siemens & Halske for penetrating urban spaces with its slogans.[158]

Between 1920 and 1950, urban transport was further diversified: automobiles, both private cars and public buses, became more popular. Subway systems were established in large cities, where they became the most efficient means of transport. In hindsight, the gradual increase of cars in this period was of utmost importance for the acceptance of the automobile before it took off after the 1950s.[159] Parallel to the slow rise of the automobile, the tram gradually started to lose its dominant position in urban transport. In particular, the tramways' public image was damaged due to the car-friendly Nazi politics of low tram investments during the 1930s, which resulted in outdated and unimproved tram technology. Nevertheless, tramways and bicycles still prevailed in German urban transport prior to 1945.[160]

Figure 2.2 Potsdamer Platz in Berlin, *c.* 1907. Bundesarchiv.

While the issue of hygiene had dominated the urban debate in the nineteenth century, the housing question was the most prominent topic of urban planning in the twentieth century. Both housing quality and quantity were an issue: first, dark homes caused sanitary problems. Additionally, the total stoppage of housing construction during the First World War led to a housing shortage in the 1920s.[161] Thus, even though central water supply and sewage systems had become standard in German cities, severe problems remained in urban housing. For contemporary critics of the living conditions of the urban poor, modern technologies held many promises. For instance, in 1899 the psychiatrist Hans Kurella protested that the working class dwelled in 'houses without clean air, without sunshine, without comfort, without the thousand aids of modern technology'.[162] Thereafter, comprehensive housing reform efforts ranged from construction and urban planning to the floor plan, furniture and technical applications.[163] However, in the early twentieth century, electrification caused further social division; only well-off homes and shops could afford the shift to electricity at that time.[164]

After a period of adopting British urban hygiene technologies in the nineteenth century, German cities became international leaders in urban planning at the turn of the twentieth century. British and American progressives took German cities as a role model for their own reform efforts.[165] This knowledge transfer was mutual and German architects were strongly influenced by the American efficiency movement. On the eve of the First World War, Martin Wagner, who would later become head of the Berlin municipal planning and building control office, adapted the latest American methods in building, namely Frank Bunker Gilbreth's findings based on time and motion studies. Due to this concept of rational construction, Wagner built a residential estate of standardized small apartments in a suburb of the northwest German town of Wilhelmshaven.

Even before the First World War, reformers had developed the concept of comprehensive urban planning. This concept sought a rational subdivision of any space, both on a city and on an apartment scale.[166] In the 1920s, positive references to the American school of scientific management were omnipresent in the writings of the German modern architects.[167] However, standards for housing had been applied in Germany since the early modern era, and to a certain degree the principles of rational urban development preceded architectural modernity.[168] The German reformers were particularly fond of the efficiency movement and standardization because these new American ideas fit well with their own traditions.

Housing reform only came into full effect during the rather short period of 1924 to 1931, when public housing had its heyday. At that time, there were many opportunities for urban reformers to create homes according to the standards of modern architecture.[169] A tax on profits from real estate property, the so-called *Hauszinssteuer*, provided the means for public housing programmes after the currency stabilization in 1924. The programme was stopped in 1932 in the context of the Great Depression, when the conservative government cut the tax and supported a shift to detached houses. The Nazis continued this policy after their 'seizure of power' in 1933.[170] In this regard, the historical turning point was not in 1933 but rather in 1932, when the short period of rationalization euphoria ended due to political change and economic pressure. Yet even during the last years of the reform period, changes had occurred due to the economic crisis and housing standards had decreased rapidly. A Frankfurt suburban housing estate for the unemployed called *Goldstein* was set on a dirt road and supplied neither gas nor water. Moreover, it was not connected to the sewer system. Instead, self-sufficiency became standard again. Thus, the technology-fuelled reform euphoria ended, and housing briefly returned to early modern standards.[171]

The 1920s housing reform is best understood as a direct reaction to the social grievances of the most infamous legacy of nineteenth-century urban growth: the rental barracks. Reformers sought to avoid the structure of high tenement blocks around interior courtyards. They were aiming for a maximum of air and sunlight and designed rows of rather low buildings 'at different angles from the street'. A common feature of the otherwise quite varying local reform projects was modern technology: 'central heating, modern plumbing, built-in kitchens, and a compact floor plan'. The large housing estates were situated on the outskirts, so the projects provided transport and the necessary infrastructure of schools, shops, and other amenities.[172] The Frankfurt housing programme, for example, was part and parcel of the municipal infrastructure policy, which contained the projected autobahn between Hamburg, Frankfurt and Berlin, the electrification of the railway line to Basel, Switzerland, and the construction of a new airport.[173]

The eminent modernist architect Walter Gropius, one of the most popular protagonists of the reform movement, pointed out that the standardization of construction was not meant to standardize individual lives. Quite the contrary, the building components were standardized to enable more people to live unique lives in an adequate housing environment. In the late 1920s, reformers built residential areas in – among other cities – Berlin, Karlsruhe, Stuttgart, Dessau, Frankfurt and Munich. Enhancing housing efficiency was an essential issue, so Gropius applied methods of prefabricated building to his construction project at the Stuttgart-Weißenhof estate.[174] However, only the first steps in the mechanization of construction sites were made due to the large investments needed. Frankfurt also aimed for an expansion in prefabricated building and to this end a municipally owned company produced pumice concrete slabs from 1926 onwards. However, only 832 of the total 15,000 houses were built with the prefabrication construction method.[175] The very idea of industrial series production in housing derived from the United States. The aforementioned Martin Wagner, a social democrat and unionist who had become head of the Berlin municipal planning and building control office, took study trips to the United States in the 1920s. At that time, the United States and the Netherlands had become international leaders of prefabricated building. After returning from his trip, Wagner became one of the most important German spokesmen for this highly efficient building method and promoted the future construction of 'new cities'. Wagner conceived of this new city as the 'perfect machine'.[176]

British reform architecture was another crucial influence on German urban reformers at the time. Wagner's counterpart in Frankfurt, the architect Ernst May, had worked for a while in England. After returning to Germany, he combined the ideas of the English garden city movement with the concepts of German Bauhaus architecture.[177] Within only six years after 1926, Frankfurt municipal authorities had organized the construction of 15,000 flats. Nearly every eleventh family living in Frankfurt moved into a newly constructed building. Floor plans as well as the interior design and furniture were standardized. The new material plywood, which originated in aeroplane construction, was applied to new furniture. The average flat size decreased due to the Frankfurt housing programme and housing became standardized and functional (dining room, bedroom, children's room), while showcase rooms, namely the bourgeois parlours, disappeared.[178]

Housing reformers were dependent on state and municipal support. As such, the actual reform efforts were limited to these famous projects. Most German architecture of this period did not conform to Bauhaus or other modernist architectural dogmas. However, the concept of efficiency won and paved the way for post-war building projects.[179] Although the 1920s reform projects 'did not reshape the existing city . . . they became the model for alternatives to the *Mietskaserne*, at least until the 1970s'.[180]

As mentioned earlier, even before the Nazi seizure of power, the conservative national government put an end to progressive reforms. Initially, the Nazis continued this policy and supported suburban small housing estates.[181] However, the earlier-dismissed modernist housing concepts had a surprising comeback in Germany from the late 1930s onwards, when Nazi housing officials became interested in Swedish housing. Then again, German modernists in the 1920s had strongly influenced the Swedish architects now admired by the Nazis. To a certain degree Weimar modernism was re-imported to Nazi Germany through Sweden.[182] In 1940, the Nazis announced the 'social housing of the Führer' programme, departing from the earlier promotion of private housing. This was a vast programme intended mainly for social housing on a very large scale after the war. Some historians consider Hitler's decree 'the beginning of mass housing in Germany'. On the one hand, this turnaround in 1940 meant a revival for the ideas of some 1920s-era proponents, although the most prominent reformers had been expelled or escaped from the Nazis.[183] The concepts of technological progress, industrial rationalization, Taylorism and Fordism also influenced the Nazi proponents of housing, who considered themselves social engineers or members of a technocratic elite. Once again standardization and efficiency were key concepts of the project, which consisted of six types of flats ranging from 62 to 82 square metres. The components too were standardized. Between 1941 and 1943, roughly 100,000 flats complying with these standards were built. In spite of these impressive numbers, it was still not series construction on a large scale.[184]

The 'thousand aids of modern technology' of Kurella's optimistic vision of future tenements in 1899 would become central to all housing reform projects of the twentieth century. Yet the outcome was quite ambivalent: on the one hand, urban technologies were a means of integration in the ever-growing cities; on the other, they also fostered social separation. While the early twentieth-century working class lived in back buildings, such as the infamous 'rental barracks' of Berlin, the middle class enjoyed the benefits of the electropolis – the metropolis equipped with the benefits of the electrical age. Running water and electric light were soon standards of middle-class apartments, while workers initially could not regularly afford the prices of electric trams.[185] The invention of technological artefacts marks only one aspect of the history of technology. The more important question is when their widespread usage began.

Most now-common electric household appliances were invented by 1911. The AEG listed, for example, the stove, hair dryer, coffee machine, vacuum cleaner, flat iron, food processor, washing machine, sewing machine and blender, among others. In spite of their early invention, these appliances were widespread only in the 1930s. Indeed, most of them spread throughout Germany as late as after the Second World War.[186] Rather slowly, household electrification was just beginning on the eve of the First World War. While most households had been connected to water supply and disposal systems, gas supply reached only half of German households. Electricity supply was even worse: before 1914, only 10 per cent of households consumed electricity. Paraffin lamps and gas lighting still prevailed. Furthermore, the traditional stove had multiple functions in the eat-in kitchen: heating the flat, heating water, cooking and waste incineration. At that time, the high installation costs were the most important reason for the slow proliferation of electricity.[187] To a certain degree, the First World War gave incentives to small-scale applications of electricity because oil was needed for military purposes. Thus, petroleum was hard to get for both craftsmen and private households.[188] Yet a high degree of social inequality initially hindered a general appropriation of electricity. At the same time, the highest representative of the German state was an early adaptor and thus promoted the general acceptance of the new technology: Emperor Wilhelm II's palace was among the first buildings in Germany to be illuminated electrically.[189]

Low prices were not the only reason for the persistence of the old technologies. The new competitor electricity gave crucial impulses for innovations in gas technology. One of these innovations, gas boilers, supported the establishment of bathrooms in standard flats. Thus, during the 1890s, private gas consumption nearly doubled.

German gas works campaigned for new uses of gas, advertising for gas geysers, stoves and central heating. Thus, even though gas lighting started losing out to electricity, new gas works settled on the outskirts of many German cities at the turn of the twentieth century. At that time, gas and electricity were complementary; each technology found its niche.[190]

Furthermore, in the late nineteenth century, gas lighting became much more efficient by the establishment of the incandescent mantle, invented in 1886 by the Austrian scientist Carl Auer von Welsbach, which was based on an alloy of rare earth metals. The gas mantle improved gas lighting quality by imitating electric light: it used the heating power of gas, not an open flame, resulting in a brighter, more regular and cheaper light. This points to another often-overlooked aspect of technological progress: while innovations prospered, old technologies were often improved in such a way that they persisted alongside the new technologies for some time. This happened to gas lighting as electricity emerged. Further back in time, candles persisted after gas lighting had emerged, because wicks were improved to burn in a cleaner way. Developments in old and new technologies were often intertwined: twelve years after his invention of the gas mantle, Auer von Welsbach introduced the osmium mantle, which was the first working metal-filament lamp. Innovations in electric lighting had first helped to improve gas lighting; later the material knowledge, which Auer von Welsbach had thereby gained, made it possible to enhance electric lighting. Only improvements like this gave electric lighting a competitive advantage. Edison's carbon filament lamp had not been brighter than gas lighting. By contrast, the tungsten lamp, which was invented on the eve of the First World War, covered the whole range of lighting levels from dim light to high brightness. Only by then had electric lighting become 'modern' in a full sense.[191]

Apart from illumination, the elevator marks the beginning of tenement electrification. Werner von Siemens presented the first electric elevator in 1880. After security issues had been solved, the new technology prevailed from the 1890s onwards. The older steam-driven or hydraulic elevators had caused problems by their loud noises and their need for large spaces. By contrast, the electric elevator fit seamlessly into tenements because its technology was invisible and silent. In general, this is often a typical signature of modern technologies: they entered everyday life without reminding users of the technological alteration of their lives. By the 1880s, the elevator had become standard in the new American apartment houses, while Paris and Berlin followed only in the 1910s. The establishment of the elevator eased public acceptance of tenements and put an end to the general criticism of high floors. Whereas some problems persisted, such as heat or stuffiness, the potential for multistorey buildings changed due to the new technology.[192]

Comprehensive household electrification evolved only slowly in Germany. The Chicago World Exhibition had introduced the electric stove in 1893, but as late as 1911, a kitchen showroom on the outskirts of Hamburg presented electric cooking for the first time.[193] The electricity companies soon realized that they had to increase demand for household electricity. The AEG heir Walther Rathenau told his father in 1907 that the producers had to promote electrical applications, which literally had to be 'forced on consumers'.[194] In spite of these intentions, the low quality of the hotplate and high costs prevented the electric stove from becoming a success before the 1930s. In 1931, only 50,000 German households owned an electric stove. From the mid-1920s onwards, however, electricity broke through. In 1928, 55 per cent of Berlin households were connected to the electricity grid, effectively doubling the figure from 1925 and a marked increase on the figure of 6.6 per cent in 1914. This was mainly due to lighting and the widespread use of flat irons, whereas the much costlier refrigerators and electric cookers were still very rare. Eighty per cent of German households were connected to the electricity grid in 1932. By now, the German electricity grid was much larger than its British counterpart, largely due to the fact that gas still prevailed in Britain. Simultaneously, stove prices began to fall in Germany. Despite these developments, owning an electric cooker was still the exception before the 1950s. Furthermore, there were vast differences between German cities. In 1936, Chemnitz had the largest share with 4.2 per cent of households owning an electric cooker, whereas Münster had the lowest score with only one per 1,000 households.[195]

Thus, when the short-lived housing reform era of the 1920s began, the 'aids of modern technology' had barely reached the average German household. The Frankfurt housing estate Römerstadt became the first 'fully electrified' German settlement. This was extraordinary because there were only 30,000 electric kitchens in all of Germany by 1929. In this context of German backwardness, the Römerstadt settlement was seen as the outcome of Americanization. As mentioned earlier, the settlement's construction was standardized and mechanized, with the flats fitted to a high technological standard. All flats were connected to the radio; each dweller only had to turn the speakers on in his or her flat because the building had one central receiver. At that time, this was a low-cost way to enjoy radio, which was still uncommon due to the usual high acquisition costs. Moreover, every flat had a separate bathroom and a fitted kitchen, the so-called Frankfurt kitchen. In particular the equipment with electric household appliances was exceptional: stove, light, water and room heating as well as electrical sockets all over the flat. At that time, this omnipresence of electricity was rare. However, there were similar, but less comprehensive, projects in Schweinfurt and some other German towns.[196]

By contrast, the Berlin municipal energy utility rejected the idea of electric cooking until 1931. Accordingly, Berlin's large housing estates did not employ comprehensive electrification comparable to Frankfurt's Römerstadt estate. Indeed, there was an economic reason for the Frankfurt energy utility to propagate household electrification, which the Berlin counterpart overlooked: luminous flux was still prevalent. Thus, in the evenings when the factories did not consume energy, the municipal provider had serious sales problems. Consequently, municipal offices decided to provide boilers in the large housing estate. However, the daily amount of hot water was sufficient only for one bath. Thereafter, family members had to change the tradition of the weekly family bathing day. One result of this hyper electrification was that the Frankfurt Social Democrats criticized the Römerstadt electrification for being simply too costly for the working class.[197]

Housing reformers gave particular attention to the issue of the kitchen, and the reformers adapted the principles of scientific management to the work of housewives. Kitchen spaces had to be designed rationally.[198] In general, the rationalization of housekeeping was a prominent political issue in the late 1920s, as municipal administrations took measures in this direction. Additionally, the national ministry of economics discussed the topic in a meeting with experts in October 1929. Many institutions all over Germany addressed efficient housekeeping technologies and practices. These efforts were widespread and more diverse than a mere focus on the famous municipal reform projects in Frankfurt, Dessau and Berlin would suggest.[199]

The Frankfurt kitchen, however, got the most attention both from contemporaries and from later architects and historians. The kitchen, which was designed by the Austrian architect Margarete Schütte-Lihotzky in 1926, was part and parcel of Ernst May's Frankfurt housing reform. With 5.6 square metres, the Frankfurt kitchen was very small, but it used space extremely efficiently. The kitchen included all furniture: stove, wall unit, work table and sink as well as an ironing board installed on the wall. Even small details were thoughtfully designed. The colour of the kitchen furniture was blue to repel flies. Although it was not the first of its kind, it became the prototype for a standardized and – with 10,000 units – mass-produced fitted kitchen.[200] However, the adaptation of Fordist principles to architectural issues, of which the Frankfurt kitchen is an eminent example, brought with it the foremost problem of Fordist industrial production, namely a lack of flexibility. The Frankfurt kitchen fitted exactly into the floor plan of the Frankfurt reform housing estate, for which it was designed, but it did not match with other floor plans. This lack of flexibility prevented mass production on a nationwide scale. Despite these practical constraints, the underlying concept became a huge success through the fitted kitchen in the post-war era.[201]

In hindsight, the Frankfurt kitchen architect Margarethe Schütte-Lihotzky stated that her idea of an efficient kitchen design had its origins in 1922 when she had first heard of Taylor's scientific management. At the same time, she had read the German translation of the American efficiency expert Christine Frederick's 1914 book *The New Housekeeping: Efficiency Studies in Home Management*. Moreover, the kitchens of American Pullman cars and Mississippi steamboats inspired the housekeeping efficiency movement in Germany. Yet Schütte-Lihotzky took this approach to a new level by transferring the principles of scientific management consistently to housing: thus, the distance between the Frankfurt kitchen and the dining table in the living room was not to exceed 3.2 metres. At the same time, old technologies were a source of inspiration to the modernist architect. One of Schütte-Lihotzky's simple, but efficient, appliances was a dish drainer, which saved the housewife from drying the dishes. The Austrian architect did not invent this modern labour-saving technology but found the traditional artefact in Italian farmhouses.[202] However, there was one essential difference between Schütte-Lihotzky and her fellow German urban reformers on the one hand, and the American school of efficient housekeeping on the other. Whereas Frederick sought to transform the kitchen into a space of consumption full of electrical gadgets (and succeeded as consumer capitalism soon afterwards transformed American everyday life), German architects were obsessed with efficient production and rather neglected aspects of consumption. Indeed, German living standards were still far from the level upon which a comprehensive consumer society could have been built.[203]

Schütte-Lihotzky was not the only architect in Germany who addressed rational kitchen design in the 1920s. The architects of the famous Weimar Bauhaus School designed a similar kitchen in 1923. They praised this efficient kitchen as 'the housewife's laboratory'. The family dining table gave way to a smaller kitchen table designed as the housewife's work table, while wall units replaced the traditional sideboard. The heights of table, stove and cupboards were likewise standardized with the goal of facilitating mass production. Furthermore, in 1924, the renowned architect Bruno Taut designed a 'new flat', of which the kitchen also was an important part. Taut was driven by similar ideas of efficiency as his colleagues. He even explicitly connected his efforts to rationalize housework with Taylor's scientific management. As did his Weimar colleagues, Taut also referred to Christine Frederick's calculations of distances inside the kitchen.[204] Yet, none of the competitors reached the production figure of the Frankfurt variant.

While focussing on the Frankfurt kitchen, most historians overlook the 'counter-model proposed by the city of Munich' at that time. The Munich kitchen offered mostly the same technical appliances but it was more flexible; indeed, it was a mere extension of the living room. Thus, it corresponded better with the traditional requirements of working-class life. Importantly, the Munich kitchen, which was an integral part of the city's social housing projects of the late 1920s, was cheaper and smaller. Subsequently, it was 'used very frequently in German social housing projects'. Political disputes did not colour the opposition between the two models, as the Social Democrats governed both cities. In fact, both kitchens were variants of modern architecture, while they signified 'competing modernities'.[205]

After the Nazi seizure of power in 1933, the reform efforts of leftist architects were rejected. The Frankfurt kitchen almost fell into oblivion because the Nazis made the eat-in kitchen a national standard by decree. However, the Frankfurt kitchen inspired architects and designers in Sweden, Switzerland and the United States during the 1930s and 1940s. Although the Weimar reform concept of the Frankfurt kitchen was internationally acclaimed, it was seldom transferred unchanged. The Swedish housewives' association preferred the larger eat-in kitchen because it would be more appropriate for families with children. Otherwise, the association feared a decrease in birth. Further research in these countries improved the heights of the kitchen furniture, and, after the war, a modified Frankfurt kitchen returned as the model of a 'Swedish' or 'American' kitchen to Germany. As such, after 1945, German kitchens became standardized to a wide degree. Wartime destruction and the subsequent post-war urban housing shortage offered a strong impetus to an increasingly efficient use of space. In these circumstances fitted kitchens proliferated.[206]

The second most important technical kitchen appliance – the refrigerator – succeeded even more slowly on the German market than the electric stove did. Two million American households owned a fridge in 1936, whereas Bosch introduced the first German fridge only in 1933. This was due to differences in national technological styles. While American companies mass-produced small fridges, German industry sought technological improvements in refrigeration. Thus, the German variants were more sophisticated but far too costly for mass consumption. In 1935, there were only 30,000 fridges in Germany; this number increased, but in 1938 it was still only 150,000. After the war, German companies changed to mass production.

Before 1945, the rather affordable electric iron was the most common electrical household appliance, which more than a quarter of German households in the mid-1930s owned.[207] In general, household electrification was a joint venture of manufacturers, middle-class housewives' associations and modernist architects as well as Social Democratic housing reformers in the Weimar period. Later, Nazi bureaucrats continued the efforts because they shared the common goal of household modernization.[208] The underlying approach, which Jeffrey Herf has termed 'reactionary modernism', will be discussed at length in Chapter 4.[209]

In the shadow of urban electrification, water supply and sewage systems were further developed and standardized. Even though central water supply and disposal provided the means for establishing bathrooms, only 10 to 15 per cent of German homes had a separate bathroom by 1900. However, housing reformers were convinced that separate bathrooms were a must due to hygienic considerations and placed the issue on the political agenda over the next several decades.[210] Meanwhile, water quality had improved as German cities adopted enhanced mechanical water filtration systems, the so-called fast filters, which American engineers had introduced in Ohio in 1890. After the Hamburg cholera epidemic of 1892 (at the latest), municipal authorities were aware of the necessity to react to the sanitary threat. Chlorine addition was seen as a sufficient and inexpensive means of water disinfection, which was also developed in the United States in the early twentieth century, and the Ruhr city of Mühlheim was the first German city to introduce this method in 1911. As late as the early 1940s, however, only 30 per cent of German waterworks used a chlorine addition, although filtration by activated carbon was often used from 1929 onwards. A multistage process of these procedures, which was established in the first half of the twentieth century, is still prevalent today. Another crucial improvement took place in pipeline construction. By the turn of the century, cast iron pipes for public water supply had replaced toxic lead pipes. In the 1930s, cast iron pipes were then superseded by asbestos cement, steel and plastic pipes.[211]

By the 1920s, German sewage pipelines, which had had diverse local standards, were also standardized on a national scale. As mentioned earlier, German engineers had originally adapted British technologies of mechanical sewage cleansing. Later, they developed new technologies such as rakes and sand traps, which had international success from the turn of the century forward. These, improved procedures of mechanical cleansing prevailed in many German cities in the early twentieth century, while the era of biological sewage treatment began in England at the same time. Activated sludge nourished with germs meant a breakthrough on the eve of the First World War, with Essen the first German city to adopt this technology in 1926. The rapid progress in sewage treatment encouraged many German experts to believe in the concept of a technological fix. The German pioneer of sewage technology, William Philipp Dunbar, declared in 1912 that modern technology would surely solve the issue of sewage treatment. By that time, septic drain fields had become less important, although this technology fitted quite perfectly to the natural conditions of German soils. However, ongoing rapid urbanization raised the problem of finding sufficient quantities of adequate fields near the cities from the late nineteenth century forward. Only the Nazis, who put self-sufficiency on top of the political agenda, revived septic drain fields to some degree. After the Second World War, this technology largely disappeared. Instead, sewage treatment plants, which combined mechanical cleansing and biological sewage treatment, prevailed. Later, chemical precipitation became established as the third treatment stage in sewage treatment plants due to new kinds of pollution, namely phosphates from detergents and nitrogen from fertilizers.[212]

Many housing reformers and urban planners sought to solve the very problems caused by modern technologies via technological means. Even if many of them harshly criticized some of the outcomes of modernity, they by no means longed for a romanticized vision of the premodern past. Quite to the contrary, many engineers, architects and municipal officials were optimistic technocrats who conceived of modern technology as a means to 'humanize' the modern city.[213] And at first glance, urban electrification and housing reform seem like success stories: streets and houses became cleaner and better illuminated, healthy residential areas at the outskirts were connected to the tramways and electricity replaced steam in the inner cities. Most of all, the urban mortality rate had decreased sharply. However, emissions were only translocated to the rural areas without solving environmental problems.[214]

The period of rapid urbanization soon came to an end, and the industrially caused urbanization had already been completed in Germany by 1914. The new industries of electric, chemical, precision and apparatus engineering established new factories and thus paved the way for the foundation of several cities at the turn of the twentieth century. Afterwards, there was no more growth of the major cities to be expected. Instead, urban growth shifted to smaller towns and cities: between 1925 and 1970, German cities of more than 50,000 inhabitants stagnated, while small and medium towns grew. The population of villages, however, decreased further.[215]

Urban reconstruction, the advent of the automotive city and the issue of ever-growing urban waste

Wartime bombing affected the large cities worst, and West German cities with more than 250,000 residents faced a disastrous housing situation after 1945. On average, 45 per cent of the housing stock was destroyed.[216] In the eastern part, which later became the GDR, large cities suffered even greater losses with two-thirds of houses destroyed.[217] The challenge of urban reconstruction turned urban planning into one of the most 'effective weapons in the Cold War'. The capitalist West and the socialist East competed in showing their competence in caring for their citizens through the politics of housing construction.[218] The urban housing concepts each side preferred clearly differed from its counterpart. While concrete prefabricated buildings shaped the appearance of East German cities, in West Germany suburban detached dwellings prevailed, although larger settlements also gained ground.

Americans held considerable influence over West German urban reconstruction. Initially, housing programmes were dependent on Marshall Plan funding and American administrators released funds only if Germans were ready to break with their 'traditional building methods'. Thus, standardized building techniques finally succeeded in Germany during the post-war era. These American building concepts, however, had some German roots. Two of the most important German housing reformers of the 1920s, Martin Wagner and Walter Gropius, had to emigrate after the Nazi seizure of power. Settling in the United States, they promoted prefabrication and standardization in housing construction. Both already owed large parts of their understanding of construction to the American efficiency movement. All in all, a long history of mutual international knowledge transfer characterizes the history of building, as does the history of technology in general. In the late 1950s, after American control of West German housing construction ended, the Germans nonetheless adhered to the now generally accepted standardized techniques. Yet they developed a particular style that differed from the small residential settlements that the Marshall Plan–sponsored programme preferred. Now, the flat design was larger, signifying the rising standard of living in Germany.[219]

Nonetheless, suburban, single-family homes were more popular in West Germany. Even in the early post-war years, detached dwellings had a larger share of new housing construction than tenements did.[220] By contrast, the East German government heavily supported prefabrication techniques. As a result of this policy, the majority of East Germans lived in a concrete prefab building, or *Plattenbau*, by 1990. This high share of prefab buildings was without precedent worldwide.[221] This figure is particularly remarkable because East Germany had a low rate of urbanization compared with west European countries. By the late 1980s, only a quarter of East German citizens lived in large cities with more than 100,000 residents. At the same time, roughly the same percentage lived in small villages of less than 2,000.[222] Thus, there were very different housing situations in urban and rural areas in East Germany. While most rural residents still lived in privately owned detached houses or duplexes, state-owned large settlements shaped the cities.[223]

The GDR government advocated the building of tenement blocks from the outset. From the mid-1950s onwards, prefabricated building techniques played an eminent role in GDR housing construction. This policy was due to the ruling Communist party's concept of the 'scientific-technical revolution', which had a strong influence on housing construction. Party leaders conceived of society as 'a complex system that could be closely regulated and automated through the use of technology'.[224]

Over time, the tenement block changed its aesthetics. Construction of the first large concrete prefab building project began in the city of Hoyerswerda in 1957. After a period of neglect during the 1960s, the new administration under Erich Honecker prioritized the housing programme.[225] The new 'Dwelling Series 70' was the most prominent outcome of this policy, which put even more emphasis on efficient housing construction and sought to 'correct some of the aesthetic problems of the earlier prefabrication systems'.[226] By 1970, improvements in concrete-slab mass-housing technology allowed efficient construction of very large settlements with relative low costs. However, technology transfer was a necessary precondition to this, and the GDR imported a factory from Finland to mass-produce concrete slabs. Among those mass-produced settlements of the 1970s and 1980s was 'the largest mass-housing project on European soil' with some 400,000 residents: Berlin-Marzahn.[227] Then again, what at first glance appears to have been one of the more successful policy areas of the GDR created some serious problems. There were hidden costs to the relatively cheap prefabricated buildings. Costly infrastructure had to be provided for satellite towns such as Marzahn.[228] Additionally, while 1.8 million new housing units were constructed through Honecker's programme, nearly the same number of tenements became uninhabitable due to a lack of refurbishment.[229] Parallel to the housing programme, there were some efforts to preserve pre-war building in East Berlin but only on a rather experimental, low scale. The government's clear priority was prefab construction of large settlements. Perhaps the most surprising and long-lasting outcome of the East German refurbishment experiments was their impact on the later West German concept of 'careful urban renewal'.[230] After reunification, German urban planners revived the old tenement blocks on a large scale following an international trend of refocusing on inner-city traditions.[231]

The East German urban housing situation thus clearly differed from the West German experience. The historian Eli Rubin has shown that housing was closely intertwined with transport policy. Those who lived in large settlements such as Marzahn did not necessarily have to own a car: most people's workplace was within walking distance to a large industrial area, while shopping centres, schools and other communal facilities were integrated parts of the settlement. Furthermore, the elevated public transport train connected the settlement easily to the city of Berlin. In short, 'Marzahn was built specifically with those who did not own cars in mind.'[232] By contrast, most West German families owned a car and lived in suburban, detached houses.

Rising automobile ownership was a general trend after the end of the Second World War: most large cities were transformed due to the concept of the automotive, or car-friendly city. Arguably, the automobile was the single technology with the strongest impact on European urban life after 1945.[233] This trend was most comprehensive in the capitalist West, but it also hit the socialist East.

Even before cities were transformed according to car-friendly concepts, the beginnings of motorization had taken place in large cities. In the early twentieth century, the centres of commerce and industry were the first places with significant, if still low, volumes of car traffic. However, German automobile development fell behind France: the car volume in Berlin was considerably lower than in Paris.[234] Nevertheless, prior to the First World War, Germany had taken important steps on its way to motorization. In most German cities, horse-drawn hackney carriages lost significant shares to motorized and electric cabs. While electric cabs remained a mere episode (before the return of the electric car in the twenty-first century), the motorized variant became prevalent by the mid-1920s.[235] In spite of these first steps, the poor economic situation and concomitant reduced purchasing power of most Germans prevented widespread access to private cars after the First World War.[236] For its part, the German government had no interest in a strong automobile industry because the German National Railway, or *Reichsbahn*, was central to the government's ability to pay reparations to the allies.

This changed after the Nazi seizure of power in 1933 when the Nazis repudiated the Treaty of Versailles and its obligations.[237] The Nazis put the 'People's Car', or *Volkswagen* on top of their traffic policy agenda and made large investments in *autobahn* construction. Yet the average German did not own a car before the 1950s.[238] For a long time, the bicycle was the most important private vehicle. In 1932, there were fifteen million bikes and only 2.2 million motor vehicles (among them only 490,000 cars, roughly 1 per cent of today's car population). German mass motorization began with the motorcycle: 730,000 motorbikes were registered by 1930.[239] This signified a particularly German path to motorization. In each year between 1926 and 1960, Germans owned more motorcycles than cars.[240] In the inner cities, tramways and bicycles were the most important means of transport before the Second World War.[241]

Regardless of its moderate diffusion in the first half of the twentieth century, the car had already inspired urban planners' imagination by the mid-1920s. The aforementioned Berlin urban planner Martin Wagner supported the vision of a car-friendly city. At that time, relatively few cars were on the streets of Berlin. A mere 42,844 cars had admission in 1929, compared with 1.15 million in 2013.[242] By the mid-1920s, the Stuttgart city manager for construction pleaded for road construction in favour of automobiles, although there was a car population of only 2,200. With automobiles signifying a modern approach to urban life, urban planning projected the coming demands of future car traffic. The city authorities discussed concepts which would become important after the Second World War, when Stuttgart was remodelled on the concept of the car-friendly city. While the financial means were missing in the interwar period, the concept of the automotive city had already convinced the municipal authorities to invest when it was possible.[243]

Immediately after the end of the war, German traffic planners were pessimistic about the economic future of their country. They believed that Germans would suffer a rather low standard of living for the near future and that public transport would account for 75 per cent of transport for a long time. Accordingly, trams seemed to be the most useful means of urban transport. Around 1950, most German cities modernized their tramway systems, following American models. The economic miracle, however, became a game changer for traffic planning. Rising mass consumption and car ownership involved new plans for the cities, meaning that the second German tramways boom was rather short-lived. Instead of trams, German politicians advocated for subways (which signified that a city had reached the status of a modern metropolis), with the goal of separating different means of transport. This gave automobiles a virtual monopoly on the roads.[244] Additionally, most small cities closed their tramlines and switched to busses. In general, between 1950 and 1970 German cities repeated a process that had transformed American cities twenty years earlier as the automobile began to dominate transport.[245]

While American traffic concepts had a great impact on European debates about car-friendly urban redevelopment, both the conditions and the outcome clearly differed. The narrow roads of European old towns did not allow planners to simply copy the American model of an automobile-centred urban redevelopment, whereas the American example fitted well for European suburbs and new development areas. German inner cities, it was thought, had to be transformed. Streets were widened at the expense of sidewalks or historic facades. In 1959, the German architect Hans Bernhard Reichow published a book on his concept of 'the car-friendly city', or *Die autogerechte Stadt*. Although his holistic concept owed more to German urban planning of the 1930s than to the contemporary American discussion, many German experts took study trips to the United States. Due to their experiences abroad, the concept of the car-friendly city became a hybrid of the German urban planning traditions and the

American example. Afterwards, 'the car-friendly city' became a catchword for the German period of comprehensive motorization from the 1960s onwards.[246]

Although traffic planners denied that they privileged cars, their plans did exactly that. Urban roads were transformed according to the needs of automobile traffic. Contrastingly, urban planners rejected the outspoken model of a car-friendly city from the late 1960s onwards while the competing expert group of traffic planners continued to refer to the underlying concept. Despite new rhetoric, which promoted a turn towards 'city friendly' traffic, the automotive city remained the standard in Germany. At present, most cities are still car-friendly.[247] This creation of car-friendly urban spaces challenged urban development from the late 1950s onwards, leading cities to establish special protected zones for people such as playgrounds, pedestrian crossings and cycle tracks.[248]

As a logical consequence of the automobile boom, public transport passenger rates decreased after the long period of growth that had begun in the nineteenth century. By the 1950s, public transport lost a significant share of total transport, and more and more commuters took to the car.[249] From the 1950s to the end of the 1980s, a reversal in the modal share of urban traffic occurred in both West and East Germany. While 80 per cent of travellers in both Munich and Dresden used public transport in the 1950s, the share decreased to 40 per cent in both cases by the end of the Cold War. This trend characterized West German urban transport in general.[250] As mentioned earlier, the East German situation was different. At least in theory, public transport was preferred for ideological reasons.[251] Furthermore, the car per capita ratio in East Germany did not even reach half of the West German ratio.[252] Nevertheless, traffic planning was surprisingly similar in both German states. In both states, the United States was a role model for German experts in transportation studies. Those engineers considered themselves neutral, non-ideological experts whose duties were solely mathematical and technical. While propagating a socialist alternative to American models of urban planning, East German urban planning was itself based upon core elements of the car-friendly city. Although public transport prevailed in most cities, multilane roads, tunnelling and elevated roads shaped the appearance of cities such as Dresden or Halle-Neustadt. Only by the 1980s did a new generation of interdisciplinary trained transport experts begin to question this general approach.[253]

The issue of urban traffic became more urgent by the 1960s when the concept of 'urban densification' evolved. A trend towards higher buildings had the consequence that more people worked in inner-city business towers while they dwelled in settlements on the outskirts.[254] In spite of ever-increasing urban traffic, the car-friendly policy continued. From the early 1970s onwards, many municipalities expanded urban light rail and subway networks while they closed tramlines. This was an expensive alternative that was inefficient in the long run. From the early 1980s onwards, subway euphoria vanished due to the everyday experience of car-crowded cities. Protesters who argued for humane transport and ecological considerations challenged the municipal traffic policy in cities such as Munich and several German cities began to expand public transport and cycle paths along with traffic calming measures. Yet most spectacular was the revival of the tramways. The latest American trends of a metropolitan public transport revival again influenced German traffic planners. Later, the German tram renaissance became a role model itself with other nations such as Great Britain following.[255] Interestingly enough, while the distance for commuters increased during the twentieth century, the expenditure of time was only slightly higher in the year 2000 than 100 years ago. Only in the immediate years after the German reunification did commuters require commensurably more time as a result of the unemployment problem in the eastern states.[256]

Increasing rates of car ownership signified rising mass consumption in post-war Germany. However, the coming of the throwaway society gave urgency to salvage considerations. Apart from the unknown masses of waste, new product materials increased the problem of waste handling. Urban areas with a high population density had to face this problem particularly. Even before the 1960s, the waste issue had been conceived of primarily as an urban problem.[257] Until then, waste consisted mostly of organic material. Thus, composting offered a simple solution to that issue in the countryside. The urban sanitary movement of the nineteenth century had followed this approach. Waste disposal was considered a mere side aspect of their tasks. Once the waste was out of town, their job had been done. Thus, most cities disposed of their waste on the urban fringe, 'either as fertiliser or landfill material to reclaim land'.[258] Even in the first decades of the twentieth century, most often urban waste was merely dumped at the outskirts.[259]

Although this seemed to be a satisfying solution to urban administrators, people living in rural areas raised serious objections to it. During the cholera epidemic of 1892, farmers violently protested the dumping of Hamburg's waste on the outskirts of the city. These protests resulted in the establishment of the first waste incineration plant in Germany, which operated in Hamburg from 1896 onwards. Hamburg followed the English example, where the first incineration plants had been established by the 1870s. In the Hamburg case, the technology transfer was quite simple. Copying the English technology functioned flawlessly because the inhabitants both of English cities and of Hamburg preferred heating with coal. By contrast, lignite-based heating was prevalent in most other German cities. However, only coal remained in sufficient amounts to guarantee an efficient operation of incineration plants. Without these coal remains, fuels had to be added. Therefore, incineration plants were established only bit by bit in Germany. On the eve of the First World War, there were only nine of these facilities, and only three new incineration plants followed in the period up to the beginning of the Second World War. Some of these proved of little worth. Only by the 1960s had this technology resumed on a wide scale in Germany.[260]

The modern history of urban waste disposal and city cleansing began in the late nineteenth century. In the mid-nineteenth century, there was still no central waste disposal in German cities. Most often, urban dwellers stored their waste in their own backyards for several months.[261] Yet even at that time city dwellers were taking the first steps towards the modern system. Townsfolk put open waste vessels besides the road where they were collected on certain days. The city of Frankfurt pioneered this method in the 1860s. First, local peasants were contracted to dispose of household waste from the city centre. Then, after 1873, the city established a regular waste collection service. The cities of Mannheim and Dortmund followed this example of a regular inner-city waste collection in the 1880s. By the turn of the twentieth century, inner-city waste collection had been established in many German cities.[262]

In 1895 a Berlin police regulation initiated a profound change that had long-lasting effects. This regulation introduced standardized, closed vessels, which provided the means for dust-free waste collection. There were two different systems: most often, the waste collectors emptied the dustbins. The standardized garbage cans were locked onto a fitting on the garbage vehicle when the cans were emptied. The alternative was a container-changing system, which had been introduced in Kiel and was adopted by several German cities, among them Dortmund. This very hygienic system had the disadvantage of high costs because twice as many cans had to be provided; thus, it slowly vanished by the mid-1950s. In both cases, initially horse-drawn carriages collected waste, then, as soon as 1911, electric driven vehicles were used by the Fürth waste collectors. Soon after, Hamburg introduced the first waste collection automobiles. During the 1920s, a comprehensive waste collection service became standard, although it was still confined to inner-city districts. At that time, technological improvements were of utmost importance for the expansion of the service. This clearly distinguished German developments from those in Britain, where waste collection service was extended at the same time but without technological changes on a large scale. In Germany, the companies of Krupp and Kuka cooperated in manufacturing waste collection vehicles. While Krupp represented German industry's past and present, Kuka would become the most important player in industrial automation as a robot manufacturer from the 1970s onwards. Krupp produced the chassis and Kuka the technology for waste compression. Later improvements point to the close connection between these business experiences and Kuka's later success in industrial robot manufacturing. From an engineering point of view, the equipment that picked up garbage cans was 'a kind of a robot already'. Nonetheless, horse-drawn waste carriages persisted in several German cities for quite a long period. It was only after 1945 that waste collection automobiles became prevalent in any German city. Despite the usage of garbage trucks, the infrastructure and technologies of waste disposal in the 1960s still resembled those of the late nineteenth century.[263] A rather small, but profound, change began in 1964: the city of Freiburg initiated the gradual replacement of sheet metal bins by plastic garbage cans in 1964. Within the next two decades, every German city followed, which meant large increases in waste collection efficiency due to the decrease of the cans' weight.[264]

At that time, however, three new problems emerged: a shortage of landfill, the transformation of city structures and a new composition of waste. First, some cities suffered a severe lack of landfill space from the mid-1950s onwards. To address this incineration technology was improved and modern incineration plants were established. The heavily industrialized cities of the Ruhr district were pioneers, while most large German cities followed from the 1960s through the 1980s.[265] Second, urban transformation continued. Central heating became standard for tenements and this put an end to the tradition of burning one's own kitchen waste at home. Additionally, the outskirts lost their character as villages and were transformed into regular suburbs. As a result, the habit of composting kitchen waste in the garden vanished too.[266] Third, as mentioned earlier, new plastic products and the rise of the throwaway society challenged the capacities of waste management.[267] With the advent of regular waste collection services, West Germans recycled far less than before. Hence, the amount of waste increased further.[268] By contrast, in East Germany salvage had a higher profile and the recycling of waste materials played a more important role in East Germans' economic considerations.[269] Some West German cities profited from the possibility of exporting their waste to East Germany, where waste disposal sites were exclusively established for the import of Western waste. As the GDR was in urgent need of foreign currency, it offered these waste disposal services to the capitalist West.[270]

Recycling had a revival in West Germany from the 1970s onwards. On the one hand, this was a necessary reaction towards the ever-increasing amount of waste and the shortage of landfill space. On the other hand, the renaissance of recycling corresponded with a change in public opinion towards environmental and waste issues. This change had a large impact on everyday lives, which will be discussed in Chapter 7. The technology of waste separation, which was the basis for industrial recycling, had a long tradition, albeit one that was not continuous. In 1907, the Berlin suburb of Charlottenburg was following the example of several American cities by introducing waste separation for its 300,000 inhabitants. The residents separated three different sorts of waste from each other. For this cause, municipal authorities distributed tripartite dustbins for ash, kitchen waste and the remaining waste material. While the ashes were used for land development in the Berlin hinterland, the kitchen waste was served as food for municipally owned hog-feeding farms. This system suffered financial losses, however, and it was cancelled soon thereafter. Only during the First World War was the Charlottenburg system revived. A similar renaissance of recycling occurred during the Second World War. In general, however, these remained mere episodes. The same was true for the early German waste sorting plants, which also resembled American models. The first of those was installed in a Munich suburb by 1898, but it was not able to operate profitably. Before the revival of recycling in the late twentieth century, landfilling, composting and incineration prevailed in waste handling.[271] By the 1980s, the first large success apart from the deposit bottle system was the recycling of glass and paper. By that time, all large cities had either installed regular waste paper collection or established collection points.[272] Yet the official infrastructures of waste handling did not cover all waste. This is also true for present issues of salvage: while Germany is acclaimed for its e-waste policies, probably more than half of the total e-waste runs through informal networks. Thus, Germany is both 'Europe's biggest e-waste producer' and a large exporter of digital waste.[273]

Another technology that would change the everyday life of Germans in the latter half of the twentieth century was initially confined to large cities: the telephone. There were several reasons why the telephone was diffused rather slowly and why it was perceived foremost as an urban technology in its first decades. Most important for its slow development were technical restrictions, the persistence of old technologies and the policy of the German Post Office. The technical restrictions to the early telephone era stemmed from the fact that the telephone had a limited transmission range, which hampered its usage apart from in high-density urban areas.[274]

Another hindrance was the reluctant strategy of the Post Office. Although the German Postmaster General supported the new technology of the telephone immediately after he first heard of Alexander Bell's invention in 1876, it was established only slowly due to a lack of demand. For most people, there seemed to be no need to invest large sums for the installation and operation of the telephone because the postman came several times a day in large cities; for instance, the postman rang up to eleven times daily in Berlin during the 1880s. Additionally, other old communication technologies such as personal visits and messenger services offered an adequate service and thus hindered the rapid success of the telephone.

Beyond the question of demand was how the telephone was seen by the Post Office, which operated the telegraph as well as the telephone. The Post Office still perceived the telegraph as an adequate technology for long-distance communication, while the telephone should only supplement the telegraph in the cities. The office had a very cautious budget policy and preferred to establish telephone networks in cities only after fifty citizens had subscribed.[275] The German telephone premiere in Berlin began with only eight subscribers in 1881. Within the next few weeks, Mulhouse and Hamburg followed with a mere seventy-two and ninety-five subscribers, respectively. At the end of the year, seven German cities had phone systems with 1,004 communication stations.[276]

By 1883, all large German cities had established local telephone networks; however, the Post Office wanted to restrict further diffusion for reasons of cost. Only urban networks promised high traffic and fast profits. In contrast, rural telephone networks meant high costs and a negligible return.[277] As such, the telephone was still foremost an urban technology by the turn of the twentieth century. In 1890, 176 million of 184 million calls were local calls. The networks only slowly expanded to rural areas.[278] Indeed, the early interregional lines were not even projected as a network. Instead, the telegraph system of city-to-city connection served as a model for the installation of single lines between cities.[279] As mentioned earlier, this model of point-to-point communication had first served for the early railways. A comprehensive telephone network was only established gradually. Initially, telephone lines were confined to the cities. Thereafter, operations began between neighbouring cities, and finally, the lines were gradually extended to a proper network, which connected the villages of a certain region.[280] This meant that telephone density increased rather slowly in the first part of the twentieth century. Before the Second World War, there were fewer than five telephone stations per hundred citizens. After the war, the figure increased to 10 per cent in West Germany in the late 1950s. In the late 1960s, it reached 20 per cent, while 50 per cent coverage was reached only by the early 1980s. In socialist East Germany, the rate was significantly lower.[281]

Conclusion

There was no specifically German path to modern urbanity. Yet, several features distinguish the development of cities in Germany from those in other Western countries. The spread of modern amenities in German cities was a piecemeal and uneven affair, affected by historically contingent developments. Rapid urban growth in the mid-nineteenth century had very different outcomes. Tenement construction, for example, in many of the growing cities was rather anarchic. As a reaction, a countertrend of a rather strict, bureaucratic regulation set in.

At the turn of the twentieth century, Germany had become a leader in urban planning. It is important to emphasize, however, that this scientific approach towards the issue of urbanization was by no means grounded in something like a German national character. This trend emerged out of contingent developments only in the nineteenth century. The same is true for municipal socialism. The Germans appropriated this idea from British examples after some less-than-convincing experiences with private sector suppliers of water, gas and electricity. Only in the late twentieth century did a neoliberal trend of privatization see private ownership return to Germany.

Another long-lasting effect of the rapid growth of German cities was the burden of the social question. Although urban issues remained controversial between various political factions, a progressive approach towards settlement policy had prevailed by the late 1920s. The reform movement was in retreat during the Great Depression, and finally the Nazis forced the reformers out. However, some of the technology-based reform concepts shaped the policy of urban growth in Germany as well as in other parts of the world during the latter half of the twentieth century. Primarily, the idea of efficient housing became widely accepted. Additionally, household electrification began to slowly take off.

Belated mass motorization is another example of a reversal of a trend. Only from the late 1950s onwards did automobile ownership become common for Germans. Nevertheless, urban planners immediately and wholeheartedly appreciated the concept of the car-friendly city. Only decades later, after several German cities were coming close to a traffic collapse, was public transport revived. German cities even became pioneers in the international trend of the tram comeback. Nonetheless, the urban automobile system had acquired technological momentum and cities continued to be arranged in a particular way to facilitate car traffic. On the other hand, urban dwellers and particularly suburban commuters had become dependent on their cars.

By and large, this chapter has shown the weakness of historical approaches that privilege the history of ideas and focus on particular, largely unrepresentative, technologies. Although some studies continue to overrate the influence of expert discourses on practical matters of how technology transformed urban life, this chapter has shown that pragmatic local middle-class interests were in most cases the most important driving force for the establishment of water supply or other kinds of urban technologies. These specific interests frequently outshone scientific or political concepts of urban development. Accordingly, public or expert discussion must not be mistaken for the history of everyday life. Although reform projects such as the Frankfurt kitchen were widely discussed, most kitchens looked very different to it at the time. Competing models of efficient kitchens were in fact more successful, without being as 'new' or 'innovative' as the Frankfurt variant. Overall it was compromise models that succeeded, meaning that household electrification was slowed down and postponed until the post-war period.

3

High tech

Traditionally, high technology has been a prominent theme both in popular narratives and historical studies. At least since David Edgerton's *The Shock of the Old* (2006) historians of technology have been aware of the fact that the notion of 'high tech' is always embedded in narratives of progress. This chapter scrutinizes such progress narratives and explores the contradictions inherent to the term 'high tech'. As will be shown, high-tech innovations were often intertwined with old technologies. In several cases, tinkerers did the pioneer work, while in later stages interdisciplinary expert teams further enhanced these technologies. Furthermore, high tech was by no means the sole preserve of well-paid specialists. Rather, in some cases, the Nazis introduced forced and slave labour, that is, archaic forms of labour, into the realm of high-tech production. At the same time, the most innovative technology proved to be complementary with reactionary ideology and even with open racism and anti-Semitism.

It is challenging to define high technology, and this chapter looks at those very advanced technologies that relied on big science. Although big science is often associated with the American Manhattan Project, which gave birth to the atomic bomb in 1945, there is a prehistory of big science. From the 1930s onwards, German aviation and rocket research was already close to the definition of big science, boasting interdisciplinary teams of scientists, technologists and skilled workers who collaborated on large-scale projects based upon heavy state support in cooperation with private companies.[1] Specifically, this chapter investigates the three examples of aviation, rocketry and nuclear power. Even before these technologies became embodied in large-scale projects, there was widespread euphoria surrounding these nascent technologies. In addition, political interest in their assumed military potential was very high. After the end of the Second World War, a pronounced ambivalence towards high technology became obvious, as the dual use of high-tech innovations for military and civil causes opened opportunities to sell these German products abroad.

Equally worthy of consideration is the connection between nationalism and high technology. After the defeat of 1918, German aviation served to fuel fantasies about the renewed rise of the German nation. In particular, right-wing politicians saw high technology as the realization of German supremacy. This idea had fatal consequences under the Nazi regime. After the war, the allies were suspicious about German military recovery, and it was only European cooperation that allowed the comeback of German big science. The tension between nationalism and international cooperation persisted, however, as did that between civilian and military usages. Even where the civil use of high tech prevailed, the military option remained omnipresent in the background.

Airships and aeroplanes

Military experiences sparked the beginning of German aviation. During the Franco-Prussian War of 1870–1 the French army employed balloons, which impressed the Germans. Even after the victory, fear of losing ground to France in military technology was widespread in Germany. In 1881, the outcome of this fear of backwardness was the establishment of the Society for the Promotion of Aeronautics, or *Deutsche Verein zur Förderung der Luftschifffahrt* in Berlin.[2] The most prominent of the society's members, Otto Lilienthal, was a gliding pioneer who inspired the American Wright brothers. In 1891, Lilienthal undertook the first controlled glider flight, although the glider was rather primitive and the wider public showed no interest at all.[3] In general, tinkerers dominated the first decades of German aviation. Some, however, shared Lilienthal's fate, who died gliding in 1896. By 1894, however, the German military developed a keen interest in aviation and established the Prussian aeronautical regiment. Furthermore, similar societies for the promotion of aeronautics were established in many German cities from the 1880s onwards. At that time, aviation was both locally based and internationally connected. It was simply unavoidable for the pioneers to seek out international cooperation because there were only very few experts in scientific aviation.[4]

Despite this, nationalistic ideas managed to conquer the field. This became most obvious with the advent of the airship. Although German inventors such as Lilienthal were important aviation pioneers, aeroplanes only reached the stage of innovation after an initial technology transfer with France. French technicians adopted and modified the German invention.[5] To a certain degree, the success of other nations in aeroplane construction explains the airship mystique in early-twentieth-century Germany. Many hoped that the airship would help the German nation catch up and close the technological gap.[6]

On a large scale, German aviation began to affect cultural and political life in August 1908, when Count Ferdinand von Zeppelin attempted a twenty-four-hour circuit over south-western Germany to demonstrate the military practicality of his airship LZ 4, or *Luftschiff Zeppelin 4*, to the German government, which was interested in purchasing the technology. Tens of thousands of spectators watched the flight and their airship euphoria provoked a wave of grassroots nationalism, which differed from the established authoritarian patriotism. This was the nationalism of the industrial middle class, proud of its products, as represented by the airship, now popularly called the 'Zeppelin'. In the era of high industrialization, this new kind of nationalism celebrated 'the technical virtuosity and the material achievements of the German people, not the German state'.

Most of Zeppelin's supporters were middle-class liberals or conservatives, but the Zeppelin cult also reached parts of the working class and even some determined Social Democrats. There were multiple aspects to this airship euphoria. The Zeppelin worked for divergent political agendas, both chauvinists and believers in technological progress.[7] By and large, from that time on, nationalism and modern technology were intertwined in Germany. Many Germans conceived the Zeppelin airships as the wonder weapons of the German Empire, while all other European countries clearly preferred the aeroplane. By the end of 1908, the Germans placed their hopes for aviation almost exclusively in the airship.[8]

There is another peculiarity to the German history of aviation. Zeppelin's twenty-four-hour demonstration in 1908 ended in a disaster, and the failure became crucial for the nascent cult of the airship. In the early hours of the morning of 5 August, shortly before the twenty-four-hour circuit would have been successfully finished, the airship had engine troubles and landed at the town of Echterdingen near Stuttgart. In Echterdingen, a sudden wind tore the airship off its moorings. Then it caught fire and was completely destroyed. It was only through good luck that no one was injured. Count Zeppelin was already somewhat used to such failures: it was the third destroyed airship in three years. Despite this, many German newspapers reported the flight and its tragic, but somewhat heroic, end positively. In the following weeks, an early form of crowd funding generated up to five million marks.[9] Inexplicably, these accidents even fostered the popularity of the airship in Germany, with airship crashes conceived as a patriotic sacrifice for the cause of technological progress.[10] Postcards and other memorabilia celebrated the Zeppelin airship and the 'heroic' disaster of Echterdingen (see Figure 3.1), while many Germans imagined the count as the 'conqueror of the air'. Zeppelin's fight against all drawbacks made him both an inventor of high technology and a heroic figure led by 'will and idealism'.[11] Thus, the myth of 'the lone genius battling obtuse bureaucracies' was established. Before the event, as historian de Syon points out, the count was actually 'but another inventor with no technical training, one of hundreds claiming to have solved the problem of dirigibility'. Despite this, the airship had a large impact on the Germans' attitude towards modern technology, with the middle class in particular celebrating the Zeppelin as the product of the superiority of German culture. While there was still widespread resentment against the political and social values of modernity among German conservatives, modern technology was now accepted as an integral part of Germanness.[12]

In addition to crowdfunding, both the German economy and the state, which hoped for military applications, supported the airship in its economic competition with the aeroplane.[13] By 1909, Zeppelin had raised sufficient funds for the completion of two new airships. A further twenty followed before the First World War. Most of these were purchased by the German army and navy, while the German Airship Company bought six for day trips. By 1914, at least 17,000 wealthy Germans had embarked on one of these expensive trips. At that time, the airships moved rather slowly at only 20 kph. In their heyday in the 1920s, however, they would be able to travel as fast as 120 kph.[14]

In the final years before the First World War, the new management of the Zeppelin company entirely changed the enterprise. Zeppelin was transformed 'from a cottage industry into a full-scale concern'. Until then, tinkering prevailed in the Zeppelin workshop. Now, it was replaced by systematic engineering. As a result, Zeppelin came to dominate German aviation before 1914. Although German aeroplanes had largely

Figure 3.1 Postcard in commemoration of the first twenty-four-hour flight of an airship, 4–5 August 1908: The ascent, flight and end of the airship *Luftschiff Zeppelin 4*. Library of Congress, public domain.

caught up in the meantime, the Zeppelin company had access to more capital than all seventeen German aeroplane manufacturers combined.[15]

In spite of the German airship euphoria, aerial warfare played only a minor role in the First World War. Besides Zeppelin, the German military used airships by other manufacturers. Among them, the wooden airships of Schütte-Lanz were most prominent. The first airships raids over Britain took place in early 1915, but the German airships were anything but wonder weapons. Although airship raids killed a total of 4,000 people during the war, they inflicted no strategic damage, while half of the German airship fleet was lost and many crewmembers were killed. Without proper radio navigation, airship captains 'usually had no idea where they were'. It even occurred that one captain attacked a place near Birmingham while he was convinced that he had bombed the far-off city of Manchester.[16] Yet, once again, many Germans praised the technological failure as a heroic act. Although it was admitted that the technology had some teething troubles, the German crews were lauded as having demonstrated the virtues of will and courage. In this way, the airship was associated with the 'aesthetic of struggle and defeat'.[17]

After the war, the Allies restricted the military use of airships. Thus, Zeppelin's only serious competitor Schütte-Lanz, which did not construct airships for commercial use, went out of business. For his part, Count Zeppelin had died in 1917 and his successor

Hugo Eckener switched the corporate strategy to passenger aviation. Due to the rapid progress in fighter aeroplane construction, the airship had in any case lost its assault potential. Nonetheless, the LZ 126, which was sent as German war reparation to the US Navy, signified the renaissance of German airship construction in 1924. Most spectacular, the airship crossed the Atlantic and served experimental causes for the US Navy for several years. From 1928 to 1929, the airship Graf Zeppelin even flew around the world attracting many spectators worldwide, including 250,000 in Berlin in November 1928.[18] After 1933, airships were used for propaganda causes. The Nazis strengthened the propagandistic ideal of national technology. For them, high tech signified the nation's greatness and fuelled their hopes for future German supremacy.[19] Only the infamous *Hindenburg* disaster in 1937 ended both transatlantic airship travel and the history of the airship as a significant technology in general. The luxury passenger airship *Zeppelin Hindenburg* had been celebrated as the largest airship of the world, before it was destroyed while landing in New Jersey. Compounding the negative effect the disaster had on public opinion was the fact that the disaster was repeated and reproduced in newsreels, radio broadcast and press photo coverage.[20]

In the shadow of the airship, aeroplane development quietly prospered in Germany from 1910 onwards. To a certain degree, the airship boom contributed to the growth of the German aeroplane industry in the long run. While Germany celebrated the *Zeppelin*, the first American air shows boomed in many European countries in 1908 signifying the triumph of the aeroplane over the airship.[21] Within a decade German aeroplane manufacturers had caught up. Before 1910, German aeroplane construction entirely depended on French and American licenses. Only ten years later, however, the German companies Junkers, Zeppelin and Fokker had become global leaders in aviation technology. One German innovation between 1910 and 1920 paved the way to modern aeroplane design that soon became the international standard, namely the monoplane with thick wings. This German particularity was a consequence of the former backwardness of the fledgling German aviation industry, which had relied on state-sponsored research. Almost accidentally, German aviation became the prototype of a science-based industry, as aerodynamics and statics prospered in Germany during this period. In its own way, the popularity of the airship in Germany was also an important factor for this scientific advantage. While the Zeppelin boom contributed to the belated development of aeroplanes in Germany, the airship attracted structural engineers to the aviation industry. Furthermore, the airship was the perfect test object for aerodynamics because of its relatively low degree of complexity compared with aeroplanes.[22]

Also not to be overlooked are the efforts of Zeppelin himself, which were important for Germany's aeronautical history. His company used its experiences with airships for further developments in aeroplane construction. Although Zeppelin preferred the airship, he established a department for aeroplanes before the First World War. During the war, the Zeppelin company became one of the German hubs for aerodynamic research.[23] Germany soon became the leader in basic research on aviation. The 'father of aerodynamics', Ludwig Prantl, and his team had already begun with aerodynamic experiments in a wind tunnel in Göttingen in 1908. Prantl's institute later became the famous Aerodynamics Research Institute, or *Aerodynamische Versuchsanstalt*. Even before the First World War, the community of practitioners elsewhere lost ground to the Göttingen scientists. Mathematical derivation replaced the older practical forms of knowledge based on observation and practice. Despite these important developments, applied research in Germany was still far behind that of the United States, Britain and France. For the next decades, these countries kept their advantage in propeller and engine technology.[24] On the other hand, German aerodynamics knowledge was transferred to the United States prior to the war, with American scientific journals reporting on Prantl's wind-tunnel experiments. After the war, this scientific exchange intensified and one of Prantl's leading disciples, Max Munk, left for the United States to work for the National Advisory Committee for Aeronautics.[25]

During the war, the German aviation industry built roughly 47,000 aeroplanes. There were three different kinds of players in the sector: first, the early aviation companies such as Fokker. Second, corporations such as Siemens and AEG established departments for aviation. And third, train carriage construction companies employed their experiences with wooden construction on aeroplanes. By contrast, the innovative companies Junkers and Dornier, which would later become leaders, remained outsiders in German aviation during the First World War because their scientific approach had the side effect of a lower level of flexibility. The military needed flexible aeroplane construction for adaptation to changing tactical needs. The military's priorities also lay with low cost, high velocity, high manoeuvrability, robustness and a high level of adaptability.

After the end of the First World War, aviation lost its remaining amateur status and some of its romance. It was transformed into a science-based industry, which was led by academically trained engineers. From 1919 onwards, Junkers had a virtual global monopoly with its F13 model. This plane performed better than any competitor until the mid-1920s.[26] In the 1920s, the history of German aviation was once again tied to a narrative of resistance. While the earlier Zeppelin myth was based on his heroic resistance against bad luck, the forces of nature and the initial indifference of the German state, the German aviation cult of the 1920s emphasized the battle against assumed international enemies. The Treaty of Versailles and following restrictions had limited the German aviation industry until the mid-1920s. The sanctions imposed maximums of velocity, load capacity, ceiling and range for German aeroplanes.[27] These Allied restrictions contributed to a 'myth of victimization' that saw German conservatives celebrate aviation's progress against fierce international hindrances. This aggressive nationalism positioned Germany as a victim of unfair international sanctions. Many right-wing Germans now considered innovative high tech as a means of resistance.[28]

Curiously, however, the restrictions resulting from the Treaty of Versailles actually helped the German aviation industry. First, the restrictions rationalized the sector. The resulting higher degree of market concentration helped the remaining aeroplane manufacturers improve their economic outlook. While companies such as AEG and Siemens closed their aeroplane departments, only the most innovative outsider firms Dornier and Junkers remained. Second, the Allied restrictions hindered German aeroplane manufacturers from participating in the international race for maximum speed. Thus, the German companies concentrated their research on slow-flying aeroplanes. This gave them a competitive advantage, especially against their British competitors, who maintained their traditional modes of aeroplane construction well into the 1930s.

Another important impact was the dissolution of the German air force prescribed by the Treaty of Versailles. Many of the former war pilots were trained engineers who had little chance of finding a job as a pilot after 1918. As a result they opened workshops for aeroplane parts or design offices. The practical knowledge of these former pilots greatly enriched the German aviation industry, which had hitherto been dominated by science. This mixture of science and practice gave a decisive competitive advantage to German aviation compared with its international competitors.[29]

Non-motorized flight was another niche that boomed during the period of Allied restrictions. In the early 1920s, gliding became the symbol of German resistance to these restrictions. In practice, gliding contributed to the German efforts in aerodynamics. Even after the restrictions had ended in 1926, gliding kept its impact on German aviation. During the Great Depression, many tinkerers constructed homemade gliders. Even in this late stage, tinkering was still important for aeronautical high tech.[30]

Despite the difficulties faced by the industry, air transport established itself in Germany from the mid-1920s onwards. In 1925, the merger of the companies of Junkers Air Transport and German Aero-Lloyd gave birth to the semi-public airline Luft Hansa, which – with a minor change in spelling: Lufthansa – is still a major airline today. The German state provided huge subsidies and guaranteed the monopoly for Luft Hansa. The new company boomed, becoming the European leader in passenger air transport by 1928. Even before the Nazi takeover, the Luft Hansa fleet could have easily been converted for military applications. Once the Allied restrictions were abandoned, the German manufacturers enjoyed an enormous upswing. Junkers and Dornier dominated the market in Sweden, Poland, Italy, the Soviet Union and South America.[31]

At that time, air shows with tens of thousands of spectators took place in Germany. The Germans had their own national star pilots while internationally renowned figures like Charles Lindbergh were widely ignored. At this time record flights caught public attention. Most spectacularly, for the first time an aeroplane, namely the Junkers machine 'Bremen', crossed the Atlantic non-stop from east to west in 1928. Two Germans accompanied by an Irish pilot made the flight. They succeeded after British and French pilots had failed in their own attempts at a non-stop Atlantic crossing. Thus, the German success was seen as a record against all odds. More than a million people welcomed the returning pilots in Berlin. Modern technology set the stage for the renaissance of the idea of heroism combined with German nationalism.[32] As historian Peter Fritzsche puts it: 'machine dreams mingled with national dreams.'[33]

This is even more true for the period from 1933 to 1945. The Nazi takeover marks a caesura for German aviation, when Nazi rearmament pushed the industry to an unknown boom. While a mere thirty-six planes were produced in Germany in 1932, 40,000 were mass-produced at the end of the war in 1944. In a certain sense, the modern German aviation industry was a child of the Third Reich. The impact of this boom was not restricted to the aviation sector. During the Nazi rearmament from the 1930s onwards, series production of planes demanded the rapid expansion of other technology sectors in Germany. Chemical industry, engineering and light metal factories all profited from the growing aviation industry. Thus, aviation filled the gap that the weak automobile industry left in Germany.[34]

In 1936, German aviation research both quantitatively and qualitatively reached the realm of big science. The German Research Institute for Aviation, or *Deutsche Versuchsanstalt für Luftfahrt*, which had been established in 1912, then had more than 2,000 employees. As important as scale were structural changes. The institute reduced the autonomy of work groups by formulating a joint objective. Similar transformations occurred at the Aerodynamics Research Institute. A modern wind tunnel was established in the mid-1930s, which was built with the Portuguese cork harvest of a whole year. Within six years of the Nazi takeover, the number of employees had risen from a mere 80 to 700.[35] Moreover, the German Aeronautical Research Institute, or *Deutsche Forschungsanstalt für Luftfahrt*, later known as the Hermann Göring Institute, was established in 1936 in Brunswick to close the gap with state of the art international-engine design. Research on the new technology of turbo jet engines was intensified, and this required different kinds of wind tunnels and engine testing institutes. Yet, despite the Nazis' efforts, American aviation research was still far ahead in 1939.[36]

By and large, German aviation research impacted many aspects of big science during the rearmament period of the mid-1930s. The research was multidisciplinary and built upon large-scale equipment. Additionally, significant financial resources were spent on aviation research. The government also set new objectives in line with military goals such as high-altitude flight, jet engines, high-velocity aerodynamics and ballistic missiles. As such, aviation research was deeply entwined with the Nazi state.[37]

Aviation research was also avant-garde research. In 1944, 10,000 employees worked in aviation research institutes all over Germany. Scientists intensified their collaboration with engineers, and the ideal of the traditional scholar was replaced by the modern concept of research teams. At the same time, large-scale equipment such as wind tunnels became ever more central to the research agenda. Although some more traditional scientists opposed these innovations as a 'mechanization of research', this big science approach prevailed.[38] This big science approach was not monolithic, however, and in many fields big science and small science coexisted. This became the rule rather than the exception. Yet, there was a specific feature of Nazism, which hindered the comprehensive establishment of big science to a certain degree: German aviation, like other high technologies, suffered from the polycratic structure of the Nazi regime, which resulted in 'the chaos of overlapping research agendas'. In the end, the Nazis failed with their approaches to coordinate the cooperation of science, state and industry.[39]

German aviation research nonetheless profited from the initial victories in the early years of the war, the blitzkrieg. The Nazis conquered eminent European aviation research institutes such as the French Etablissement d'Experiences Techniques des Chalais Meudon in Paris and the Dutch National Luchtvaartlaboratorium near Amsterdam. While the Germans forced the Dutch and French specialists to cooperate, they annexed the Czech Aviation Research Institute at Prague.[40] Thus, the preconditions for the mass production of high-tech warplanes seemed to have been met. Indeed, the production growth in aviation was remarkable. Yet, as historian Adam Tooze has shown, too many historians have taken at face value the Nazi rhetoric of a singular 'armaments miracle'. While significant, the output rise was 'far from miraculous'. Most of all, it was a 'deliberate sacrifice of quality for an immediate increase in quantity'.[41] In addition, the Nazis' preference for the huge and spectacular, a reaction to American advances in big science, saw them begin numerous competing construction projects in rocketry and aviation that never were being completed.[42] In this context, German aviation research lost its focus after 1943 and became ever more unsystematic. In the final stages of the war it was mostly based on trial and error.[43]

In the entire period between 1919 and 1945, there was an inherent tension between the production of a small number of high-quality transport aeroplanes and the mass production of short-lived fighter planes. The German aeroplane companies managed to overcome this contradiction by pursuing their long-term interests in air traffic even in times of war production. This corresponded with the personal character of the German aviation pioneers: all of the leaders of German aviation – Junkers, Heinkel, Messerschmitt and Dornier – were both entrepreneurs and inventors. They never privileged modern management over their inventor spirit. Until 1945, their factories more resembled laboratories than modern factories.[44]

At first glance, the German jet engine programme seemed to have been a result of the successful establishment of mass production. The Nazis decided only in 1944, and thus later than the United States and Britain, to manufacture jet aircraft. Nevertheless, within only six months, Junkers produced more jet engines than Ford in Britain.[45] Yet, there is a controversy among historians, as to whether this programme was really the success it seemed to be. From a technical point of view, Budraß correctly states 'that the German turbojet was a remarkable technical achievement'.[46] However, a different picture emerges if one considers the particular historical context. As Hermione Giffard points out, the impressive quantities obscure the fact that the programme was a failure because the engines lacked quality. The Germans produced the largest number of jets during the war, but their jets crashed far more often than their Allied counterparts. Giffard's argument goes even further: from the beginning, the jets were anything but wonder weapons. Instead, it would be more appropriate to label them 'engines of desperation'. The jet engine programme was the result of the failure of the German piston engine programme which were far more expensive than jet engines. The jet engines were thus a kind of 'ersatz technology', a substitute after the plan to develop innovative and fast piston engine aircraft had failed.[47] In addition, both Nazi leaders and entrepreneurs were convinced that turbojets were superior anyway, while the conventional piston engines were aerodynamically inefficient,[48] and that jet engines were a 'simpler, cheaper substitute for extremely complex and expensive piston engines'. Thus, the invention of the German turbojet depended on the specific conditions of production in Nazi Germany during the last months of the war.[49]

The history of the German jet engine has a lot in common with the rocket programme, which will be discussed further later. In both cases, production facilities were relocated underground. In fact, the Junkers jet engine factory was based in the same tunnels near the town of Nordhausen as the V-2 rocket programme. The Junkers jet engines followed a relatively simple design in response to the shortage both of material and of manpower in 1944. Assembly was designed in an undemanding way to allow production to continue while only unskilled labour and forced labour was available. At the same time, 'the large-scale production of jet-engines saved strategic materials', in particular nickel.[50] The brutal exploitation of forced labour was less murderous at the Junkers plant than at the neighbouring rocket facility in the same tunnels; nevertheless, 6,000 forced labourers mostly from East European countries had to work under harsh conditions in the so-called North Works, or *Nordwerk*. Slave labourers from the nearby concentration camp Mittelbau-Dora had earlier prepared the tunnel system for industrial production. The harsh working conditions killed many of these workers. Later on, concentration camp inmates had to work in rocket production at the Central Works, or *Mittelwerk*. While the working and living conditions of forced workers were terrible, the circumstances for concentration camp prisoners were even worse.[51]

The transfer of German high-tech expertise after the war is well known. Less well known is the fact that neither rocket scientists nor nuclear experts were the most sought-after group of German specialists, although many of those went to Allied countries after 1945. The single largest group of scientists and engineers employed in the United States, however, were aviation experts. In particular, the US Air Force was interested in German expertise in high-velocity aerodynamics, which was due to the extraordinary widespread employment of wind tunnels in Nazi Germany.[52] Thus, besides German experts, the Americans also transferred test sections and instrumentation from Peenemünde wind tunnels to the United States, where they were used for experimental aircraft test and aerospace experiments.[53] While the US Air Force made use of the German experts, these were not imported geniuses but rather complementary additions to American expert teams. In comparison to other sectors, transfer of high tech was rather the exception after 1945. The most important cases of technology transfer from Germany to the United States took place in industries such as the chemical sector.[54]

Although some specialists went abroad, German aviation recovered step by step. By the late 1940s, West German aviation research was enjoying a comeback because contemporaries believed aviation to be a pacemaker for general technological and industrial development.[55] To a certain degree, military aviation endeavours abroad paved the way for the re-establishment of the aviation industry in West Germany. The West German government supported the famous inventor and entrepreneur Messerschmitt in developing turbojets for Francoist Spain from 1951 onwards.[56] Soon afterwards, in the 1950s, the West German aviation industry began with the series production of aeroplanes. However, state funding for aviation was first offered in 1963; and the subsidies were only modest. The weekly news magazine *Der Spiegel* ridiculed this West German research policy in 1969. According to the magazine, the West German state had funded the production of skimmed milk at a rate ten times higher than aviation.

West German aviation became a large industry only after the establishment of the European Airbus project. The first Airbus had its maiden flight in 1972 and became a huge economic success breaking the virtual American monopoly in this sector.[57] In the meantime, Spain stopped the Messerschmitt programme, although it was quite successful, due to high costs and sold it to Egypt in 1960. There, however, Messerschmitt built only very few of his supersonic jets because the Soviet Union provided Egypt with much cheaper MiG fighter aircraft.[58] A general problem to such high-tech transfers was the lack of domestic know-how and industrial infrastructure. In the late 1940s, the German engineer Kurt Tank had developed a jet fighter in Argentina. The prototype's production was successful, as was the first test flight in 1951. Yet series production never began due to the absence of technological and industrial networks. German high-tech expertise remained in demand and Tank and his team moved further to India.[59]

Lacking comparable European cooperation, the East German aviation industry had only little success. The forced exodus of specialists to the Soviet Union hit the GDR harder than its West German counterpart. Yet, after the return of most experts in 1954, aircraft construction enjoyed a modest revival. The GDR depended heavily on this manpower because among the repatriates were not only scientists and engineers but also skilled labourers who had been badly missed. 'Soviet assistance' in technical matters and matters of material procurement allowed the establishment of an East German aviation industry.[60] Due to the problems already outlined in Chapter 1, the GDR was not able to maintain different fields of high technology. Thus, the GDR cancelled aeroplane construction in the early 1960s as a result of enormous costs.[61]

Military rockets and imaginary spaceships

There are many parallels between aviation and aerospace history. Initially spaceships were even conceived of as advanced airships. Even prior to the First World War many Germans had begun to develop a certain interest in spaceflight. Most exceptionally, the inventor Hermann Ganswindt claimed that he found the solution for the problem of spaceflight in 1891. Despite the fact that Ganswindt remained an outsider, ideas about spaceflight were canvassed prior to 1914. Space fiction changed from merely fantastic stories to speculation about the technological possibility of spaceflight. Some of the most important German rocket pioneers of the 1920s to the 1940s were heavily influenced by the space fiction of the pre-war era.[62] Beyond these cultural influences, the academic basis for the later aerospace research was provided when the institute for applied physics was established at the Technical University of Berlin after the First World War. Its graduates would later play important roles in German rocketry.[63]

Weimar Germany experienced a modest 'spaceflight fad' that was set off by the physicist Hermann Oberth's book *The Rocket into Interplanetary Space* (*Die Rakete zu den Planetenräumen*) in 1923. This book was not an immediate bestseller, but it influenced the right people – promising rocket scientists, engineers and other space enthusiasts; some of whom would meet Oberth in person when collaborating in an amateur Berlin outfit in 1930, and later at the Peenemünde Army Research Centre for the Nazi rocket programme. By 1929, rocket enthusiasm had hit popular culture with the release of Fritz Lang's film *The Woman in the Moon* (*Die Frau im Mond*), for which Oberth was hired as scientific adviser.[64] Besides this popular film, several rocket events attracted widespread attention in Germany in the late 1920s. Although they did little to solve the scientific or engineering problems of rocketry, popular figures such as Max Valier and Fritz von Opel fuelled the collective imaginary of spaceflight. They organized test events with black powder rockets attached to race cars, train cars or bicycles that attracted thousands of spectators. Although these experiments had no scientific value, they did a lot for rocketry's public relations.[65]

Second in their enthusiasm only to the Soviets, Germans were especially fond of rockets. The rocket fad of the 1920s profited from, and overlapped with, the widespread aviation enthusiasm. Both fields of innovation raised public hopes for overcoming defeat in war. This imagined supremacy in a nascent high-tech field fed nationalistic hopes for a rebirth of the German nation.[66] Most explicitly, Valier appropriated the language of aviation and connected high-tech development to the concept of national renewal.[67] His collaborator Opel did the same. He told a newspaper interviewer in 1928 that he expected the first spaceship to launch in less than six years. This spaceship would be named *Deutschland* and celebrate the comeback of the German nation after its wartime defeat. The spaceflight fad was, however, by no means limited to the nationalistic right. Instead, it reached the whole spectre of Weimar politics. Another collaborator of Valier, the more serious science writer Willy Ley, wrote two nearly identical articles on spaceflight in May 1928: one was published in the Nazi paper *Völkischer Beobachter*, the other in the Social Democratic *Vorwärts*.[68] However, the enormous value of Valier's public relations became obvious after he died experimenting on a new engine in 1930. Thereafter the Weimar spaceflight fad faded out, having lost 'its most public champion'.[69]

Besides these sensation-seeking demonstrations, Valier also contributed to serious rocketry in establishing the Society for Space Travel (*Verein für Raumschiffahrt*) in 1927. Among its members were Oberth and some of the later rocket pioneers of Peenemünde such as Wernher von Braun.[70] Those formed the most important of several amateur groups of rocket tinkerers, named Rocketport Berlin (*Raketenflugplatz Berlin*). In 1930, Army Ordnance secretly funded this group. However, these tinkerers underestimated the complexity of rocket technology and pinned their hopes on a rich investor to succeed. As Michael Neufeld has pointed out, these hopes were illusive because only a big science approach with heavy state sponsoring of a military-industrial complex was able to realize such a high-tech project. The army's interest was serious, although the first funding was rather modest. In 1929, the Ministry of Defence formally entered rocket science and sponsored a solid-fuel rocket programme. After the Rocketport Berlin group failed with a demonstration in April 1932, Army Ordnance decided to establish a liquid-fuel rocket programme on its own behalf, employing some of the group's most promising members such as von Braun. Von Braun began working for Army Ordnance at the test stand of Kummersdorf near Berlin in December 1932. Only five years later, he was the head of the large research facility at Peenemünde.

After the Nazi takeover and the establishment of the Kummersdorf test stand, rocketry disappeared from the public. The programme was top secret and by then the rocket fad was more or less over. In late 1934, rocket tests with aggregate 2 (*A2*) were promising. Two rockets reached roughly 1,700 meters in altitude. This test proved that it was possible to launch a liquid-fuel rocket with a range of a few hundred kilometres.[71]

After these successful tests, it took some fifteen more months until development began on the rocket that later became known as the infamous weapon V-2. The

Peenemünde engineers began their development of this aggregate 4, or A4 rocket only in 1936. The propaganda name V-2, or *Vergeltungswaffe 2*, suggested that it was a mere weapon of retribution for the Allied bombings of German cities. This ballistic missile was, however, to be a terror weapon to be aimed at civilians from a distance of 250 kilometres. The Peenemünde tests succeeded on 3 October 1942. The rocket sped up to 5,600 kph and reached an altitude of eighty-four kilometres.[72] In the following weeks, it became obvious that the war had turned against Germany. In response, Hitler gave the highest priority to the rocket programme in late 1942 after successive defeats in Stalingrad and North Africa.[73] Failed rocket test launches were still common, however, in 1943.

The programme was severely challenged in the summer of 1943. The Royal Air Force raids over Peenemünde in mid-August 1943 gave the impetus to the relocation of the A4 assembly into the underground facilities of the Kohnstein Mountain near Nordhausen. The first group of 107 concentration camp prisoners from Buchenwald was transported to the tunnels only ten days after the RAF raids. They had to dig out further tunnels and transform the underground system into factory halls. More than 3,000 prisoners died before the barracks of camp Dora were erected in early 1944.[74] Dora was established as a satellite camp of Buchenwald concentration camp in August 1943 and for many weeks the prisoners had to work and sleep in the tunnels. By autumn 1944, Dora developed into the centre of a new camp system under the name Mittelbau concentration camp. To catch the whole camp period between 1943 and 1945, historians have used the term Mittelbau-Dora.[75]

Mass production began in the underground facilities of the Dora concentration camp in January 1944. Until the end of the war, Nazi Germany shot more than 2,000 rockets on London, Paris and Antwerp.[76] The engineers actively participated in the murderous system of slave labour, with von Braun going at least once to the concentration camp Buchenwald to select prisoners for rocket assembly. Even after rocket assembly had begun, however, most Dora prisoners still had to work on construction sites for further underground armament facilities. Most of those were never completed. Only 10 per cent of the total 60,000 inmates who were imprisoned in concentration camp Mittelbau-Dora worked on the assembly of the V-2 rocket or the V-1 cruise missile, while the overwhelming majority had to do low-tech work on construction sites. Thus, Mittelbau-Dora was less an armaments camp than a construction camp.[77] While the assembly line work on the V-2 was harsh and brutal enough, being downgraded to construction work meant almost certain death for thousands of prisoners. Compared to the unbearable conditions of most concentration camp prisoners, the assembly line workers received higher rations because the SS management needed a steady workforce with rather low rates of turnover. As Michael Thad Allen puts it, the 'rocket, as a technological system, was not compatible with arbitrary murder'. However, the whole concentration camp system built around the assembly line did in fact mean murder to more than 20,000 prisoners. The permanent threat of being degraded to construction work meant a macabre, yet 'powerful incentive' to assembly-line workers.[78]

In this context, the concurrence of high-tech assembly lines with low-tech construction work is a remarkable, yet often overlooked fact. Some concentration camp inmates began to work on the rocket assembly line, but were dismissed to construction work if they became weak and were replaced by new prisoners. For construction workers, the living conditions were much worse. The type of work was also different. Low-tech machines such as hand-held drills, hammers and pneumatic shovels were used for the labour-intensive tunnelling.[79] This work, undertaken with rather primitive tools, laid the foundations for the Nazi prestige high-tech project: the rocket.

It was the combination of high-tech enthusiasm and the desperate war situation in mid-1943 that paved the way for the programme's megalomaniac plans and ruthless exploitation of slave labour. It was also this particular 'atmosphere of high tech desperation that drew the SS into the project'.[80] To a certain degree, the SS turned the rocket programme inside out. After the Minister of Armaments, Albert Speer, and Heinrich Himmler's SS had taken over the rocket programme in 1943, a new generation of managers was employed and slave labour entered the rocket assembly on a larger scale. As a result of the polycratic structure, however, competing authorities made plans that became ever more unrealistic both in schedule and in scale.[81] Initially the new managers redesigned the rocket assembly. While the rocket scientists of Peenemünde knew little about industrial engineering, the SS delivered both know-how of modern shop-floor management and slave labour. The newly formed Special Committee A4 ended the Peenemünders' tinkering and introduced accurate scheduling and most of all 'broke down manufacture of the entire rocket into 20,000 components'. Subsequently, concentration camp inmates had to do less complex work at the assembly line.[82] The far-reaching hopes of the SS were soon disappointed, however, and the illusion of high tech through slave labour crumbled. Where the original plans had foreseen a relation of eight slave labourers working at the side of one German worker, the relation turned out to be merely 2:1 in April 1944.[83]

In summer 1944, the rocket programme lost the enthusiastic support of the Reich's leadership that it had initially enjoyed. Both production rates and the rockets' war performance had proved disappointing. As such, the Fighter Staff claimed the northern part of the tunnels for their own project, namely the Junkers jet engine. In the end, most tunnels were employed for fighter aircraft manufacturing, while only a small part of the tunnel system remained for the rocket programme. In early 1944, the Nazi administration even discussed halting the entire rocket programme. It was continued, albeit on a lower scale, while the fighter aircraft programme was prioritized. More broadly, it became obvious that the Nazis had completely lost their sense of reality. Even in the last months of war, they continued to establish new construction grounds for more and more underground armament facilities.[84]

When the Allied troops reached the German rocket production sites, both the Soviets and the Americans were interested in harvesting rocket material and immaterial expert knowledge. The German rocket specialists profited from this headhunting situation. The Allied transfer of high tech offered attractive career opportunities to the Peenemünde experts. In particular, the competition between the superpowers in the context of the nascent Cold War proved to benefit this group and their Nazi past did usually not threaten their post-war careers. Both superpowers established similar operations in 1946 to employ German know-how for their own rocket development programmes.[85] There were nonetheless important differences. Today, it is well known that the secret US Operation Paperclip (and its predecessor Operation Overcast) paved the way for many Peenemünders to immigrate to the United States, where they were soon integrated.[86]

Wernher von Braun and some of his most prominent collaborators on the V-2 rocket project made impressive careers in the post-war United States. Von Braun became the director of the Marshal Space Centre in Huntsville, while Arthur Rudolph led the Saturn V programme. For his part, Kurt Debus became the director of the Kennedy Space Centre. Although the former Peenemünde engineers generally succeeded in obscuring their Nazi past and became honoured for their work on US aerospace programmes,[87] criticism did not entirely disappear. During the aerospace boom in the mid-1960s, the popular singer-songwriter Tom Lehrer wrote a satirical song on von Braun. Von Braun always maintained that he had never been a Nazi, but had been merely pursued scientific research and thus had nothing to do with warfare. Lehrer ridiculed this defence strategy in his song: '"Once the rockets are up, who cares where they come down? That's not my department," says Wernher von Braun'. The song ends by revealing von Braun to be the ultimate opportunist, only interested in rocket countdowns irrespective of the political cause: '"In German, oder English, I know how to count down. Und I'm learning Chinese now!" says Wernher von Braun'.[88] More seriously, Arthur Rudolph had to face his Nazi past, but only after his career had ended. He renounced his US citizenship in 1984 and returned to Germany after a non-prosecution agreement with the US authorities, which had investigated him for war crimes.[89]

The US Army liberated the Mittelbau concentration camp and secured the most interesting technological artefacts of the rocket project for its own use. In addition, most Peenemünders, who had left for south Germany offered their service to the United States. Yet, after the US Army retreated in line with the agreed division of Germany into occupation zones, the Soviets, who had already captured a complete V-2 rocket in Peenemünde, also benefitted from both rocket parts found at the former Mittelbau camps and the expertise of some of the German rocket specialists.[90] Immediately after the war, the Soviets ordered German engineers to reconstruct the A4 rocket at the sites of the former Mittelbau satellite camps Kleinbodungen and Bleicherode.[91] At the end of 1945, 1,200 German specialists worked on the Soviet rocket programme in these newly established Soviet research centres in East Germany. Although these were not the most prominent members of von Braun's team, they still made good progress.[92]

Yet, to complete the high-tech projects, roughly 1,600 scientists and engineers as well as 1,300 skilled workers, all accompanied by their family members, were moved to the Soviet Union according to the plans of the Operation Osoaviakhim in October 1946.[93] This operation included – along with aviation experts and nuclear scientists – about 300 Germans linked to the rocket programme.[94] Most were forced to go, but 'even those who had gone voluntarily, felt like prisoners at least part of the time'.[95] By October 1947, the Soviet V-2 tests had enjoyed some success: the rocket flew roughly 200 kilometres, but missed the goal. By contrast to the integration strategy of the United States, the Soviets sent the German experts back after the technological gap was closed in the early 1950s. Later, Soviet propaganda obscured the German origin of the rocket for nationalistic reasons.[96]

Less well known than the American post-war careers of some Peenemünders is how the French and British rocket projects initially made use of German rocket specialists. Those contributed to the French Véronique rocket and the British Blue Streak. Even beyond the Western World, German rocket engineers were in great demand. However, the Argentine and Egyptian projects with tactical missiles failed due to a lack of industrial infrastructure. The first truly successful outcome of German contributions to post-war aerospace took place in the Soviet Union. The resulting launcher R 14 was the processor of the SS 6, which launched the world's first satellite, *Sputnik*.[97]

Except for this early participation, East Germany did not play any role in aerospace development. In contrast to aviation, where cooperation with the Soviet research programme was quite common, the GDR contributed little to the Soviet satellite programme, except for optical devices and measurement technology. In the 1950s, a group of rocket enthusiasts was quite active and even established the German Astronautic Society (*Deutsche Astronautische Gesellschaft*) in 1960. This society maintained a certain distance to the socialist government before becoming infiltrated in the early 1970s. However, its members were mere amateurs. None of them was a trained aerospace engineer. Except for the East German army's modest experiments with sounding rockets from 1970 onwards, there was no East German rocket research programme.[98] Nevertheless, the first German in space was the GDR citizen Sigmund Jähn, who was a cosmonaut on board the Soviet spacecraft *Soyuz 31* in 1978.[99]

In West Germany, private associations kept rocket research alive while state-financed big science was banned until the mid-1950s. Strictly speaking, however, it was largely a matter of mere rocket tinkering, which nonetheless provided the basis for a certain continuity in German rocketry. In 1951, the Northwest German Association for Aerospace was established. The Allies even permitted the testing of two rockets in 1952.[100] This informal period of semi-legal rocket tinkering ended in 1954 when the first official West German rocketry institute was established in Stuttgart. This was in itself remarkable, because its establishment came before the Allied ban of military research had ended. Two years before the first nuclear research institutes were established, this Stuttgart rocketry institute marks the modest beginning of West German big science conducted as a collaboration of government and science.[101] This facility was camouflaged as the 'Institute for the Physics of Jet Propulsion' (*Forschungsinstitut für Physik der Strahlantriebe*) to alleviate international distrust of German rocket science.[102] Formally, the establishment of the military alliance Western European Union in 1954 paved the way for the return of German aerospace research, even if limited by a range of constraints. International cooperation was the only way for German rocket research to be revived. After the Second World War, the international community had good reasons to distrust German high-tech military endeavours.[103]

Initially, the Ministry of Transport was responsible for rocket research. This provided German rocketry with a new image, desperately needed after the V-2 attacks of the Second World War. Now rocketry was depicted as a means of transport. Later on, the Ministry of Transport lost its impact on big science, as the Ministry of Defence established its own aerospace industry in the latter half of the 1950s. From the early 1960s onwards, the West German government supported fledgling discussions on pan-European rocketry, not least to overcome international fears of a new German solo effort towards military rocketry.[104] European aerospace development also gained momentum from Cold War developments. In particular, the Soviets' success in launching the *Sputnik* satellite in 1957 gave impetus not only to US aerospace endeavours but also to the Europeans. In the early 1960s, the leading European countries, including Germany, were deeply interested in expanding European big science because they feared being overrun by the two superpowers and becoming entirely dependent on their ally, the United States. Some voices such as the German conservative politician Franz Josef Strauß preferred cooperation with the United States, but the vision of Europeanization prevailed. Thus, the idea of a European aerospace programme emerged, built on the model of the European Organization for Nuclear Research, or CERN, in Switzerland. In 1961, the leading European states formalized their cooperation in aerospace research with the establishment of the European Space Research Organization, or ESRO. Thus, in the early 1960s, West Germany became both a member of the ESRO and the European Launcher Development Organization, or ELDO.[105]

Tensions between unilateral actions driven by national vested interests on one hand and the political goal of Europeanization on the other hand nevertheless continued. Consequently, West German national ambitions were not entirely stifled. In 1960, the minister of defence commissioned the German Institute for Aviation, or *Deutsche Versuchsanstalt für Luftfahrt* (DVL), to coordinate German efforts in aerospace. This research policy stemmed from a desire to merge aviation and aerospace research. Until today, this unity of aviation and aerospace is a German peculiarity. The German path to aerospace depended on scientific research in another way that was almost unique in the West.[106] In general, large West German technologies such as aerospace were based upon the cooperation of state-sponsored big-science institutions with virtually state-owned companies.[107]

In 1962–3, the German Commission for Aerospace proposed the national development of a satellite, which was supported by the government. Urged to decide between the national and the European option of satellite development, the West German government privileged the national project, which was more generously funded.[108] German scientists and industrialists were also more interested in this national programme than in supranational projects. In the end, however, this first German satellite Azur, which had been the most important part of the national aerospace agenda, turned out to be a total failure.[109] On the other hand, supranational endeavours were important for the German government as proof that Germany had become equal to France and Great Britain in high-tech industries again. The ELDO was also a total failure, however, and ceased its efforts in 1973. It failed in its aim to develop a launcher for satellites, and the rocket 'Europa' never had a successful launch. Wernher von Braun had foreseen this failure in 1965, doubting that the European countries would be able to overcome their national rivalries. Von Braun had long been convinced that only a common European endeavour held promise.[110] However, even after the establishment of the European Space Agency in 1975, the single European states still followed their own agendas and failed to develop a common aerospace strategy.[111]

For some decades, the United States was both a partner and a model for European aerospace projects. It had been virtually impossible for West Germany to re-establish aerospace in the 1960s without American cooperation. For the United States, the cooperation with the European countries represented a chance to control European aerospace endeavours and thus avoid unwanted competition. From the late 1960s onwards, the European aerospace project had some success with satellites, albeit dependent on American launchers. Today, US aerospace remains dominant, but not to the extent it was during the 1960s. European countries have to some extent caught up. This became obvious in 1981 with the success of the Ariane rocket. Although developed in Europe, this rocket nonetheless still depended on essential components imported from the United States.[112]

There were also national hindrances to aerospace success. The ministry for atom issues, which would become the research ministry in 1963, led German aviation and aerospace politics, but two other ministries had an important say on these questions. As such, overlapping competencies hindered German aerospace research. Only in 1967 would the establishment of the German Test and Research Institute for Aviation and Space Flight, or *DFVLR* (which became the German Research Institute for Aviation and Space Flight/*DLR* in 1990 and finally the German Centre for Aviation and Space Flight in 1997), paved the way for efficient restructuring. Thereafter, the research ministry was solely responsible for aerospace affairs.[113] A further institutional change also favoured German aerospace development. On one hand, the establishment of the German Space Agency or *Deutsche Agentur für Weltraumangelegenheiten* (DARA) in 1990 enabled Germany to participate more easily in international big science cooperation. On the other hand, legislative control of aerospace policy was severely limited. Now, parliament had only a limited say, once international cooperation had started. The dilemma was obvious: big science to a certain degree relied on a certain loss of democratic control. Additionally, as the sociologist of technology Johannes Weyer has argued, the international high-tech race is to some degree a product of large technology companies.[114]

Nuclear power

There has been much speculation as to why the Nazis did not succeed in developing an atomic bomb. Despite its outcome, the sheer menace of Nazi nuclear weapons alone had driven their war opponents to intensify their own nuclear research programmes. The world's first comprehensive big science project, the American Manhattan Project, was initially motivated by the 'spectre of Nazi nuclear weapons'.[115] In August 1945, however, the US nuclear attack on Hiroshima was not least a demonstration of American military power to the then US ally and future Cold War opponent, the Soviet Union.[116] The effort to develop the bomb was enormous. The Manhattan Project had a budget of more than a billion dollars and roughly 250,000 employees in various research institutions throughout the United States between 1942 and 1945.[117] By contrast, Nazi Germany 'squandered huge resources on its rocket project'. While the V-2 project was smaller in scale, it was nonetheless somewhat 'comparable to what the Americans spent on the Manhattan Project'.[118] On the other hand, the Nazi government did not give the highest priority to its nuclear programme. The following section will explore the reasons for this decision.

The starting conditions for German nuclear research were near perfect. Weimar Germany was 'the mecca for physicists' attracting promising international scientists such as Robert Oppenheimer who would become the leading physicist of the Manhattan Project. The Nazi takeover saw some of Germany's best scientists expelled, but the country still had world-renowned physicists.[119] However, theoretical physicists faced some severe ideological attacks: some dedicated Nazis tried to establish 'German physics' condemning all kinds of theoretical physics as being 'Jewish in spirit'. The most prominent case was the Nobel laureate Johannes Stark who had been a long-term Hitler follower. Stark published an anonymous article in the SS propaganda weekly *Das schwarze Korps* in 1937 attacking his opponent Werner Heisenberg – also a Nobel Prize winner – for being a partisan of Einstein and thus a 'white Jew'.[120] This intrigue failed, and after these attacks, Heisenberg and his followers, who had a certain distance to the Nazis, were driven 'into the arms of National Socialist patrons more amenable to modern science'.[121] Thus, their research in theoretical physics continued.

In early 1939, the German physicists Otto Hahn and Fritz Strassmann published an article in the journal *Die Naturwissenschaften* describing their experimental demonstration that nuclear fission was possible. Soon afterwards, scientists in different European countries calculated that nuclear fission would produce enormous energy. And already in late 1939, French, German, British, Japanese, Soviet and American research on the military potential of nuclear fission began. A few months earlier, German Army Ordnance had started to consult physicists about nuclear fission in preparation for the war. This was for two main reasons: first, they wanted to avoid a situation in which the enemy would surprise them with any new weapon technology. Second, if nuclear fission should prove to be of military use, they had to ensure that it was Germany that developed this decisive weapon. For German physicists, the military interest in their research meant the chance to accelerate their careers. Most of the scientists were nationalists anyway, even if most of them were not dedicated Nazis.[122]

Soon, it became clear that there were two realistic paths to nuclear fission. In February 1940, Heisenberg calculated that nuclear explosives could be produced through a uranium machine on the basis of natural uranium with heavy water as a moderator. The alternative path would be isotope separation of uranium resulting in enriched uranium; in this case, light water would serve as a moderator. In summer 1940, German physicists explored the dual use of uranium machines: while slow chain reactions were of economic interest for heat production, fast chain reactions resulted in a potential weapon. In practice, any research on both paths to nuclear fission always implied research on its military potential.[123] Both alternatives were hard to realize, however, due to the scarcity of resources. Nonetheless, the Allies had serious fears about a German atomic bomb. By spring 1940, the Germans seemed to have it all: high-profile physicists and control over both the Belgian uranium mines and the world's largest heavy-water plant in Norway.[124]

By late 1941, German experts had prioritized the heavy-water path to nuclear fission due to failures in producing enriched uranium. This also seemed to be a realistic path with the Norwegian heavy-water plant under German control. The

Norwegian company Hydro had already begun to produce heavy water soon after American researchers had developed a procedure to separate heavy water from light water in 1932. Under the German occupation, the German corporation IG Farben took over Hydro and increased production. Yet, even this plant did not provide a sufficient amount of heavy water. Whereas five tons were needed for a single uranium machine, IG Farben only delivered 150 kilogrammes. As such, Army Ordnance changed priorities in 1941. Although they knew about the military potential of nuclear fission, they did not believe in the realization of an atomic bomb before the end of the war. The initial German victories seemed to be proof that the war would be won before the physicists could develop nuclear explosives. While nuclear research still received important resources, it was not the highest priority. Even some postgraduate students who were important researchers in the field were sent to the front. At that time, the famous Heisenberg was a mere advisor in the background while some of Carl Friedrich von Weizsäcker's doctoral candidates continued the practical work on the uranium machine.[125]

Even after the war had reached its turning point in winter 1941–2 when it became clear that the Russian campaign would not lead to fast victory, the nuclear programme was not upscaled. Historian Mark Walker points out that at that time the American and the German nuclear fission projects were 'evenly matched'. The only difference was expectations. The German physicists (correctly) informed Army Ordnance that they probably needed several years until success. Whatever the outcome of war would be, the German Army was sure that it would be over before nuclear weapons could be provided. Thus, nuclear research continued in a civilian institution. By contrast, their American opponents took the decision to dedicate vast resources into the big science Manhattan Project, even if for the wrong reasons, expecting as they did the war would continue for four or five more years. Given this time horizon, they hoped that the United States could successfully develop nuclear weapons first.[126]

This development sheds light on the failure of the Nazi nuclear project, which remains controversial. Paul Lawrence Rose has argued that Heisenberg fundamentally misunderstood 'the scientific principles of an atomic bomb'.[127] Against this, Thomas Powers has maintained that leading German physicists 'had no desire to make a bomb for Hitler'.[128] Nevertheless, both Rose and Powers miss the important point that, although the German researchers made good progress, 'they were never in a position in which they had to decide whether they should help make atomic bombs for Hitler'.[129] What is clear is that they did not show any moral concerns about working on weapons of mass destruction. It was only much later in a British POW camp that the Heisenberg group established 'the apologetic myth of resistance to Hitler by denying him nuclear weapons'.[130] In fact, the issue of success or failure in nuclear fission was less a matter of scientific genius but rather of technology, organization and production. Unlike the Manhattan Project, the German nuclear programme 'was simply not boosted up to the industrial scale'.[131] It all depended on the big science approach, large-scale industrial facilities, adequate hierarchical structures and, last but not least, staggering levels of funding. Although in theory Germany may have had the industrial resources, it is most unlikely that the Nazis had any chance to succeed with nuclear research before the end of the war even if they would not have privileged rocket and jet propulsion projects.[132]

Despite this, at the end of the war, the myth of 'Nazi Germany's scientific and technological superiority' was widespread.[133] While it was true that German experts were highly skilled, their former wartime opponents exaggerated the Germans' high-tech expertise. The Soviet nuclear programme benefitted far more from espionage in Los Alamos than from the expertise of German scientists deported to the Soviet Union after the war.[134] Nonetheless, German high-tech experts enjoyed a high reputation after 1945. In the early 1950s, the Argentine government fell for the fraud of German physicist Ronald Richter, who became head of a nuclear research centre exclusively staffed with German emigrants, but never even got close to the promised controlled thermonuclear reaction. The fraud was obvious but at the time this internationally known project gave a strong impetus both to the United States and to the Soviet Union to intensify their respective nuclear research.[135]

After the war West Germany was in need of international agreements for restarting any kind of high-tech research with military potential. The Treaty of Paris in 1952, which foresaw the establishment of the European Defence Community, allowed Germany to begin nuclear research even if on a rather modest scale. Although the treaty failed in 1954 because it was not ratified by the French parliament, it marked the beginning of West German post-war nuclear research. The German Research Foundation, or *DFG*, had established a commission for nuclear physics chaired by Heisenberg by 1952. Three years later, Heisenberg resigned because the research reactor was established in Karlsruhe and not at his preferred location of Munich. In any case, physicists lost their dominant role in reactor development in the following years. By that time, US practice had demonstrated that although physicists were good in the atomic bomb programme they had failed to construct a reactor for industrial use. Instead, engineers took over in close cooperation with the scientists of the Max Planck Society.[136]

In general, the history of post-war nuclear development is most adequately told when putting less emphasis on luminaries such as Heisenberg. Instead, the 'formation of the interest triangle among state, industry and science' should be central to our understanding.[137] Without doubt, NATO membership in 1955 paved the way for West German nuclear research. The nuclear research centres at Karlsruhe and Jülich established in the latter half of the 1950s received large amounts of state funding from the atom ministry, which developed itself into the research ministry. From the 1960s onwards, the ministry funded other fields of high tech such as aerospace and data processing on the model of nuclear research.[138] As such, nuclear research became a model for big science in general.

There has been a controversy among historians about the West German government's decision in favour of heavy-water reactors: did the specifically military potential of this kind of nuclear technology play a decisive role for the choice? The decision for heavy-water reactors meant potential autonomy while the light-water path would have resulted in reliance on imports of enriched uranium from the United States. In that case, the United States could have controlled the amount of delivered enriched uranium to hinder any German plans of military usage. The chosen path of heavy-water reactors, by contrast, promised vast production of plutonium. Officially, plutonium served as fuel for future fast breeders. On the other hand, plutonium was a dual-use material perfectly fitting for nuclear weapons.[139] Radkau has answered this question by explaining that the West German decision as a result of a kind of path dependency: he argues for a 'German line' of nuclear plans from the Third Reich to post-war West Germany. According to Radkau, Heisenberg's research in 1940 leads directly to the decision in favour of the heavy-water reactor, culminating in 1961, when the Karlsruhe heavy-water reactor FR 2 achieved criticality or nuclear fission chain reaction.[140] In hindsight, this explanation of heavy-water continuity makes sense. During the 1950s, however, West German experts seriously discussed graphite as an alternative moderator.[141] Against Radkau, Hanel and Hård have argued that only concrete political targets around 1960 can explain the heavy-water decision. For them, the 'German line' of this nuclear technology from 1940 onwards was not sufficient for the decision. According to Hanel and Hård, the multiple interests of different players were decisive: autarky, cheap energy or plutonium, be it for fast breeders or atomic bombs.[142]

It seems more appropriate to expand analysis beyond the intentions of politicians and interest groups. In general, dual use was inherent to any kind of nuclear power; and all protagonists understood this. The military use was pre-structured in both uranium enrichment and nuclear reprocessing, by which plutonium was extracted. Both were keys to the atomic bomb. Whenever the West German government was interested in reactor export to the Global South, it had to acknowledge the fact that these countries were mostly interested in the bomb option.[143] In the 1950s, nuclear fuel autarky was a dominant consideration and heavy-water reactors opened the door to nuclear technology without reliance on US imported enriched uranium. While the military option was crucial to the German government, this was not only an issue of the possibility of developing domestic nuclear weapons. Economic considerations pertaining to the possibility of exporting German reactors were probably more important.

In fact, there is no doubt that there was political interest in the bomb option during the 1950s. For a short period of time when he doubted the protective efficacy of the US nuclear shield, West German chancellor Konrad Adenauer fancied the option of a German nuclear weapon. Accordingly, in 1956 Adenauer conceived the establishment of the European Atomic Energy Community or Euratom as a path leading to national nuclear power. Despite this plan, Adenauer was not interested in producing a German atomic bomb. Rather, the mere national option of a nuclear bomb project was meant to serve as a political weapon for negotiations with the Western allies. Thus, it was a technological means of returning as an international player, even if the supremacy of the superpowers was accepted. In the event, Adenauer's point of view was rather isolated and soon thereafter, trust in US nuclear protection prevailed in any case.[144] In addition, the leading German experts including Hahn, Heisenberg and von Weizsäcker denounced the prospect of German nuclear armament in the famous 'Göttingen Manifesto' of 1957.[145] In hindsight, Siegfried Balke, who served as West German atom minister from 1956 to 1962, also acknowledged that the country's early nuclear politics had little to do with energy issues. By contrast, the military potential played a crucial role in the decision to invest significant amounts in this kind of technology. It was particularly important for export. The shah of Persia, for example, invited Balke to talks, even though Persia was an oil-producing country with no need for nuclear energy. It was the military option that had aroused his interest.[146]

In addition, Heisenberg pointed out that the first German nuclear export was due to his original concept of a plutonium-generating technology. Argentina hired a German company to construct a nuclear power plant in 1968 because this reactor was based on Heisenberg's war plans. The combination of natural uranium, heavy water and the production of plutonium was still attractive for both military and civil purposes.[147] Nevertheless, after the light-water variant broke in the 1960s because US nuclear technology had made great advances and promised efficient energy generation, the West German nuclear sector adapted to the new situation. In any case, light-water reactors also proved to be demanded export goods. In this context, even spectacular failures were helpful. The failed innovation of the *Otto Hahn* nuclear ship initially gained a lot of publicity in the late 1960s and early 1970s. This prototype ship took trips to Brazil, but it never reached the stage of series production. Brazil was nonetheless interested and, despite the final failure of the German nuclear ship project, the mere publicity it generated helped to sell German light-water reactors to Brazil. While the German government ensured that it was a mere civilian technology, the military option was of interest to Brazil with regard to the ongoing political tensions with its neighbour Argentina.[148]

To capture the whole picture of West German nuclear policy in the post-war era, it is crucial to analyse international relations. The 'German line' played a certain role, but the broader situation was decisive for the outcome. In the mid-1950s, the German government had to decide between the British and the American model of nuclear power. The American light-water reactors were combined with cumbersome uranium enrichment plants, which also offered a dual-use option. By contrast, the British model, which was quite advanced on its way to reactor series production, fitted better with the German traditions. In general, the German specialists already had the know-how for heavy-water reactors, while Germany would have relied on American know-how transfer for light-water reactors. Most of all, it was a decision based on the strategic goal of fuel autarky: the natural uranium path was based upon fast breeders and reprocessing plants, but avoided the most complex enrichments plants. The alternative of importing American enriched uranium was also rejected because it would have fostered the reliance upon the United States, too. In hindsight, it is surprising, how unimportant safety and economic issues had been.[149] At that time, safety issues were 'the stepchild of nuclear power developments' both in West and in East Germany. This changed only later, when safety also meant reliable energy supply and the US and Soviet safety regimes were adopted.[150]

Although a 'German line' of nuclear research explains some particularities of German nuclear history, in general, the American model influenced post-war nuclear development. Even the Europeanization of nuclear policy was the outcome of US impact, with the US government supporting the establishment of Euratom as a counterbalance to the British nuclear industry, which was the main competitor to US nuclear technology. To this end, the United States provided the means for a common European nuclear research policy and rejected German wishes for its own bilateral uranium deals.[151] In addition, the American 'Atoms for Peace' exhibition travelled around the world to propagate nuclear power and American leadership. In Germany, the exhibition toured Frankfurt and Berlin in 1955. Afterwards, nuclear technology transfer deepened Germany's reliance upon the United States.[152] Finally in 1963–4, the US light-water reactors became a huge commercial success and the particularly German emphasis on heavy water and plutonium became a dead end.[153] The first commercial light-water reactors started operation in the 1960s, while the only German commercial heavy-water reactor in south German Niederaichbach operated from 1972 to 1974. It was a complete failure with permanent technical and economic problems.

Initially, West German electric utilities had been reluctant towards nuclear technology. These companies embraced the light-water reactor, however, due to its improved energy efficiency.[154] From the mid-1970s, commercial reactors were established, with energy generation the prevalent motive for nuclear plants. Until the end of the 1980s, the number of West German commercial nuclear plants rose to thirty-one. West German nuclear technology was largely Americanized. Light water became the standard moderator, and US safety techniques were adopted. This adaptation was due to both the high standards of American commercial reactors and the West German 'desire to please its most important ally'.[155] The Kraftwerk Union Company – owned by Siemens and AEG – was the only producer of nuclear reactors in Germany. While the commercial use upstaged the military option, nuclear plants lost the former confidence of the public. Despite this rising protest scene, there was important political backing for nuclear plants. Before the 1986 Chernobyl catastrophe, all major German political parties except for the Greens supported nuclear energy.[156] Due to the energy crisis and the recession of the 1970s, economic and technological modernization became even more central to the political agenda, while the German provinces maintained a strong interest in the persistence of nuclear energy.[157] As Langdon Winner has argued in a well-known article, the history of nuclear plants proves that artefacts have politics. After the establishment of nuclear power the 'adaptation of social life to technical requirements' was easily justified. By contrast, 'those who cannot accept the hard requirements and imperatives will be dismissed as dreamers and fools.'[158]

The history of East German nuclear research is rather short. By the early 1960s, most of its expensive nuclear programme had ceased. Afterwards, the Soviet Union simply delivered power plants to the GDR. As was the case in West Germany, considerations about the military potential played an important role. The Soviets only supplied light-water reactors because they did not want East Germany to have plutonium-producing nuclear plants. Besides hardware, the GDR was also dependent upon knowledge transfers. For their part, the Soviets were reluctant to deliver both the latest reactor models and ongoing expertise. This resulted in severe security issues: for instance, the GDR reactors were not equipped with improved containment structures.

From the 1970s onwards, the GDR nuclear plants increasingly followed Western models and imitated Western technologies.[159] However, fuel autarky was never reached. The GDR relied upon energy imports from the West in the late 1980s. In the last months before the state's downfall, the GDR even considered Western support in repairing its nuclear plants and considered buying a Western plant. For safety reasons, reunified Germany shut down all East German nuclear plants.[160]

At that time, the West German nuclear industry also lagged behind the political plans of the early 1970s. In 1986, it produced 34 per cent less electricity than had been calculated thirteen years before. This stagnation too was a German particularity, especially when compared to its neighbour France, where atomic energy prospered. As is well known, Germany is currently a pioneer on its way to shutting down nuclear power after the Fukushima catastrophe in 2011. However, this is the outcome of a long history of scepticism and protest, which will be discussed by large in Chapter 8. In 2015, only nine reactors remained, generating 15.8 per cent of the total energy produced in Germany.[161] In 2020, there were a mere six nuclear reactors.

Conclusion

High tech and big science were always both national and international. On one hand, big science relied upon heavy state support. On the other hand, scientific progress was almost always the result of transnational cooperation. In many regards, the history of high tech in twentieth-century Germany is similar to many other industrial countries. There had been remarkable endeavours before 1945, but during the Cold War only the superpowers had adequate resources for big science projects such as aerospace and nuclear power. By contrast, Germany depended on European or American cooperation.

Similarly, the technological nationalism of the first part of the twentieth century was not a German particularity. Often, high tech was the projection screen for fantasies of national supremacy. However, at least after the First World War, high-tech nationalism became an important topic in German politics and culture in a particular way. While high-tech nationalism was also quite common in the Western colonial countries, many Germans conceived of their nation as having been treated unfairly by the Allied restriction and perceived high tech as a means of resistance.[162] According to this worldview, high tech was the result of a special German national character that would pave the way for national rebirth. This unholy alliance of self-victimization and high-tech chauvinism after 1918 was the perfect breeding ground for German fascism.

For scientists and engineers, the substantial financial support offered by the German state with an assumed military application for high tech in mind proved to be compelling, most often overcoming any moral qualms where there were any. Many middle-class experts, however, were convinced nationalists who believed in the rightfulness of their cause. In particular, during the Second World War, delusions of high-tech wonder weapons opened unknown opportunities for engineers and scientists to follow their own ends. The hubristic promise of decisive weapons gave the resources – both material and manpower – to realize their most ambitious projects. In the case of the V-2 rocket, however, this proved to be just another vehicle of senseless Nazi mass murder – both for the civilians in bombed cities and for the slave labourers of the Nordhausen tunnels – while it had no strategic use at that time.

In times of peace, or to be more precise Cold War competition, corporations and engineers benefitted from the promises of both the space age and the nuclear age. Big science funding reached unknown amounts for risky projects, and only a minority of these were successful. In general, politicians justified these projects with military spin-off myths while in fact it was the hopes for later civilian uses of military inventions that were the least realistic.[163] Leaving aside cost-effectiveness criteria, however, the investments in high tech resulted in a certain intensification of technological efforts, from which the economy and research institutes benefitted.

4

Visions of progress

There is more to the history of technology than inventions, innovations and usage of technology. To a certain degree, past storytelling is an important aspect of this history. From the Enlightenment and the industrial revolution onwards, narratives of progress sought to make sense of the multitude of innovations. Only these narratives have made them appear to be part and parcel of 'identifiable historical patterns of progress'.[1] In these narratives, 'progress' implied a concept of historical development, which went far beyond the mere idea of improvement. Historian Robert Friedel distinguishes between a more general 'culture of improvement' and this specific 'faith in progress'. According to Friedel, the culture of improvement characterizes Western history at large; it is the belief in the possibility to do things better through 'small, gradual improvements' mostly contributed by 'ordinary, anonymous workers and tinkerers'. By contrast, faith in progress is based upon the teleological belief that technology 'is moving linearly to a divine end'.[2] This divine end has, of course, been secularized. Throughout the nineteenth and the twentieth centuries, the notion of progress served manifold ends: be it civilization as such (with racist and colonial undertones), the glory of the nation or the overcoming of capitalism.

This chapter investigates the multiple visions of progress in German history. Since the latter half of the nineteenth century, both Marxist and liberal thinkers viewed technological innovation as a vehicle for social progress, but its application remained contested. After the First World War, technocracy and social engineering became the most influential agents of progress. Designing central aspects of modern life and work, many engineers were convinced that they could deliver an antidote to social unrests. With the rise of the National Socialists, alternative visions of technological progress emerged that did not comply with the liberal or socialist visions that the modernists had believed in. Competing liberal and socialist visions of progress gained both strength and support during the Cold War's economic boom. In the socialist GDR, the concept of the 'scientific-technological revolution' was crucial to technological and economic policy until the serious economic crisis of the 1980s. It was believed that further innovations in productive technologies, in particular automation, would lead socialism to win the Cold War economically. In capitalist West Germany, the popular perception of ongoing technological change – most prominently nuclear power and computerization – was more ambivalent. Yet, even critical minds often employed visions of progress to their own oppositional ends.

In 1917, the famous sociologist Max Weber gave a lecture entitled 'Science as a Vocation' that has influenced assessments of modern technology ever since. Best known is Weber's phrase that 'the world is disenchanted' through the progress of science and technology. According to Weber, however, scientific and technological progress did not 'indicate an increased and general knowledge of the conditions under which one lives'. Thus, contemporary Germans did not understand more about the conditions of their own lives than their ancestors or non-Western people did. The crucial point was that they did 'not need to know . . . Unless he is a physicist, one who rides on the streetcar has no idea how the car happened to get into motion'. Yet, the passenger could count on the modern transport technology to function. Furthermore, he or she knew or believed 'that if one but wished one *could* learn' about the technical matters. In principle, anything was calculable. For modern Westerners, there were no 'mysterious incalculable forces' anymore.[3]

Ordinarily Weber's lecture is taken to represent modern thinkers' trust in the progress of rationality, science and technology. By this account, pre-modern mysterious forces were replaced by the 'disenchanted' modern world of science and technology. Bernhard Rieger has criticized Weber's concept of disenchantment for neglecting crucial aspects of technology assessments in the nineteenth and early twentieth centuries. According to Rieger, innovative technological artefacts astonished most of Weber's contemporaries, not least by their sheer size. Many conceived of these novel technologies such as railways, bridges and electricity as 'modern wonders'. This perception resulted in two different reactions: 'both public euphoria and technophobia'.[4] In the long run, however, even this ambivalence fostered the acceptance of technological developments. To a certain degree, public observers admired the new technological artefacts because they did not totally grasp how they functioned. Thus, the sublime character of modern technologies was based upon a combination of astonishment, admiration and even fear. One misses the point when neglecting the emotional aspects of visions of technological progress.[5]

Even if Rieger's point is convincing, he misunderstands Weber. As has been shown, the eminent sociologist stated that contemporaries did *not* understand technology. Instead, they merely conceived of spectacular and wonderful technologies as outcomes of modern science and technology, which they did not have to understand in detail but put their trust in nonetheless. This chapter explores this very combination of emotion and rationality, of modernity and archaic beliefs because visions of progress have been built upon both.

Technological innovation as a vehicle for social progress

In the mid-twentieth century Alexander Gerschenkron pointed out the emotional components of progress narratives. In his controversial study on 'economic backwardness', Germany figured as a prime example of a 'backward country' in the nineteenth century, which had experienced neither political revolution nor national unification. According to Gerschenkron, rational arguments dominated the debates about industrialization in 'advanced' countries, whereas 'a New Deal in emotions' was needed to convince the public in backward countries. Thus, the famous German promoter of industrial progress Friedrich List employed a language of 'nationalist sentiment' for his cause.[6] It is very doubtful if this emotional aspect in narratives of progress has really been a particularity to 'backward' countries – leaving aside that research findings have rejected the assumption of a general German backwardness (see Chapter 1). Nevertheless, Gerschenkron hit the nail on the head when pointing to the relevance of emotions and in particular national sentiments.

Although nineteenth-century Germany was not backward in general, German officials did worry about being backwards for the first time. The industrial revolution had brought an unknown boom to the English economy, and the emerging concept of progress was a game changer. Before that time, economies were conceived of as either weak or strong. Now, 'people started talking about advanced and backward economies', and the concept of progress had reached the realms of technology and economy. This was a 'new way of thinking' about industrial development. Furthermore, a mutual relationship between nationalism and the concept of progress was asserted. German unity offered the opportunity to strengthen the political economy with combined forces and to close the technological gap to England.[7]

This was not, however, a monodirectional development. Sometimes, the pacesetters of German industrialization had visions that clearly differed from the later outcome. The first generation of Prussian bureaucrats who pushed the technological and industrial development in the first half of the nineteenth century, for example, had visions of an alternative modernity that never eventuated. Leading officials like Peter Beuth envisioned a 'rural, aesthetic industrialization' for Prussia. In the end, their own policy of business development led to mass construction of railroads and an enormous increase of heavy industry.[8]

By the mid-nineteenth century, common people also shared the fascination with technological progress. Industrial exhibitions were among the most important popularizers of modern technology and the very notion of progress. These exhibitions established themselves in the 1830s and peaked in the late nineteenth century, attracting millions of visitors all over the country. They proudly presented scientific and technological products that had been 'made in Germany', reinforcing the idea that technological progress and national strength were indissolubly intertwined. 'German work' manufactured products in 'German style', which led the newly established empire towards economic and political strength in the late nineteenth century.[9] These industrial exhibitions built a bridge between visions of the future and representations of the past. In this context progress became tangible and the future was predictable. Watching the exhibitions, visitors had the impression that they were taking their first look into the future. At the same time, the aesthetics of the exhibitions owed much to traditional middle-class high culture. Thus, the visitors saw novel technologies, but not in opposition to traditional values. On the contrary, the old and the new were presented as fitting perfectly together. The exhibitions presented progress as an evolutionary process that would conserve national traditions without challenging the social and political order.[10]

Belief in technological progress was omnipresent in nineteenth-century Germany beyond the boundaries of class and politics. In particular, Karl Marx's account of technological progress had a lasting effect on the German labour movement and the political left in general. Interestingly enough, Marx's critical analysis of the industrial age largely subscribed to a description given by the British liberal advocate of the capitalist factory system, Andrew Ure, in 1835. Ure painted a picture of a near future in which the automated organization of factory labour would gradually make skilled labour redundant. In this automated factory of the future, skilled workers would be replaced by workers who merely minded the machine.[11] Marx likewise followed this vision in the first volume of *Capital* in 1867, contradicting only Ure's political assessment. Just like Ure, Marx himself expected an 'automatic system of machinery' in the none-too-distant future that would largely do without human labour and require 'only attendance' from the worker.[12]

Besides this vision of automation as a process that had already begun, Marx's most sustainable impact on left-wing technology policy was the assertion that technology was neutral in principle. The workers only had to learn to direct their protest against 'the capitalistic employment of machinery', not against machinery as such.[13] The machine breakers known as Luddites in early nineteenth-century Britain had failed to make this distinction. But several decades of work experience in capitalist factories had seen the industrial workforce learn.[14] If the ownership of the means of production was changed after a socialist revolution, technology in itself would serve the cause of the workers:

> Since therefore machinery, considered alone, shortens the hours of labour, but, when in the service of capital, lengthens them; since in itself it lightens labour, but when employed by capital, heightens the intensity of labour; since in itself it is a victory of man over the forces of Nature, but in the hands of capital, makes man the slave of those forces; since in itself it increases the wealth of the producers, but in the hands of capital, makes them paupers.[15]

Following Marx's analysis, socialists and leftists alike were convinced that technology as such was neutral and that it was a matter of politics to design technology in specific ways to foster social progress. This belief gave the German labour movement a long-term goal that was crucial to its political strategy. By contrast, the labour movements in Britain and France were more interested in short-term objectives to improve workers' conditions. In these countries, a rejection of technological change was more common, whereas the German left embraced technological progress in general.[16] This had practical political consequences. At the party level, Social Democrats supported, for instance, the establishment of the Imperial Institute for Physics and Technology in the late nineteenth century, whereas many conservative members of parliament opposed it. While liberals were in favour of the institute due to the economic benefits of science-based industry, Social Democrats joined in the notion of progress, but gave it a different twist, hoping for a future of scientific socialism.[17]

At the grassroots level, initially, two different interpretations of technological progress competed. For example, the metal workers' union of Solingen conceived of factories per se as representing progress. Their perspective was based on a teleological understanding of history, which expected an inevitable economic and technological development resulting in social progress. Thus, the union's goal was bargaining, and it neglected the concrete methods of production. The union conceived of 'technological progress' as the workers' natural ally against capitalism because technological development by itself would dissolve capitalist production. This union policy caused criticism from two sides. First, grinders working in cottage industries criticized the union for its naïve conception of technological progress. The union had missed that the introduction of partially mechanized grinding had meant no real progress because it was far below the quality standards of handicraft. Second, the anti-union workers' weekly newspaper had a Marxist opinion towards technological progress. They looked down on both Luddites who blindly opposed and unionists who blindly celebrated capitalist machinery. As late as on the eve of the First World War, some unionists realized that rationalization at least partially downgraded the conditions of work and was anyways a means of enhancing management's power. However, these unionists belonged to the internal opposition and did not represent the union's official position.[18]

There was broad public consensus in the nineteenth century that embraced the notion of technological progress. Even those parts of society that had been initially sceptical towards technological change endorsed novel technologies in the latter part of the century. After the establishment of the German Empire in 1871 (at the latest), the overwhelming majority of Germans were quite enthusiastic about technology. It was the one aspect of modernity that Germans embraced the most.[19] Contemporaries had witnessed rapid technological change (as described in Chapters 1 and 2), most prominently the coming of electricity and new technologies of communication and mobility. Due to this experience, technological change seemingly promised further improvements in social life in the future.[20] Even before the breakthrough of electricity, the hopes associated with this new technology attracted more than one million visitors to the 1891 International Electricity Exhibition in Frankfurt.[21] This exhibition followed the example of US electricity exhibitions that had presented the new technologies as 'man-made wonders' to masses of astonished visitors by the 1880s.[22]

Furthermore, most Germans also associated the rise of their nation with technological development, described by one observer as 'the most wonderful technical progress that humanity had ever seen'. This widespread enthusiasm for technology would last well into the twentieth century.[23]

In the late nineteenth century, even conservatives made their peace with modern technology. Technology design that combined the old and the new eased the conservative acceptance of innovations. The first automobiles resembled carriages, train stations were built in the neo-Gothic style and historic facades decorated modern factory buildings. Thus, conservatives embraced these modern artefacts because they still signified traditional cultural values while somewhat hiding their functional aspects.[24]

Nevertheless, at the turn of the twentieth century, most German observers were aware of the fact that 'culture itself had metamorphosed' due to the triumph of modern technologies. By then, railways, telegraphs, airships and electric tramways had become essential components of German culture. As a result, German national identity was increasingly built upon the international comparison both with its 'civilized' European rivals and those non-Western states, which represented 'old culture'.[25] This conception of technological progress as a global competition of nations had an impact on colonialism. Rivalry with other colonizing states like Britain motivated the German government to sponsor colonial technologies such as the unsuccessful AEG endeavour with wireless telegraphy to counteract the British submarine cable monopoly.[26] Furthermore, colonial racism was to a certain degree based upon assumptions of technological progress. For instance, the former fighter pilot Gunther Plüschow depicted his adventure stories of the colony Qingdao in the language of technological supremacy. According to the German pilot, the Chinese 'children of nature' had been totally bewildered by Germans' modern technology. The self-image of white supremacy was, at that time, not entirely based on a biological racism but rather on a hybrid form of racism that emphasized the achievements of German technology and culture.[27]

In the colonial metropole too, German engineers established a self-image as 'bearers of culture', to follow an expression of the famous engineering professor Franz Reuleaux.[28] Many middle-class engineers asserted that in the long run, modern technologies would increase the nation's cultural life.[29] Around the turn of the twentieth century, most German engineers believed in the quasi-unlimited progress of technology. Their optimism was nearly total, and they conceived of their own profession as social engineers who could solve any social problems by technological means.[30] These proto-technocrats camouflaged their political agenda as altruism or service to the nation while in fact following a middle-class agenda.[31]

Several historians have asserted that there was a crisis of modernity around 1900. Hänseroth, however, argues that this assumption was wrongly based on privileging contemporary voices of the humanities and the fine arts who in fact often criticized some of the outcomes of modern technology. Although many disliked certain aspects of modern technology, most middle-class Germans were optimistic about technological development and its social consequences, with the overwhelming majority looking forward to further developments.[32] In this context of widespread technological optimism, even technological catastrophes such as air crashes seemed to make sense. Those victims were perceived as collateral damage or even unavoidable sacrifices for technological progress. These accidents were also seen as bringing important future safety improvements.[33]

For many, aviation symbolized the possibilities of technological progress par excellence. Indeed, early aviation combined myth and high tech, with older dreams of flying now realized through modern technological means. While the world became disenchanted by technology, the very same technology fascinated the masses and became a myth in itself. In the light of the enormous technological achievements of the late nineteenth and early twentieth centuries, the idea of general feasibility became widespread. In a particular mélange of creative conception and rational planning capability, engineers conceived of themselves as wanderers between the spheres of art and science.[34]

Even those who criticized modern civilization around 1900 were not reactionary per se. Most of those critics embraced modern technology, but dreamt of an alternative modernity. Later, the Nazis followed this approach, conceiving of modern technologies as a means to ends very different to the visions of a liberal modernity.[35]

The rise of alternative modernisms after the First World War

After the First World War, concepts of feasibility increased further. Thanks to modern technology and social engineering, many social and political issues had already been improved, for instance, in urban planning and public health (see Chapter 2).

To a certain degree, the 'dream of perfectibility' was based on such successes. As will be shown, this dream 'that anything can be done' dominated modernists' thinking during the first decades of the twentieth century.[36] In this context, visions of progress peaked. As Peter Fritzsche has argued, 'the social experimentation of the nineteenth century was only a preamble to the hyperimaginative technologies of the twentieth'.[37] In general, this development was continuous, although the German defeat in 1918 was disruptive to some degree. A specific nationalist science fiction literature, filled with imaginary wonder weapons, emerged in Germany. These bestselling books once again mingled technological progress with dreams of national rebirth.[38] Although there was a similar debate about technology and the nation in Britain at the same time, it differed profoundly from its German counterpart. Whereas in Britain, technological development was meant to defend the status quo, many Germans hoped for future technologies that would aggressively transform the global order.[39] That is, the political context was of utmost importance to the meaning inscribed to technological artefacts. While in the mid-nineteenth century US, 'sublime technological objects were assumed to be active forces working for democracy',[40] similar artefacts became forces of nationalist and anti-democratic ambitions in Weimar Germany. To be sure, the political meaning of technology was always contested. In the 1920s, the Social Democrats praised Zeppelin's newest gigantic airships as 'progress of human civilization', even though their predecessors had been symbols of German aggressive nationalism (see Chapter 3).[41]

Assessments of technology that embraced the universalist values of the Enlightenment were prevalent in Germany until the Great Depression put an end to any kind of global optimism. Before then, Germans had celebrated communication and transport technologies for opening up the world. Suddenly, former faraway regions seemed to be very close. New technologies provided the means for global commerce and European, or – as many hoped – German dominance. In some fields, globalists and nationalists shared certain goals, although their political opinions clearly differed. More generally, however, the visions of Germany's comeback as a global player relied on a racist concept of technological supremacy.[42] In particular, German engineers had long assumed a nationalist synthesis of technology and culture. At the latest, the crisis of the late 1920s brought these traditions to the fore and 'also opened the door for Nazification'.[43] Crucially, modern technology and modern society had many 'radically differing potentials'.[44] As such, the 'reactionary modernism' that Jeffry Herf described for the Weimar Republic and the Third Reich was but one possible outcome of modernity that saw the Weimar right-wing anti-democrats and the Nazis blend 'anti-modern and modern elements', embracing modern technology but rejecting the values of humanism and democracy.[45] Strictly speaking, Germany did not fail to modernize. Instead, it had become 'a nation of troubling modernity'.[46]

In general, a strained combination of convictions characterized modernity: on one side, visions of progress made many contemporaries believe in the enormous potential of technology – almost anything was feasible. On the other side, modern life, and not least modern technology, raised fears of 'permanent crisis'.[47] In Weimar Germany too, the discussion about the future of industry moved between these extremes: fear of the redundancy of human beings on one side (discussed in Chapter 8), and the hope of liberation through technological progress on the other. The belief in progress is exemplified in a 1930 publication by bestselling popular science author Hanns Günther about 'automatons': in his view, technological progress was already on the brink of engendering the 'self-conditioned automaton' that would relieve humans of the most difficult tasks and thus become the 'liberator' of humankind.[48] In addition, many engineers on both sides of the Atlantic believed in the transformative power of technology: they conceived of themselves as social engineers who would create new kinds of workers.[49]

There is some evidence for two particularities that underpinned German visions of progress. For one, Weimar political tensions had become a fertile ground for fears associated with modernity. Second, 'reactionary modernists' conceived of technology as 'an externalization of the will to power'.[50] Thus, 'danger and energy' were both associated with the future of technology and modern society.[51] Oswald Spengler's 1931 book *Man and Technics* (*Mensch und Technik*) was essential to this discourse. In particular, he introduced Friedrich Nietzsche's term 'will to power' into the realm of technology. Spengler, famous for his bestselling *The Decline of the West* (*Der Untergang des Abendlandes*), presented a history of technology that was more or less identical to the general history of mankind. In general, inventors followed the 'beast-of-prey nature of man'. Consequently, their inventions did not seek social usefulness. Instead, innovative technologies represented the personalities of their inventors who were solely driven by their will to power: in this case, the wish to overcome old technologies.[52] For Spengler, progress was a mere 'war on Nature' conducted by these men.[53] This 'cult of feasibility' was significant for both Spengler and the Nazis.[54]

Spengler's genius-centred narrative of technological development was combined with a deep-rooted cultural pessimism. He mocked his contemporaries who were driven by a 'rose-coloured progress-optimism'. By contrast, Spengler was convinced that he was witnessing the beginning of the decline of Western culture, or 'machine culture'.[55] In the near future, the former 'lord of the World', the 'Nordic man', would be degraded to a mere 'slave of the Machine'.[56] Afterwards, the 'Faustian man' would vanish and thus mechanical engineering would disappear from world history. His pessimism was based upon a racist narrative of progress that saw Spengler agree with most historians that the harnessing of coal was the most important resource for the rise of the West. According to his view of white supremacy, only the 'white' engineer had the competence to produce coal in large amounts, although many parts of the world had large coal reservoirs.[57]

Leaving aside his racism, Spengler offers a reasonably accurate account of the exploitive forces of colonialism. This, for him, had exacerbated the decline of the West because it had engendered a fatal 'treason to technics': the transfer of technical knowledge to the colonies, the 'dissemination of industry'. Due to this transfer, the Western countries had lost 'their greatest asset' forever. Over the long run, he argued, the 'coloured people' would be using the new technologies as skilfully as their former masters. Thus, white labour would soon be redundant. This would be the revenge of the 'exploited world' against 'its lords'.[58] This pessimism distinguished Spengler's technological racism from its counterparts in the West: white American's self-image around 1900 was similar but had a different imagined outcome. The descendants of European settlers were convinced that they had replaced the Native Americans because they had been 'technologically backward'.[59] By contrast, Spengler envisioned the revenge of oppressed peoples.

Spengler was a convinced anti-democrat but not a Nazi. Nevertheless, the Nazis admired his writings, but clearly privileged the 'cult of feasibility' over his inherent pessimism. Ultimately, its combination of 'danger and possibility' gave Nazi technology policy a particular form which revealed its most aggressive nature during the war.[60] To overcome any dangers of decline, it stated, the will to power had to become even more ruthless. In this context, the Nazis conceived of the most impressive technological wonders as the result of a uniquely German will to power.[61] This idea was combined with the adoption of some elements of the American technocracy movement, like the strong belief in feasibility.[62] To quote the words of Fritzsche, the outcome of these developments was a specific and 'horrifying' version of modernism: 'The Nazis were modernists because they made the acknowledgement of the radical instability of twentieth-century life the premise of relentless experimentation.'[63]

Technology euphoria was an important component to this Nazi variant of modernism, with Hitler designating himself as a 'fool for technology', or *Techniknarr*.[64] Most important to Nazism were technologies of mobility. Hitler predicted in 1933 that the roads constructed for automobiles would measure the future civilization of a nation. Once again, racism and notions of technological supremacy were intertwined: for the Nazis, German technological products proved their innate supremacy. The decisive point, however, had been that the 'German race' knew 'how to use mobile technologies in spiritually productive ways'.[65]

In addition, visions of progress still rested on hopes for further increases in industrial efficiency. In a very counter-cyclical manner, visions of automation peaked in 1944 when the end of the Nazi empire was near. An article titled 'Factories without People', or *Menschenleere Fabriken*, published in the official paper of the Nazi Party, *Völkischer Beobachter*, for example, presented an image of looming automation. It was authored by Helmut Stein, an engineer and plant manager. Stein generally favoured rationalization, but he believed the current level of mechanization, shaped by mass production, entailed the 'danger' of a 'degeneration of work'. His proposed solution was twofold: for one, Stein called for more 'distraction and relaxation' in workers' free time as compensation. At the same time, he proposed resolving the technology-induced problem through further mechanization: only 'full automation' could lead to the 'liberation of humans from monotonous and soulless work as such'. This concept sought, paradoxically, to salvage skilled labour: although all of the new tasks would 'be simple and require less training time', workers could (in terms of consciousness and self-validation) consider themselves to be the 'master of the machine forces', thereby experiencing 'spiritual satisfaction'. In this sense, they would remain 'skilled workers' and continue to sit enthroned, 'regulating and supervising', above the machine world. This vision of the 'factory of the future – the factory without people' was based on the expectation that this future would become reality very soon: that its realization was imminent.[66]

The Cold War and the 'scientific-technological revolution'

Such visions of automatic factories became pervasive after computer technology had reached an advanced level from the 1960s onward. Right after the war, novel technologies were even more central to contemporary imaginations of the future. Until the 1960s, dreams of the space age and peaceful nuclear power were the most efficacious visions of technology.[67] Both technologies played crucial parts in the Cold War and marked the end of any dreams of German autarkic high-tech development (see Chapter 3). In general, the experience of German war crimes by means of modern technology had shaken both Europeans' and Germans' belief in progress and technological modernity as such.[68] Notwithstanding this fundamental questioning, the article on 'progress' in the essential German encyclopaedia *Brockhaus* stated in the 1950s that advances in science and technology had still continued to develop and to contribute to rising standards of living.[69]

After the Nazi techno-future dreams which turned into a nightmare for millions of Europeans had fatally ended, humanistic visions of technological progress made a comeback in Germany. Above all, many placed their hopes in nuclear energy. Although technology was a main battlefield in Cold War competition, initially East and West Germany shared 'futuristic, utopian visions of the Atomic Age as an era of peace and progress for all humankind'. Modern technology, represented by the consensual force of nuclear power, meant the chance to overcome the shadows of the Nazi past and 'become modern, forward-looking nations'.[70] Prominent Social Democrats and Marxists alike envisioned lasting improvements for humankind through the 'Atomic Age'. Some even hoped for planned climate change by nuclear fission to improve infertile world regions. But liberals and conservatives also joined in these future dreams.[71] Despite this atomic euphoria, coal continued to be the most important energy source. In the 1950s, even more mines opened, and new mining technology made coal look like a future technology too.[72] Thus, as late as 1968, the *Ruhrkohle* corporation, a merger of several mining enterprises, proudly advertised that coal energy represented the future (Figure 4.1).

Initially, intellectual observers showed more ambivalence towards technological progress than politicians. While in post-war East Germany the socialist tradition of utopian visions of technology survived and was upgraded to the level of state doctrine, West German visions of technology were different. Cultural pessimism was more widespread because modern technology was associated with the Nazis' politics of genocide and the murderous war. To a certain degree, fears of technological determinism shaped West-German post-war debates on modern technology. Yet already by the 1950s, more optimistic voices began to set the tone. Intellectuals such as Arnold Gehlen, Helmut Schelsky and Jürgen Habermas believed in the possibility of a humane rationality and rejected the hitherto prevalent argument that modern technology had inherent totalitarian characteristics.[73]

In the 1960s, Habermas – beginning to become Germany's most important philosopher and public intellectual – made his point clear. He rejected both liberal and conservative interpretations of technological progress. Regarding the liberal point of view, he agreed that the possibilities of technology facilitated work and life for

Figure 4.1 'With energy towards the future'. Advertisement poster of the Ruhrkohle corp., 1968. Haus der Geschichte, Bonn.

many. But he also disagreed with liberals because he was well aware of the menace that technological progress left to its own devices or vested interests implied. With regard to conservative intellectuals such as Gehlen and Schelsky, Habermas similarly rejected their opinion that inherent necessities had driven technological progress for good. According to Habermas, this technocratic point of view had become the dominant ideology in both West and East. By contrast, Habermas pleaded for a concept of scientific-technological progress that emphasized the 'innocence of technology': technological development should become subject to deliberative discourse. Following this concept, political decision-making was to decide the concrete form that technological progress would take.[74]

At the end of the 1960s, the positive outlook towards technological change even turned into a wave of planning euphoria. The cult of feasibility had a certain revival in West Germany, while it had always been dominant in the socialist East.[75] At the same time, the aerospace and nuclear energy euphoria slowly faded away. Yet, these technologies had already fuelled visions of an all-automatic year 2000.[76] During Cold War competition, this race towards industrial automation was essential both for economic reasons and for the self-conception of the competing blocs.

From the 1950s onwards, the communist party of East Germany turned 'scientific-technological progress' into a central component of its programme and the Five-Year Plans. The goal was to win the Cold War by means of rationalization and automation, resulting in economic supremacy of the Eastern bloc, or, as the party conference of 1956 proclaimed, to 'catch up with and surpass capitalism in terms of technology'.[77] Although similar phrases were legion in socialist rhetoric, the East German communists were really convinced that the planned economy of socialism was 'peculiarly suited to modern technology, unlike chaotic, cutthroat capitalism'.[78] In hindsight, the communist functionaries were quite right about the importance of modern industrial technologies. In the end, the failure of the GDR to develop automation and computerization was one of the reasons for the economic difficulties of the system.[79] Furthermore, technology itself was central for the social and cultural life of the GDR. Visions of technology were crucial both for the 'conception of a socialist modernity' and for the construction of a 'national identity of East Germans'. The coming era of automation, or the 'scientific-technological revolution', turned technological knowledge into an important part of the 'socialist personality'. Thus, the factory was of utmost importance beyond the mere sphere of industrial production: it was the very place 'where the new socialist man and new socialist woman were forged'.[80]

The discussion of industrial automation was equally intense in West and East Germany. In both states it was claimed (albeit with differing emphasis) that the imminently anticipated technological change would spark revolutionizing transformations. The debate turned particularly euphoric between the mid-1950s and early 1970s.[81] In West Germany, the discussion among experts was quite differentiated: engineers expected the onset of full automation, that is, factories without people, only in large-scale mass-producing factories, while the target set for small- and medium-sized enterprises was partial automation.[82] The West German trade unions, too, were prominent participants in the debate surrounding automation early on. Trade union actors consistently considered both sides of automation during this debate: on the one hand, there was a fear of unemployment, and, on the other, there were the hopes associated with the social and political effects of technological progress that the labour movement had harboured ever since its inception.[83]

In East Germany, it was emphasized that fully automated factories could be both a promise and a threat depending on the state of production conditions. Correspondingly, in 1960, the popular nonfiction book for young readers *Unsere Welt von morgen* (*Our World of Tomorrow*) stated that automation in capitalism signified above all the danger of unemployment, while in socialism it would lead to the upskilling of workers because it would require significantly more and new knowledge about the production process.[84] What seemed clear, at any rate, was that fully automated factories would 'determine the face of future industrial work'. The illustration accompanying this vision of the future is almost devoid of humans; in this draft version of the factory 'of tomorrow', there are only two people standing in the control room (see Figure 4.2). It would still be a 'major step' to reach this point, but the book referenced an experimental automatic factory that already existed in the Soviet Union.[85] Finally, however, the failure of the Eastern

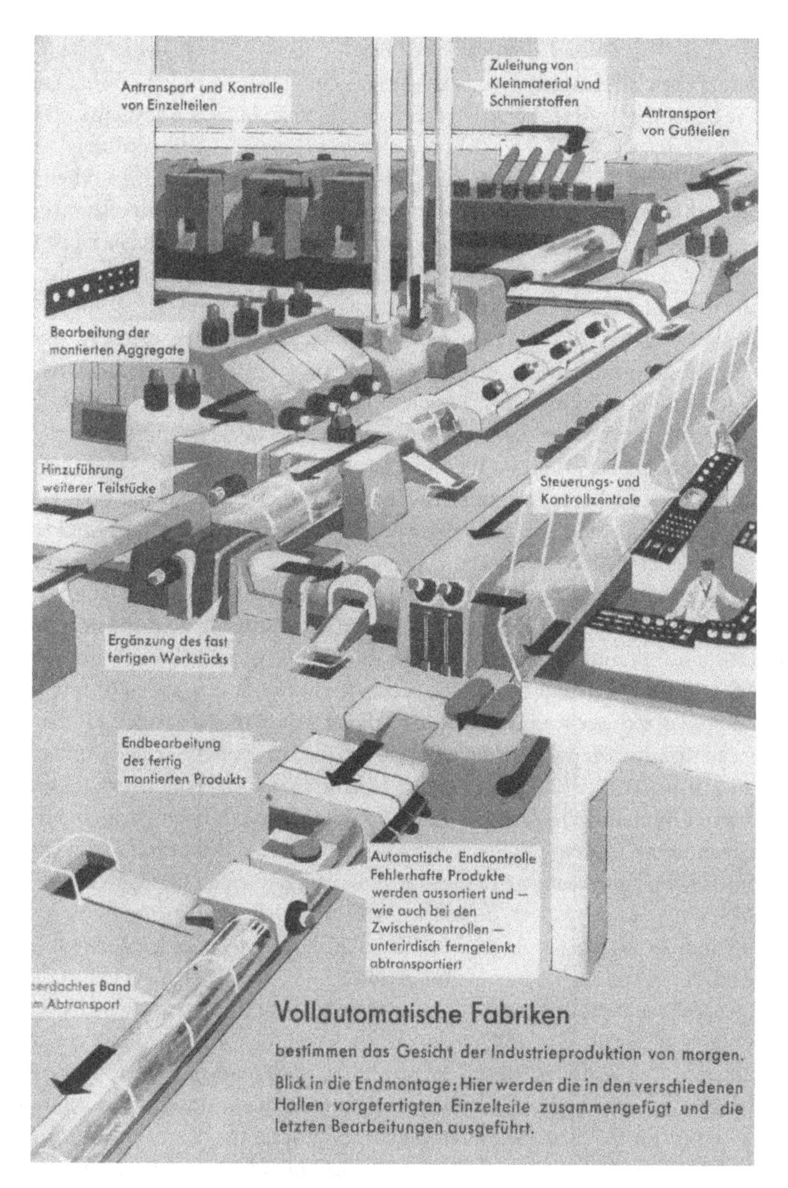

Figure 4.2 Fully automated factories. Illustration by Eberhard Binder-Staßfurt, in: Karl Böhm/Rolf Dörge: Unsere Welt von morgen (Berlin: Neues Leben, 1960).

bloc's microelectronic programme put an end to any hopes of a socialist 'scientific-technological revolution' (see Chapter 1).

In West Germany, industrial automation reached a new stage with computerization. One industry that was confronted with the more concrete advent of computer technology early on was the printing industry. During the second half of the 1970s, the first print shops in West Germany switched to computerized phototypesetting, and in 1978, there were fierce contract negotiations around the new 'computerized text systems'. The unions' political strategies for facing the challenge of computer typesetting drew on their historical experience: strong faith in union power combined with the belief that technology was neutral in principle. Indeed, during the dispute in 1978, the German printing union's board member Detlef Hensche continued to insist on Marx's dictum that 'the new technology – by itself – is neutral'.[86] In other words, the union thought that it was possible to cope with technological innovations: it had no fear of new technologies becoming entrepreneurs' newest weapons against labour interests. Instead, unionists were quite optimistic that their organization had the power to shape the social effects of technological change.

Histories from below show that the workers were also very receptive to visions of progress. Traditionally, in the printing industry, workers' self-perception relied on immense pride in working with the most recent technology. Thus, even in the late 1970s, as computerization continued to emerge, the workers still shared a strong belief in technological progress. The main base for this optimism was their belief that computerization had certain limits. The workers' experience and the trade's history suggested that skilled labour had never been replaced, and human skill would always be irreplaceable, at least to a certain degree. In 1977, a printer being interviewed pointed out that he and his fellow workers did not worry because they were like 'machine-men'. They had so much sense of the machines that it would be impossible to replace them with unskilled workers.[87] At a different company, a compositor admitted to another interviewer that he and his co-workers were in fact scared of future work in a computerized environment. Yet, one thought provided consolation: they were convinced that the computer would never do better work than the skilled worker sitting in front of it.[88] This idea demonstrates that most workers understood technology as an imitation of human skills: technology might get near the human original but could never be supreme. Moreover, most workers were convinced that technological progress was only gradual and that most innovations had already been implemented. Even in the late 1970s – in retrospect, the start of a decade that fundamentally changed the printing industry – it was common to think that further technological developments would affect only minor aspects of production.[89]

In general, the computerization debates of the 1970s and 1980s made people initially worry about job security. In 1982, in an article titled 'The Factory without People Will Be a Reality in Ten Years', the West German *Frankfurter Rundschau* newspaper reported on a study conducted by a market research institute that considered the technological preconditions for the factory without people, or at least the 'factory with hardly any humans', as given. It was expected that by 1990 half of industrial assembly processes in West Germany would be automated.[90] In the long run, however, from the mid-1980s onwards, optimistic projections of a future society enhanced by novel information and communication technologies prevailed.[91] Often overlooked, planning euphoria and technocracy experienced a comeback at that time. Based upon the unknown possibilities of advanced computer technology, belief in technological progress was even greater than in the post-war era.[92]

In hindsight, it is obvious that the increased criticism of naïve belief in progress expressed by many in the 1970s does not mean that concepts of progress had vanished. Having said that, the dark side of technological progress became obvious to many at that time.[93] As will be shown in Chapter 8, this was not for the first time in history. To a certain degree, criticism and scepticism were actually important components of the very notion of progress. Although the emerging environmental movement harshly criticized industrial societies for the squandering of natural resources, this criticism did not mean an end to visions of progress. Instead, new concepts of progress occurred. Best known was the Club of Rome 1972 report *The Limits to Growth* that was meant to bring an end to the fixation on quantitative growth. Yet, concepts of qualitative growth filled the void resulting in a new understanding of progress centred on human and ecological issues.[94]

In addition to these new visions of progress, the old path of visions of industrial automation continues through to the present. From 2011 onwards, the term 'Fourth Industrial Revolution' became a buzzword in German economic debates concerning the digitalization of the world of work. The essence of the phrase is the notion of a linear development. On the one hand, the most recent digital developments in industry are embedded in a long history of industrial progress that began in late-eighteenth-century England. Phase two and three are meant to represent mass production – the computerization of the 1960s to 1980s, respectively. On the other hand, the fourth stage is said to mark the advent of a new quality, a sea change in the industrial system.[95]

Conclusion

To what degree was there a particularly German vision of technological progress? Certainly, Germans experienced technological and social change in the late nineteenth century at a speed that had been even more rapid than in most industrializing and urbanizing countries. For this reason, many expected the past developments to continue in the future. Progress was not a mere idea but very tangible in everyday life. So was the concept of feasibility. Social change seemed to be plannable to a certain degree. In any case, engineers, urban planners and the like seemed to be able to affect the outcome of the future significantly. These processes were associated with nation-building and the rising international significance of German economic and political power.

The defeat in 1918 gave an impetus to German rightists to reconfigure the established bonding of technology and the nation. The tone of debates surrounding technology became even more aggressive than in Imperial Germany and its direction had changed, expressing the desires of a defeated nation seeking the means for a national revival. Technology seemed to offer those means. The concept of feasibility was further radicalized, leading to a hypostatization of the will to power. The resulting right-wing visions of progress omitted the values of the Enlightenment that had informed earlier discussions. Technological progress was to serve the growing strength of the reborn German nation. After the seizure of power, the Nazis combined pre-existing projections of automation with a new vision of modernity without humanism.

On the political left, there was another German particularity. Although Marx's writings inspired labour movements all over the world, the German communists, social democrats and unionists alike picked up Marx's concept of the neutrality of technology. This had long-lasting effects on left-wing political strategy. The German labour movement strongly believed in the possibility of shaping technological progress in favour of the working class. Thus, notions of technological progress generated confidence in innovations. In general, technological change was accepted because it promised social progress.

As with industrialization and urbanization, however, Germany was part of a broader Western history and many seemingly 'German' developments were transnationally intertwined. This was the case not only for technological artefacts but also for discourses on progress. Indeed, by the late twentieth century, German visions of technological progress had largely converted to the Western model.

Part II

New directions

5

The human body in a highly technified environment

The Latin term *industrius* demonstrates to what extent the history of the body is linked to the history of industry. The translation 'industrious' is most common, but 'hard-working' and 'diligent' also capture the essence of the Latin adjective. Research findings have widely acknowledged that hard-working bodies played an important role in the history of industrialization. Yet, the perspective of the history of the body is most often peripheral to these studies. This chapter explores the concept of technified bodies in modern German history. Industrialization changed the ways people perceived and experienced their bodies, at work as well as in all facets of life. According to historian Anson Rabinbach, the new technologies of the industrial age provided a new image of the human body as the analogy of the human machine became increasingly popular.[1] The long process of industrialization consistently challenged workers' bodies. Factory discipline told them when to move, when to stop or when to be silent. Scientific management and time-and-motion studies further resulted in new ways to control workers by establishing 'the one best way' to do a job (see Chapter 1). Nevertheless, workers' embodied knowledge remained important in automated work in the late twentieth century.

The factory was the first place for large sections of the population to interact with machines.[2] In hindsight, this was only the beginning of a vast transformation of work and life. The second part of this chapter explores the ongoing proliferation of technology in everyday life that meant severe changes for the experience of one's own body. Beyond the workplace, transport was arguably the most important arena of body–technology interactions. From the perspective of the history of the body, means of transport were 'mobility machines' that made people experience new ways of enhanced mobility.[3] This was true both for public transport and for individual ways of travelling and for motorized and non-motorized transport (with the most prominent example of the bicycle). While mobility machines represented one essential aspect of body–technology interactions, namely enhancement, there was also the counterpart; the fear of being replaced through ongoing technological progress.

The artificial bodies of both earlymodern android automata and twentieth-century robots oscillated between these two poles: on one side the utopia of enhancement, on the other side the threat of replacement. In the late eighteenth century, android automata had become very popular all over Europe. Most spectacularly, some of these mechanic automata played music. For the effect of their performance, it was of utmost importance that they looked and acted like humans. For example, the organist manufactured by the Swiss clock-making brothers Jaquet-Droz as well as the dulcimer player of the German cabinetmaker David Roentgen and his partner, the clockmaker Peter Kinzing, imitated the physical movements of human musicians. The dulcimer player moved her eyes to the tempo of the music, while the organist turned her head and imitated breathing by making her chest expand. At first glance, this had nothing to do with the automata's real task, the mechanical performance of music. Yet, it was an essential component of the shows, which toured Germany and other European countries. The automata's bodies were meant to present 'human' emotions to the audience.[4]

By around 1800, however, interest in automata was waning. In 1805, the famous German writer Johann Wolfgang von Goethe reported visiting a friend in Helmstedt. His host was storing some of the once-famous automata of the Frenchman Jaques de Vaucanson. Goethe commented that 'in an old garden-house the flute-player sat in very unimposing clothes, but his playing-days were past.'[5] As will be shown, however, the metaphor of the automaton had an important afterlife in nineteenth century, inspiring observers to analyse the novel factories and the special relationship between man and machine in industry.

While the early modern android automata were mere simulations of human physical attributes, the robots of the twentieth and twenty-first centuries were meant to enhance human potential.[6] Both threatened, it seemed, to replace humans.

Figure 5.1 'Machine-man' Robot presented by his British inventor W. H. Richards, Berlin 1930. Photograph by Georg Pahl. Bundesarchiv.

Where automata had little success in becoming indistinguishable from humans in any respect due to the technological limitations of these mechanical artefacts, robots seemed to embody a potential to overcome humanity. Before industrial robots proliferated in the 1970s, there had been much speculation about the forthcoming breakthrough of robotics in the first part of the twentieth century. From the end of the 1920s onwards, so-called robots toured technology exhibitions throughout the Western World. In fact, those artefacts were mere simulations of robots without any practical applications.[7]

One of the best-known robots, 'Eric', was presented by the British engineer W. H. Richards in London in 1928 and starred at a Berlin exhibition in 1930 (Figure 5.1). This 'man of tin', as the inventor called him, seemed even to be speaking to the audience. Yet, it was of course only the pretension of mechanical speech, while in fact the words were transmitted by radio.[8] 'Eric' had drawn attention in Germany even before he visited the country. The German metal workers union's weekly paper featured a somewhat scary article on 'machine-men' in 1929. In the context of Taylorism and Fordism, the coming of artificial men seemed only to be the next step in the unstoppable progress of technology. The author reported on one of Eric's shows in New York, claiming that Eric resembled the human body most closely of all robots. Even if he was only a 'man without soul', he nevertheless looked 'scarily human' for an artificial mechanism.[9]

By contrast, the industrial robots of the late twentieth century had a modern design: form followed function. Most often, those robots actually were only robot arms constructed for lifting or welding tasks. Thus, they did not resemble human body forms at all. The design changed only with the advent of service robots. While robots operating among other robots ordinarily had a purely functional design, there was an incentive to give robots operating among humans a humanoid look.[10] The ambivalent relationship between human and artificial bodies, as well as the experience of body enhancement, will be discussed later in this chapter.

More than discipline: Productive bodies

From the late twentieth century onwards, many workers, in particular unskilled labourers, described their repetitive-work routine with metaphoric comparisons to robots. Interviews by sociologists with several women workers employed in the West German metal processing industry in the late 1980s exemplify this self-perception. The workers complained about the simple, repetitive task they had to fulfil at a highly automated workplace with computer numerical control (CNC) machines. They felt like they had ceased to be human. Their bosses, they felt, wanted them 'to be like a robot' without making any complaints about the monotonous tasks. In the end, they felt like robots and lost their former joy in work. The management of this plant had a similar point of view, with one production manager volunteering that he too perceived each woman as 'a multipurpose robot'. It was only for cost reasons that industrial robots did not replace women. For this reason, the manager admitted, 'at the present time, we cannot do without women workers.'[11]

Already in the nineteenth century, critics of industrial capitalism described the situation of the factory worker metaphorically as 'both slave and automaton'.[12] Well known is the harsh criticism of Karl Marx. As already shown in Chapter 4, Marx echoed the words of the British advocate of industrial capitalism Andrew Ure but gave them an entirely new political meaning. According to Ure, the tasks of the plant manager consisted 'in training human beings to renounce their desultory habits of work, and to identify themselves with the unvarying regularity of the complex automaton'. This 'factory discipline' had been necessary for efficient production. Following Ure, both plant owners and workers benefitted from this system.[13] Thus, proponents and opponents of industrial capitalism shared a common view of the necessity of a certain dehumanization that took place at the factory.

Marx took Ure's description as a starting point, but went much further. In his account working human bodies became virtually part of the machinery. According to Marx, the factory and its 'barrack discipline' had turned the traditional relationship between man and machine upside down. Whereas, he argued, 'in handicrafts and manufacture, the workman makes use of a tool, in the factory, the machine makes use of him.' Thus, the worker had become the machine's 'living appendage'. For Marx, this had an anthropological dimension, with workers and their bodies transformed through lifelong discipline. The worker was 'taught from childhood, in order that he may learn to adapt his own movements to the uniform and unceasing motion of an automaton'. In the end, the worker was transformed 'into a part of a detail-machine'.[14] Furthermore, there were severe effects on bodies. According to Marx, industrial labour exhausted the nervous system, while the 'many-sided play of the muscles' deceased. By and large, factory work had 'confiscated every atom of freedom, both in bodily and intellectual activity'.[15] It had converted the labourer's 'whole body into the automatic, specialised implement of that operation'.[16]

Marx's account of factory discipline remains very influential to the present day. Labour historians and historians of the body alike tend to view Marx's critique as an accurate historical description. Thus, Harris and Robb are convinced that many workers understood 'their bodies as machines'.[17] As the previous statements of women workers comparing themselves to robots have demonstrated, there is some truth to this assertion. Yet, as will be shown later, this mechanistic conception of the body is incomplete. Above all, it misses the self-reliance of workers that allowed times of relative freedom even in settings of 'barrack discipline'. In this context, humans were 'machines with self-reliance'.[18] In addition, there is a widespread misunderstanding of management's intention. Managers and factory owners alike did not primarily seek disciplined bodies at the workplace as an end in itself. Instead, discipline was just one of several different means leading to the higher end of creating productive bodies. As will be shown later, management's focus was not only on how to discipline workers, it was also increasingly focused on how to utilize their physical and mental capacity for corporate interests.

In the mid-nineteenth century, care for working bodies emerged with the interest in productive bodies – if only to a certain degree by that time. The first impulses to care for workers' bodies and health did not come from the plant owners but from the outside. Diseases and accidents caused by working conditions turned the workers' bodies or, more precisely, certain workers' bodies into political objects: for example, the Prussian military complained about child labour because it reduced the fitness of future soldiers. Thus, the Prussian state banned child labour in 1839, albeit only for children younger than nine years. In addition, the working day for older children up to the age of sixteen was reduced to a maximum of ten hours. In practice, however, little changed due to the limited capacities of local authorities to enforce these measures. Only from 1853 onwards did factory inspectors gradually take over. Their reports caused little immediate improvement, but they at least attracted further attention to the issue. Eventually, industrial child labour decreased mostly due to technological developments that made many ancillary tasks performed by children redundant. Until the 1880s, occupational safety measures concentrated on adolescents and women, two groups that were perceived as particularly vulnerable. In the case of women, there were widespread fears about 'a profound crisis of working-class motherhood in Germany'. Horror stories highlighting the ostensible moral and physical decay of women caused by factory work predicted a decline in the birth rate as well as child neglect. Only the increasing number of industrial accidents, combined with the new risks facing the rising number of chemical workers, inspired general change through the amendment of the industrial code in 1891. By then, the bodies of working men were also acknowledged as deserving some state protection.[19]

By and large, men benefitted most from the gendered assumptions regarding working bodies. Once a decision was made about who was to fulfil which task, a certain momentum built that gendered the task for the near future. For their part, engineers designed machines with the bodies of those supposed to work them in mind. As such, the height and size of the machine as well as the position of levers and buttons discriminated between male, female and child labour.[20] In any industrialized country, 'cultural attitudes about the differing abilities of women and men' paved the way for the gendered division of labour. Male physical strength was associated with the power to master 'heavy and fast-moving machines'.[21] This association of machines with male bodies had long-lasting effects, even if in many case physical strength had nothing to do with machine handling. If men and women worked at the same site, assumptions about gendered bodies and their suitability for working with technology led to a gendered hierarchy of skilled and unskilled labour. For instance, in the laundry centres of nineteenth-century Germany, male workers were responsible for the repair and maintenance of machines, whereas their female colleagues actually did the laundry. Only the male work at the machines was considered to be skilled labour and paid as such. In this regard, technological development fostered the discrimination between men and women at work.[22]

While alleged male capacities qualified them for skilled labour with machines, the counterpart of supposedly female qualities fostered the gender hierarchy at work. Most often, women were assumed to be particular dexterous and patient. In fact, many women had been used to needlework since childhood and thus had developed these qualities. As a result, repetitive industrial work soon became the realm of women as long as neither physical strength nor artisanal skills were needed. Furthermore, women were considered naturally more resistant to monotony than men. The assertion of this ostensibly specific female quality prevailed in the German science of work until the 1970s. This meant that many women were hired for assembly-line work from the 1920s onwards.[23] Due to the lack of manpower, female labour also became essential to German industry during the Second World War. Experts recommended playing music from the radio to the women at the assembly line as a means of imposing the rhythm of the music on the pace of work if women workers were there.[24] The other Nazi solution to the lack of manpower, namely forced labour, was justified in similar, if harsher, terms. Convinced of German supremacy, the Nazis perceived Eastern European forced labourers as inherently suited for the monotonous tasks that male German skilled workers could not bear.[25] In this context, most obviously, sexism and racism were intertwined. In twentieth-century Germany, rationalization meant good chances for skilled factory work or even the promotion to a white-collar job for German men, while women made up the bulk of unskilled workers who had to fulfil repetitive tasks. From the 1960s onwards, migrant workers joined in this new Fordist proletariat.[26] At the same time, archaic forms of male labour did not entirely cease in the twentieth century. Often overlooked, heavy labour remained important, in particular for early-twentieth-century modern infrastructure projects such as building canals or bridges.[27]

In general, however, the last decades of the nineteenth century marked a new approach towards body and health issues at work, as medical experts began to take interest in workers' bodies. Miners' bodies, for instance, became the 'object of systematic research and biopolitical action'. A new hygienic regime was set up to control, regulate and monitor the miners, especially their health and productiveness.[28] In this context, the emerging science of work turned understandings of the correct handling of the working body into the basis for a utopian vision of industrial society. In contrast to its French counterpart that was mostly interested in artisanal work, the German science of work laid more emphasis on industrial labour. As Anson Rabinbach has shown, German scientists of work initially focussed on the adjustment of bodies to the needs of machines. The 'rhythms of the machine' were the starting point with which the working bodies had to be 'harmonized'.[29] After the First World War, the followers of Taylor's scientific management shared this central interest in the body of the worker, and in the Weimar Republic, industrial rationalization was based to a certain degree upon the 'rationalization of the body'. The mechanistic ideal was a human body without fatigue.[30]

Technological solutions to the issue of fatigue were most prominent in the late 1920s. An exhibition on efficient worker seating toured Germany between 1929 and 1932. The exhibition presented a 'mechanical form of discipline'.[31] The underlying concept foresaw comprehensive control of the workers' body movements, even if they were only unintentional. Historian Jennifer Karns Alexander asserts that this concept was to a certain degree a German peculiarity. German experts had emphasized the idea of overcoming 'workers' personal autonomy and individuality'.[32] However, Alexander overlooks the fact that managerial interest in productive bodies took many different forms besides mere exterior discipline. To some extent, external discipline was increasingly replaced by the application of workers' self-discipline.

Even more importantly, in the early twentieth century, many German experts sought ways to adjust the work environment to the needs of workers considering both their bodies and their souls. The human factor of production was no longer reduced to a problem of mechanized production. Rather, engineers and managers began to conceive of workers as human resources to be cultivated for the firm's benefit. The gender question thereby played an important role in two ways: first, several workplace designers and experts of work were considerate of female labour. They believed women workers needed a special setting in order to work at a factory. This engagement with the problems of the work environment fostered considerations regarding the whole labour force, men and women alike, as experts emphasized the humanity and corporeality of the workers. This marked an important shift: it was no longer regarded as sufficient to overcome certain limitations of the human motor. Rather, a human-friendly setting had to be created at the factory to make the most of the human resources working there.[33] Not least, physical exercise programmes at the workplace were perceived as being antidotes to work monotony.[34]

Even non-skilled labour was to some extent perceived as part of the human resources that were of some value to the companies. Thus, companies came to care for the bodies of their workers. For example, around 1908, the Hamburg-based cocoa factory Reichardt published a postcard series of fifteen pictures showing both the working conditions in the colonies and the modern German factory. Side-by-side the series displayed the slave-like circumstances of hard physical labour in the Cameroonian plantation and the social benefits of modern industry in Germany (Figures 5.2 and 5.3). In stark contrast to the coercive labour regime that the Germans had established in the colonies, at home there seemed to be concerns about the human factor of production. Of course, the postcards were meant for public relations, but nonetheless the difference between the colonized and the colonizers was evident. The German factory housed a workers' canteen and even a swimming pool for female

Figure 5.2 Transport of cacao at the 'Viktoria' plantation in colonial Cameroon. Postcard, *c.* 1908.

Figure 5.3 Swimming pool for female workers at the Reichardt cacao factory. Postcard, *c.* 1908.

workers.[35] Even the overwhelmingly nonskilled workers at the German cocoa factory represented their potential future capacities for higher levels of production to some extent. Accordingly, staff rooms were meant to create conditions increasing workers' morale and taking care of their bodily well-being. This multidimensional relationship between efficiency and humanization of work has been discussed at large in Chapter 1.

In general, the imposition of discipline at modern factories was never entirely successful. As Alf Lüdtke has demonstrated, the workers had a certain self-reliance, or *Eigen-Sinn*, that helped them withstand any attempts at total discipline. This self-reliance should not be confused with resistance to management. Rather, these everyday practices of 'momentarily slipping away' or 'walking around and talking' seemed to the workers as perfectly normal. Many of these practices consisted of contact with other workmates' bodies, in forms of horseplay or even sometimes physical violence.[36]

In the context of industrial labour, prosthetics became of utmost importance during the First World War when the technology of prostheses made remarkable progress. Due to the lack of manpower, the rehabilitation of injured veterans and their rapid reintegration into the workforce were urgent. Thus, German rehabilitation centres consisted of training workshops to test the prostheses practicability at industrial workplaces.[37] By the 1920s, the conceptualization of the prosthesis gradually changed. Initially, from the late nineteenth century onwards, prostheses had been perceived as spare parts for the human body, analogous to the spare parts of machines. Engineers conceived of the human body as being replicable.[38] In 1916, for example, when the number of war invalids reached its first peak, the plant manager of the Dresden camera manufacturer ICA proclaimed that prostheses were the missing link between man and machine. Factory machines were to be equipped with prostheses that matched the veterans' prostheses. While the machine's prosthesis would become a part of the human body, the veteran's prosthesis would be part of the machine. Thus, man and machine were becoming increasingly complementary. According to this manager, this development heralded an increased fusion of man and machine.[39]

The most commonly used prosthesis for veterans was the prosthetic arm of the manufacturer Siemens-Schuckert (see Figure 5.4). In practice, however, the experts' hopes for the occupational reintegration of the disabled veterans were to a certain degree disappointed. The full integration of injured veterans failed because employers preferred to keep the women and elderly men who had replaced the soldiers in the first place.[40] Furthermore, surveys demonstrated that only a minority of veterans wore their prostheses at work. Significantly, most considered the prosthesis as a foreign object. Obviously, the engineers' optimism had missed the psychological and emotional aspects of prostheses. The human body was not actually a machine, and its parts could not easily be replaced. Instead, disabled veterans experienced prostheses as a psychological and physical burden.[41] Unfortunately, the prosthesis designers' approach had been merely mechanical. They were convinced that restoring the harmed bodies' work ability by technological means would automatically benefit the veterans' souls.[42] Orthopaedists who collaborated with engineers similarly prioritized the ideal of industrial efficiency. Consequently, function dictated the design of prostheses. This resulted in class differences as well. While the tasks of white-collar workers needed

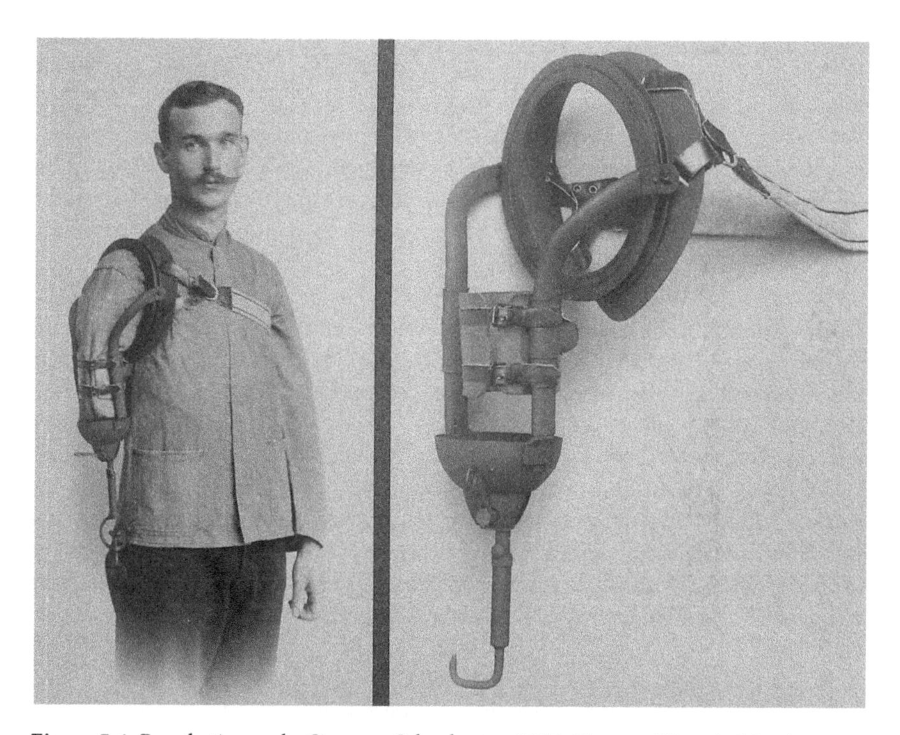

Figure 5.4 Prosthetic arm by Siemens-Schuckert, *c.* 1916. Siemens Historical Institute.

more sensitive control and thus elaborate prostheses, blue-collar workers' prostheses usually were less refined and cheaper. As a result, they appeared less natural.[43]

On the conceptual level, progress in prosthetics inspired visions of a harmonious body–machine relationship at the industrial workplace. Most German experts were convinced that the reintegration of disabled veterans by means of prosthetics perfectly fitted with modern industry following Taylor's principles of scientific management. Standardized prostheses provided the means to adapt the harmed bodies to industrial machinery. Yet the experts proclaimed that Taylorism meant more than this mere adjustment of bodies to the needs of machines. Rather, the adaptation of machinery to workers' bodies had been essential to Taylorism and modern industry as such. It was asserted that the prostheses of disabled veterans were proof of this win-win situation. Additionally, physical strength had already lost its essential status in some industries due to the modifications that were necessary to integrate women into the workforce during the war.[44]

Initially, the restoration of work ability was much more important to First World War prosthetics than aesthetic issues or the imitation of human anatomy.[45] This changed gradually in the following decades. In the 1920s, utopian visions of prosthetics as enhancements of natural bodies emerged, although the technological lag meant this was in practice unachievable. After the Second World War, when prostheses were widely needed for injured veterans again, the expert discourse returned to more realistic and immediate concerns regarding the simple rehabilitation of bodily functions. By then, disability had turned into a sign of a supreme will. The technically restored bodies needed the 'iron will' of their owners to function under difficult circumstances.[46] Only by the late twentieth century did the once-clear boundary between human body and technology began to blur. While the prostheses of the early twentieth century had not really challenged this very boundary, new technologies such as portable dialysis machines and heart pacemakers opened up new perspectives on the man–machine relationship.[47] From today's view, the prosthesis is about to become the model of a new upgrade culture. Permanent enhancement has become the ideal for the human body, which is perceived as naturally imperfect.[48]

In general, post-war industry was characterized by the promise of automation 'to liberate work from the materiality and physicality – muscles, nerves, energy – of the body'.[49] As has been shown earlier, unskilled CNC workers had different experiences of tedious tasks at computerized workplaces. Yet the case of their skilled workmates also proved that bodily functions remained important in automated factories, even if in novel forms. Most of all, the informal skills, dubbed by Michael Polanyi as 'tacit knowledge',[50] persisted. Thus, sensory skills and even an emotional relationship to the machine and the material were important to computerized workplaces. During the 1990s, German CNC operators emphasized that the sound of the milling cutter started to become physically annoying if anything was wrong. When the noise hurt their ears, it signified that something would break soon. Thus, the sensory reaction due to tacit knowledge was important for CNC operating.[51]

Besides the most prominent example of CNC, the introduction of information technology had altered industrial labour more broadly since the 1960s. The new workplaces not only challenged the qualification profile of labourers but also fundamentally transformed everyday work experience. The German printing industry gives an early example for these processes caused by information technology. Computer composition challenged the compositors' embodied knowledge. But as interviews from that period show, even under transformed working conditions the workers still imagined themselves as being 'machine-men'. They were even proud of this self-image because it implied that they instinctively knew how to handle their machines.[52] With the challenge of computerization, skilled workers desperately fought a battle to defend their tacit skills. In the end, desktop publishing took over and the employees of the printing industries became part of the growing service sector. Due to these changes, both the traditional formal and the embodied informal knowledge of blue-collar workers were lost forever.[53]

Similar transformations occurred in smokestack industries as hard manual work largely disappeared, even in industries such as steel. Beginning in the 1960s, the grafters who used to be essential to iron and steel were mostly replaced by operators, although some residual tasks still required hard physical work.[54] With this, another core problem of early industrialization disappeared bit by bit, namely workplace safety. Yet, even in 1970, West Germany continued to have an issue with occupational safety. Its rate of accidents at work was only topped by Italy globally. This changed in the following years with important reforms by the social-liberal government.[55]

Mobility machines

As has been shown, industrial labour has never been entirely dominated by the 'idea of mechanical discipline' and the corresponding concept of the 'human machine', although both were widespread in early-twentieth-century Germany.[56] Beyond the workplace, the notion of the body as a machine was clearly secondary to more individualistic body concepts.[57] The following section explores these concepts through the example of mobility machines. The dominance of the individual approach becomes most obvious with the triumph of individual transport, and with even the designers of public transport considering the physical needs of passengers to a certain degree.

The railway revolutionized passengers' perception of speed and space. In this regard, not the early decades of railways but the late nineteenth century was a period of transformation. Previously unimaginable travelling speed became commonplace within only two decades. While the train ride from Munich to Berlin took eighteen hours with an average speed of 36 kph in 1875, only seventeen years later a Prussian high-speed locomotive peaked 100 kph due to improvements in brake technology. Another fifteen years after that, the record was more than 150 kph. More speed meant more vibrations, however, and also more nervous tension for the railway staff and passengers alike.[58] While the physical strain seemed to be the same for all passengers, distinctions of class saw them have very different experiences of rail journeys. Usually, the lower-class compartments were not equipped with lavatories, restaurants or heating in the late nineteenth century. At least the third-class wooden seats were ergonomically designed, hinting that the companies and the designers had a certain sense for the basic corporeal needs of the passengers. Nonetheless, the differences with the first luxury railway trips that became established in the 1870s were tremendous. These trips offered all the travel comforts the passengers could wish for.[59] However, as Wolfgang Schivelbusch has pointed out, even the upper-class travellers had lost the feeling of pre-industrial travelling, although their compartments strongly resembled horse-drawn coaches. 'All the upholstery in the world' could not make the traveller forget that he or she had become 'the object of industrial process', who was 'being confined in a fast-moving piece of machinery.'[60]

A very different feeling of subjection was experienced by migrants from Eastern Europe on German tracks around the same time, from the 1890s to the First World War. They first had to pass through Germany on their way to one of the Western European harbour cities where they were to board a ship to America. In harsh contrast to the upper-class passengers, not the least comfort was provided to them during the journey. Instead, the bodies of the travelling migrants were perceived as possibly infectious and thus a menace to public health. Accordingly, they had to travel in sealed trains and stay in quarantine when changing. Furthermore, they were disinfected at hubs such as Berlin and Hamburg.[61]

All railway passengers enjoyed technologically enhanced speed, yet they were passively enduring the machine's motion. Conversely, individual transport offered new options to rearrange the body–technology relationship. The first individual transport technology that reached the masses, the bicycle, demanded unknown corporeal skills from its users: riders had to learn to keep their balance while moving. Hence, many observers in the late nineteenth century were very impressed by the bicycle that they conceived as an 'extension of man'.[62] The 'first-time users' of the bicycle in the late nineteenth century had a really hard time keeping their balance while riding. The bicycle even became a training device at cadet schools (see Figure 5.5). In particular, the physical challenge triggered the popularity of the bicycle, as one could prove one's own competence by overcoming any difficulties with cycling. The multitasking of simultaneously being the rider and the engine of the bike was a fascinating challenge for the early cyclists around 1900. Both muscle power and a certain sensitivity towards the machine were obligatory. At the same time, cyclists had an 'obligation to elegance': although they actually were the human motor of the bike, they had to perform a certain effortlessness to avoid making a fool of themselves.[63]

Thus, the bicycle was a prime example of a technology that paved the way for self-staging. It provided the means for the cyclist to present his or her body and thereby present certain images of himself or herself.[64] This is of utmost interest for gender history. On the one hand, supporters of the novel technology saw cycling as a 'school of masculinity'. This view was shared both by the socialist fighter for homosexual rights Eduard Bertz, who authored the first book on the *Philosophy of the Bicycle* (*Philosophie des Fahrrads*) in 1900, and by more conservative cyclists. On the other hand, the famous socialist feminist Lily Braun described the bicycle as 'a strong emancipator' in 1901. In hindsight, cycling women experienced previously unknown degrees of individuality and independence. Cycling also confirmed that women had the necessary physical skills. At that time, only middle-class women could afford a bicycle. Cyclists, whether men or women, mastered both their own body and a complex machine of mobility. While doing so, the body–technology relationship was altered to a certain degree.

Figure 5.5 Keeping the balance on the bicycle. Cadet school at Lichterfelde, *c.* 1900–14. Bundesarchiv.

Cycling was clearly self-centred; however, the relationship remained ambivalent. Although the cyclist experienced control of the machine, the human body itself became part of the machine and was conceived of as being a machine.[65] In this context, medical doctors and physiologists alike were fascinated by cycling because the cyclist could learn so much from the machine about his body functions, most of all rational breathing. According to the medical experts, the rhythm of cycling could be helpful in learning a steady rhythm of breathing.[66]

Even if many cyclists were convinced that the bike improved their agency, there were public fears about the machine taking command over the rider's body. This became evident in the debate on 'cycling rage', or *Radfahrwut*, which presumably infected both men and women.[67] According to some critics, cyclists were at risk of being tempted to ride their bikes too fast. The critical observers worried about cyclists being seduced by the unknown power of acceleration to overestimate their own physical strength and thus to overaccelerate.[68] In fact, their contemporaries did fall for the novel experience of acceleration, and not only on bicycles. The experience of speed was of utmost importance at amusement parks and sports. The roller coaster in particular made people feel the effects of acceleration on their bodies.[69] In both cases, the acceleration was directly felt due to the lack of encapsulation. The cyclist's body was, like that of the drivers of automobiles and pilots of airplanes in the early period, exposed to weather conditions. While the users of the novel mobility machines enjoyed the bodily experience, and, perhaps even more, the public presentation of their bodies, social acceptance for these practices prevailed only incrementally.[70]

The relationship between the user's body and the mobility machine is most evident in the case of the bicycle. Nevertheless, the body also played an important role in the political dispute about public and individual transport in the early twentieth century. In 1902, the psychologist Willy Hellpach, who would later become a liberal politician and minister of education in the federal state of Baden, complained about the discomfort of trams. The daily experience of passengers was characterized by missing the tram and having to wait for the next one. When finally entering the tram, it was hard to find a seat in a crowded tramcar. Additionally, the rides were usually very bumpy. Hardly surprisingly, Hellpach considered the automobile superior to the railway because it would calm the nerves. He hoped for future traffic to be dominated by individual automobile traffic, replacing the collectivist 'communist tram'.[71]

An everyday twentieth-century urban incidence like an unpleasant tram ride had different social meanings for people of different social strata. A middle-class man like Hellpach lost social status when he had to use the same means of transport as the working class. By contrast, for some workers, the electric tram meant a gradual enhancement of living conditions, namely the opportunity to leave the squalid tenement barracks of the inner city in favour of the suburbs. For women of all classes, new opportunities for mobility arose. However, from the perspective of the history of the body, regardless of social stratification, they all suffered the same disadvantages of trams: annoying waiting times for delayed trams, often in bad weather, involuntary body contacts in crowded cars, and being exposed to the sometimes-unpleasant smell of their fellow passengers.

Individual transport, in particular the automobile, provided an alternative, even if only for the upper and middle classes before it became more common in the 1960s. The experience differed markedly from public transport, with the car becoming an extension of the driver's body. According to historian Kurt Möser, German drivers resisted the process of encapsulating cars more vehemently than early motorists in other countries did. It seemed that German motorists enjoyed the physical stress of driving open cars. Automobiles enjoyed a sporty image due to the fact that the speed was palpable and the early users were not looking for more comfort but for the roughness of driving. In hindsight, however, only enhanced driving comfort appealed to new target groups and transformed cars into a product of mass consumption in the long run. The relationship between the driver and the car was particularly close; the artefact increased the self-image of the user. On the downside, the 'body-car-entanglement' became threatening in emergency situations. Even if the accident could be avoided, the driver's body reacted by increased adrenalin secretion.[72]

Apart from exceptional situations, the very conduct of driving a car conditioned the driver's body. He or she had simultaneously to fulfil four different tasks: handling the machine, driving, participating in traffic and navigating. Both the driver's sensory perception and body movements were subordinated to these functions. However, was everyone fit to drive a car? Or rather, how could the state ensure that a driver did not endanger the other road users? In the early twentieth century, when automobiles were still very rare, the belief in a driving disposition prevailed.

Only during the First World War, with the military experiencing a desperate need for drivers, were aptitude tests established. Due to this innovation, the belief in an innate sense for driving was replaced by the concept of the learnability of driving skills. The first driving schools had already been established around 1900, but now they became obligatory. This obligation was later cancelled by the Nazis in 1933 and re-established as late as 1957.[73] During the 1920s, the Technical University of Berlin offered aptitude tests that scientifically measured the presence of mind, the power of concentration, the steering sensibility and the reaction speed of the candidates (see Figure 5.6).

The debate regarding driving aptitude was gendered, with sexist sentiments about women not being gifted enough for driving common. Even high state officials shared these prejudices. In 1926, the Bavarian minister of the interior decreed that women should be tested particularly intensely during driving licence examinations. This was due to his 'various doubts about the aptitude of women as drivers'.[74] Accident statistics, however, proved the converse, and the national government also opposed gendered testing, leading to the decree being revised in 1927.[75]

Mass motorization in Germany began with the motorcycle boom of the 1920s (see Chapter 2). Women motorcyclists were even more openly rejected than women drivers. For many conservative male observers, the image of the motorcycle was not compatible with the gender norms of femininity.[76] By contrast, motorcycling was seen as the perfect stage for the performance of manliness. To a certain degree, motorcycling marked the first beginnings of the modern era of consumer society in Weimar Germany. The consumption of mobility machines was closely linked to the

Figure 5.6 Aptitude test for drivers at the psycho-technological department of the Technical University of Berlin. Photograph by Georg Pahl, 1928. Bundesarchiv.

consumers' self-image, which was in turn based upon gendered body assumptions. Differing masculine identities emerged, as motorcyclists could choose between rather rough versions to suave images of 'the new man': the 'Leather Jacket', the 'Motorcycle Apollo', the 'technically savvy cosmopolitan' or the 'true sportsman'.[77] Like other artefacts used for sports, the motorcycle had a twofold relationship with the body of the athlete: on the one hand, the technology was adapted to the needs of the human body. On the other hand, the body was modified – most prominently through muscle growth – to the needs of the sports equipment.[78]

At first glance, one would not expect physical aspects to be as important for more sophisticated technologies as they were for the sportive activities of cycling and motorcycling. But even in the high technology field of combat aircrafts, the body of the pilot was crucial. In practice, piloting needed immediate and thus intuitive reactions. Therefore, the habitualization of relevant movements was essential. Technology design anticipated this close relationship between body and machine: the design of the cockpit was based on the findings of ergonomics, anthropometrics and the feedback of pilots.[79] On the other hand, the design was made to influence the deportment of a fighter pilot's body: the heightened position of his seat urged the pilot to perceive his situation as threatening. Thus, he was more likely to react aggressively.[80] The physical senses were not only viewed as essential for combat, but also for flight safety. As with automobiles, in aviation discussions about encapsulation also took place. Even in the 1930s, pilots argued against capsulated cockpits: they wanted to feel and hear the air stream. The pilots assumed that the sensual part of flying an aeroplane was vital for safety issues.[81]

In the interwar period, the combination of combat aircraft, the masculine bodies of heroic pilots and the possibility of a national renewal fascinated many German right-wing observers. According to the psychoanalytical scholar Klaus Theweleit, German nationalists were far more fascinated by combat aircraft than by passenger aeroplanes because they fantasized about linking their bodies to these powerful machines of destruction.[82] The difference in the tone of the German debate becomes apparent when compared to its British counterpart. While Britons were soberer about the pilots merely performing their duty, the German cult of fighter pilots was dominated by the figure of the masculine hero.[83] For some German right-wing observers of the 1920s, the fighter pilot represented the renewal of mankind as such. They conceived of him as a hybrid of medieval and contemporary body ideals, a 'composite elite' of the knight and the soldier of industrial warfare. Thus, aviation embodied the vision of overcoming the man–machine dualism. In particular, the fighter pilot represented a novel unity of human body and technological artefact.[84] Some, such as General Erich Ludendorff and writer Ernst Jünger, shared a view of technological determinism: in industrial warfare, the soldiers' conduct had been largely determined by machinery. By using machines, they had even become part of a higher order of machinery.[85]

Although the heroic bodies of male flying aces dominated the public imagination, there was a spectacular countertrend, both internationally and in Germany: female star pilots. In Germany, there were at least eighty women pilots and many hundreds of female gliders before 1945 (see Figure 5.7). As women were excluded from commercial

Figure 5.7 The aviation student! The German University of Physical Exercise offered gliding courses for female students. Photograph by Georg Pahl, 1930. Bundesarchiv.

flying, becoming a stunt pilot or a brand ambassador for aircraft manufacturers was their only chance to make a living. In consequence, some of these women became flying aces of international fame: the 'flying frauleins'. Nevertheless, women made up only 1 per cent of the total number of German pilots at that time.[86] Opponents of female aviation evoked fears of women pilots becoming masculinized. These anti-feminists asserted that women lacked the physical, mental and emotional requirements of being a pilot. If a woman pilot became popular, like Hanna Reitsch, she served as an example of a woman who succeeded in overcoming her natural restriction by hard training. Yet opponents claimed that those women sacrificed their natural vocation of being a wife and mother due to the loss of femininity.[87]

After the Second World War, the body ceased to be central to debates about motorized transport. Only in the context of competitive transport, such as the bicycle or the motorcycle, was the corporeality of the rider still of utmost interest. Many functions of automobiles and aeroplanes became automatized, while driving instruction was transformed from a technical into a pedagogical task. There ceased to be any particular bodily requirements for driving a car and learning to drive became commonplace.[88] As a consequence, apart from illegal urban car and motorcycles races, sports and fairs remained the refuge for physical experiences with mobility machines. In particular, the looping rollercoasters of the late twentieth century produced the playful impression of risk. The passengers experienced ecstatic body sensations due to the playful staging of reaching the limitations of their physical capacities.[89]

Conclusion

Looking at the different examples of the body–technology relationship in modern German history, two questions emerge: first, to what extent has the beginning of the modern age altered this relationship? Second, are there any particular characteristics of the German approach to technified bodies? Beginning with the first question, even in premodern times, body modifications had occurred. Yet only from the nineteenth century onwards did new knowledge about how the human body functions pave the way for new approaches to body modifications in the Western World.[90] Since then, the human body and technological artefacts have become increasingly enmeshed: nowadays, it is hard to define where the natural body ends and the technological body modification begins.

Economic historian Andrew Pickering describes the industrialization of the nineteenth century as both an 'attack on the body' and an 'attack on the mind'. Due to his account, this was a comprehensive 'devaluation of human agency in relation to the agency of the new machines'.[91] The problem with this approach, however, is that it romanticizes the pre-industrial past. There is little evidence that there was more human agency at manufactories or cottage industries before the industrial revolution than in the later modern factories. As has been demonstrated, mechanization reduced neither the importance of human bodies nor the value of the human mind in the industrial workplace. Instead, the creative interaction of humans with technology promised to optimize productivity and required both the workers' bodies and minds. As such, the whole process of industrialization was not so much about discipline of body and mind but rather about embedding both within the goals of productivity.

With regard to the second question, German industry was characterized by some national peculiarities. Due to the importance of skilled labour to German industry, a characteristically German relationship between workers, their bodies and technology evolved. Many male German workers were proud of their close bodily relationship with the machines at work, and even in times of automation these skilled workers were convinced that they had specific embodied knowledge. They could detect by ear any unwelcome changes in how machinery functioned, they could see if something was going to go wrong and they could even feel if the machines were not working correctly. As historical research has shown, these self-images were based upon the prioritization of German quality work that emerged in the nineteenth century.[92]

Apart from any national peculiarities, distinctions of race, class and gender proved to be crucial for both the social effects and the cultural perception of technified bodies. An individual's position within these social hierarchies determined if one had the chance to improve one's work routine through working with one's body in a highly technified environment. The same was true for mobility machines: initially, access was often gendered and restricted to the middle and upper classes. Even if different classes used the same means of transport, such as the railway, the physical experience differed due to the dissimilar conditions in the different compartment classes. The experience of leisure time mobility nonetheless shows fewer national characteristics than was the case in industrial labour. Within the industrialized world, the experiences of everyday life seemed to converge internationally.

6

Rural technologies

In contrast to urban technologies, historians – with the exception of environmental historians – have often neglected-rural technologies. However, as has been shown in Chapter 1, decades before industrial cities dominated the German economy, industries in the countryside established the latest technologies, often imported from Britain. Until well into the nineteenth century, a large proportion of German industry was rural and still dependent on hydropower. The gradual changes and modifications of old technologies were often economically more successful and socially more important than large-scale innovations in growing cities. This chapter explores the impact of old, small and rural technologies that were at least as important for the shaping of modern Germany as their more prominent and often more sophisticated urban counterparts. It takes a close look at three areas of technological change. First, technologies of water and wood are examined as crucial examples of the will to dominate nature. The technological modifications of waterways and forests since 1800 illustrate how radical these changes to rural nature have been. By the nineteenth century onwards, both forests and rivers were anything but intact nature.

Second, agriculture saw many improvements to old technologies that pushed productivity to previously unknown heights in the nineteenth century. It was only with the twentieth century, however, that new technologies radically transformed German agriculture. In both post-war German states, industrial farming was established, and modernity ultimately arrived at German villages. While the paths were different both the GDR's planned economy and West Germany's market economy saw the state play a crucial role in enforcing technological innovations to agriculture. The development culminated in the 1970s, when the relative capital investment in West German agriculture rose above that of industry for the first time.[1] Technology was crucial in the form of machinery and fertilizers as well as for the transformation of dairying and livestock farming. Automation took place in cowsheds even before most factories were modified, introducing milking machines, milk pipelines and bulk milk-cooling tanks. Another aspect of agricultural automation is the human–animal relationship, which points to significant transformations during the twentieth and early twenty-first centuries.

Third, these agricultural improvements, as well as the growing demands of industry and the emerging consumer society gave rise to a peculiar section of industry based in rural areas. The ever-growing transport infrastructure opened new opportunities for marketing rural goods from the mid-nineteenth century onwards. In this context, several farmers became factory owners, for instance, establishing tinned food or beet-sugar refineries. All of these developments contributed to the economic persistence of rural Germany in modern times. In contrast to the prevailing notion of migration from the rural areas to the growing cities, the bulk of German migratory movements in the nineteenth century took place between different rural territories. There was no general rural exodus in the nineteenth century, neither in Germany nor in Western Europe. Rural districts also had growing populations, a point scholars often miss.[2] Rural areas persisted, but they changed their face. This transition was mainly due to their relationship to cities. The growing demands of urban areas led to transitions in rural regions. City dwellers consumed more and more food and energy as well as natural resources in general, all of which were primarily produced in the countryside.[3]

Dominating nature: Technologies of water and wood

There is a long history of technologies for cultivating wetlands. The colonization and drainage of moors were established in Europe in early modern times. Yet it was another technology – dam building – that became a major symbol of modernity, signifying 'human mastery over nature' more drastically than mere river regulation or moor drainage. From the nineteenth century onwards, iconic dams were built, often sparking contention between different players. Dams, of course, were hardly modern, and the global history of dams began in ancient Egypt. Even in Germany, advanced dam building was underway in the late fifteenth century, with itinerant Dutch technicians transferring the necessary know-how. The generated waterpower also helped to drain mine shafts and drove stamping mills. Until the late eighteenth century, numerous sophisticated dams were built in the German mountain areas.[4]

At the turn of the nineteenth century, a large squad of engineers was available in Germany for the project of the 'conquest of nature', among them hydraulic engineers and foresters (explored further later).[5] In particular, the new artefacts built by hydraulic engineers – sometimes unintentionally – helped old technologies to persist. The traditional energy regime had favoured rural regions before the advent of steam power. Rivers in hilly regions offered waterpower for traditional crafts and also for emerging small industrial businesses (see Chapter 1). Steam became a successful competitor to waterpower only once there was an apparent lack of labour in rural areas. The supply of waterpower was sufficient, reliable and cheap, but steam allowed factories to be built where an abundance of workers lived: in the fast-growing cities.[6] This disparity changed again later, when agricultural productivity rose strongly over the course of the nineteenth century. Thus, labour power was set free and paved the way for novel types of rural industries to emerge (as will be discussed further in this chapter's third section).

In the first place, dam building helped rural industries to cope with the problem of lower water levels due to the high water consumption of heavy industries. As French dam engineers took the European lead, the renaissance of German dam building began in the Franco-German Alsace region in the 1880s. Even before the Alsace had fallen to the German Empire after the Franco-Prussian War, French engineers had made plans for the dams from the mid-nineteenth century onwards. Their goal was to balance the water level for the benefit of both agricultural field watering and waterwheels. Notwithstanding the establishment of steam engines, water power still produced the lion's share of energy in some German regions as late as the 1870s. From the late nineteenth century onwards, dam building followed the Alsatian model in many mountainous regions of Germany. Paradoxically, the novel technology of dams saved the old technology of waterwheels – for the time being – contrary to the initial pioneering intention of dam engineers.[7]

In this context, a spatial and functional differentiation was established. While in the Ruhr area, the large technological system of mining and heavy industries prevailed; the small industrial entrepreneurs of the neighbouring highlands' valleys chose a different strategy. They built dams that provided the means for the retention of the traditional waterwheel power system. These waterwheels survived not only the challenge of steam but also the forthcoming technology of water turbines. Small hydropower installations prevailed in many German regions well into the twentieth century. Modern turbines did not succeed in replacing the old technology of waterwheels at once. Often, small meal mills or sawmills still used a waterwheel in the first decades of the twentieth century. During the working day, they used the power for their industry whereas outside the operating hours, the wheel powered an electric generator, which supplied their own business and the surrounding farms with electricity. In fact, the small waterwheels, neglected by the engineering journals that focused on modern turbines, still generated a large proportion of the total amount of hydropower during the interwar period. Many businesses in the German mountain regions kept the old technologies that they knew so well: they had invested in the waterwheels, they had the skills to maintain and repair them, and most of the time the wheels simply worked pretty well.[8] Only gradually did they disappear over the course of the twentieth century.

The manifold interests in dam building demonstrate the ambivalence of any linear notion of technological progress. At the turn of the twentieth century, engineers proudly emphasized the 'social utility of dams'. In a society of divergent interests, however, the question of who in fact profited from these artefacts was always contested. Contemporary observers already realized this. The dark side of the technological mastery of waterways was that there was ever more room for conflicts.[9] Agricultural irrigation, which had been crucial to the Alsatian dam-building projects of the 1880s, ceased to be a major topic by the turn of the twentieth century. Thereafter, conflicting industrial interests dominated disputes, with inland navigation having very different interests to hydropower companies. In this context, the state was both initiator of technological projects and arbiter of conflicting interests.[10]

In the long run, it was largely corporate interests that prevailed. Concerning the issue of energy, rivers remained crucial whatever the changing mode of power generation was. They generated hydropower, transported coal and were also essential for cooling nuclear power stations. In a certain sense, border-crossing rivers, such as the famous Rhine, and multinational energy corporations contributed considerably to the establishment of a European electricity system.[11] Thus, the Rhine, which was central to German national myths that were becoming more and more popular through youth literature and Richard Wagner's operas in the nineteenth and twentieth centuries, was obviously transformed from a mythical site of natural beauty into a crucial mediator of modern technology. In the present time, the Rhine 'more resembles a canal'. It has been shortened by more than 8 per cent. In fact, the Rhine-Main-Danube canal, completed in 1992 following a more ecological approach, 'looks more natural' than the original rivers.[12] Theodore Schatzki has shown that replacing the notion of 'nature' with the conception of socionatural sites is more adequate for the description of the entanglement of human history, nature and technology.[13] This relationship becomes clear when we take a closer look at another site of nature celebrated by Germans romanticists, nationalists and environmentalists alike: the German forest.

After a period of heavy wood consumption combined with a popular fear of a wood shortage, the forest area increased gradually from the mid-nineteenth century onwards. Until the present, wooded areas have accounted for a third of Germany's total area. Often overlooked, this steadiness in the extent of forested areas was not the outcome of a reduced demand for wood in the age of coal and steel. Demand had actually increased, and it was only new technologies of forest management that secured sufficient wood production. The establishment of railways facilitated the transportation of forest resources, whereas river navigation had been essential for the marketing of wood before that time. Despite forest management techniques, the demand for timber still vastly exceeded national resources to the extent that Germany has been a net importer of timber since 1864.[14] Often forgotten is the fact that in the latter half of the nineteenth century, the iconic new technologies of the era of high industrialization – railways, electricity and telephone lines – needed large amounts of wood for railway sleepers and pylons. In addition, the rising paper and packaging industries were large buyers of wood. On the eve of the First World War, more workers were employed in wood processing and paper industries than in the chemical sector.[15]

In this period of a prospering forestry industry, German 'foresters developed new techniques to increase production' that soon became a global model for modern forestry. Employing statistical models, the foresters entirely changed the composition of the forests. Even before efficiency became an industrial buzzword, wood experts had used scientific knowledge to maximize production. Most of all, quick-growing pine monocultures replaced the traditional mixed deciduous and coniferous forests. In the course of these developments, workers were deskilled and woodcutting was standardized.[16] After the Second World War, the trend to conifers increased rapidly, while in present times, spruces and pines dominate wooded areas in Germany.[17] Despite the search for ever-increased efficiency, however, nature still sets some limits. German forest soils were not fit for a transfer of the large timber harvest machines that were successes in Northern America and Scandinavia. As such, the transfer of mechanized woodcutting failed. Instead, workers with chainsaws proved to be an efficient method of timber harvest.[18] At the same time, traditional wooded areas increased moderately, while green areas and parks rose strongly in West Germany. Although the forest had at that time a long history of cultivation, it still symbolized the natural counterpart to industrial modernity for many Germans. This was especially true for the environmental movement, which grew stronger from the 1970s onwards (see Chapter 8). There was, nonetheless, an inherent contradiction in environmental perspectives that praised the forest both as an example of pristine nature and of cultivation.[19]

Agriculture: From peasant agriculture to factory farming

The exact relationship between agricultural change and the industrial revolution is hard to define. As historian Jan de Vries points out, researchers have proclaimed 'a large number of agricultural revolutions'. There is, however, agreement about one thing: important changes had occurred among peasant farmers well before the industrial revolution.[20] Yet, agricultural productivity growth was not dependent on novel technologies before the late nineteenth century. Occasionally, new technologies were invented, but none of them prevailed. For example, the practitioner Peter Krezschmer propagated a new ploughing method in 1748 that provided the means to go deeper into the soil and promised enhanced yields. Yet it only worked on specific soils – not on sandy ground, for instance.[21]

By and large, agrarian productivity rose from the late eighteenth century onwards. Initially, this increase had little to do with technological innovation. There were some important innovations at that time such as the seed drill, but only wealthy farmers could afford them. Instead, the rise in output was mainly due to developments similar to those that de Vries has described for cottage industries, where before the advent of the industrial revolution, an industrious revolution took place (see Chapter 1). New arable crops such as potatoes, beets and rapeseed needed more intensive soil cultivation mostly done with hoes. In addition, important changes occurred in livestock farming. Stable feeding became more common, causing more work but providing worthwhile fertilizers that in turn increased crop yields and thus provided more fodder for livestock. Furthermore, meliorations and in particular soil drainage increased the total amount of agricultural land. Thus, even before the agrarian reforms that occurred after 1806, agriculture had already been gradually intensified with more livestock and enhanced crop yields. The first beginnings of a consumer society, namely peasants' growing interest in novel furniture, gave incentives to this industrious revolution. Yet, it was a long process until innovations such as stable feeding were widely accepted. For economic reasons, it was a rational choice for most peasants to make the transition gradually.[22]

This development was not linear but a highly volatile process. At the end of the eighteenth century, traditional agriculture was in a severe crisis. Agricultural yields fell due to decades of exploitation.[23] At the same time, livestock decreased. Agricultural reformers had overestimated the yield increases, and there was no steady supply of food or fodder.[24] Although the eighteenth century brought increased agricultural yields in general, famines were still an issue. Even in the early nineteenth century, southern German regions suffered from famines; the last instance of mass starvation before the First World War took place in 1846–7.[25] More sustainable change was needed. Given the strength of feudalism in Germany, institutional change was necessary before labour was set free for the emerging industries. The agrarian reforms did not mark the beginnings of agricultural prosperity, but they were the necessary legal frameworks for a long-lasting process of agrarian growth to set in. The reforms privatized much of the commons. Now, by and large, food supply was secured both for rural people and for the growing urban population.[26]

At the same time, modern agronomy was becoming established. In 1769, the German pioneer of technical sciences Johann Beckmann published a book on the *Principles of German Agriculture* and thus paved the way for modern agronomy based upon economics, science and technology. Beckmann sent a student to England who imported seeds, grains and fertilizers.[27] Thus, even before German business owners studied the emerging English industrial technology, German agricultural experts had already imported English knowledge and resources. In the early nineteenth century, the concept of rational agriculture broke through. Albrecht Thaer established this approach with a publication in 1812, which was translated into English in 1844 as *The Principles of Agriculture*. Thaer built his agricultural principles upon his knowledge from study trips to Britain around 1800. Among other practices, stable feeding and crop rotation proliferated more and more in Germany during the early nineteenth century. In addition, peasants began to perceive themselves as merchants. Visions of technological progress inspired agricultural change, although diffusion of innovative technologies and external fertilizers was rather slow. Yet for the first time, Thaer and his predecessors conceptualized manual skills and technical knowledge in an agronomical account.[28]

Between 1800 and 1850, the amount of agricultural land doubled in Germany. In particular, the two decades after 1830 were a period of agricultural take-off and increased yields. This increase was mostly due to meliorations, embankments, drainage and reclamation. These techniques had been established for a long time, but in the nineteenth century, they were significantly intensified. A new generation of technical officials conceived of themselves as legitimate modifiers of nature. Their domination over water helped with land reclamation. In addition, the ongoing privatization of the commons and the reduction of wasteland contributed to this trend. Traditional equipment was significantly enhanced even before new machines spread more comprehensively through the countryside from the late nineteenth century onwards. Even before 1850, steel ploughs were introduced that went deeper into the soil; the scythe replaced the sickle; and equally importantly, prospering new industries cheapened the agricultural equipment. In addition, rising crop prices due to growing urban demand transformed farming into an entirely market-oriented business. This transformation furthered the increase in production. While in 1815, four farmers produced merely enough food for themselves and one further person, productivity multiplied by 1865. Now, the output of four farmers nourished themselves and four more people. It seems appropriate for historians to speak of an agricultural revolution.[29]

Market-integration and enhanced transport technology were the most important forces of change by that time. The transport revolution of the nineteenth century paved the way for supra-regional commerce. First, chausses and canals helped to reach the growing markets of the emerging urban centres, and after 1840, the railways paved the way for the establishment of a European agrarian market. In addition, the establishment of the German Customs Union and the lowering of transport costs with the expansion of the railways had a significant impact on agricultural production due to increased marketing options.[30] In retrospect, there was a mutual interplay between industrialization and agricultural change. Even before the industrial revolution reached Germany, agricultural yields had increased. Furthermore, as historian

Verena Lehmbrock has shown, the school of cameralism had established principles of agronomical rationality before the advent of industrialization in Germany. Thus, the essential contribution of the industrial revolution was to open new markets by improved transport. Only later did novel machines and artificial fertilizers provide the means for previously unknown yield increase.[31]

Although the need for new machines had already been felt by the mid-nineteenth century, their application was only gradual. Nonetheless, the institutional framework of technological change was established at that time. From 1839 onwards, an association in the central German town of Magdeburg promoted exhibitions in agricultural engineering. In 1867, a testing institute in nearby Halle began to inspect agricultural machines. However, only large farmers could afford new machinery by then. Still, technological innovations could not compensate for the ongoing issue of labour shortage. By that time, meliorations and improved livestock keeping were the most important measures for production increase.[32]

This low rate of mechanization did not significantly change until the late nineteenth century. Compared to the United States and Britain, German agriculture was mechanized rather late due to a high rate of land fragmentation resulting in many low-tech small farms. As late as the 1870s, the most important agricultural machines were still ploughs, harrows, rollers and cultivators. Most peasants did not make any more use of machinery than their elders had done. The only exceptions were the threshing machines that were used on most medium-size farms. Horses or men powered these early threshing machines. Yet large-scale farms soon changed to steam power. In the late nineteenth century, contractors with threshing machines already travelled from village to village. For most peasants, this was the primordial experience of modern technology. While small peasants who owned less than ten hectares of land had hardly any machines, even a remarkable 20 per cent of large-scale farmers (i.e. working more than 100 hectares) went without any employment of modern machines. Even modest changes like the transition from oxen to horses were gradual. Half of the draught animals on Westphalian farms were still cows and oxen at that time. What was more, iron appliances replaced wooden devices only little by little. Yet, even gradual improvements of the old technologies yielded increases in productivity.[33]

In hindsight, the wait-and-see behaviour of German peasants, which persisted into the first half of the twentieth century, was not based upon a general rejection of technological innovation. Instead, it was the manifestation of a rational and pragmatic attitude.[34] Economic historians argue that there was a discernible 'advantage to backwardness'. The peasants of Westphalia were for a long time less advanced than their Western European competitors, but they soon became the most productive market players after employing more sophisticated technologies in the late nineteenth century.[35] Bit by bit, from the end of the nineteenth century onwards, farming became more capital-intensive and mechanization transformed German agriculture. Besides machines, farmers invested in seeds, fertilizers and energy.[36] In particular, the globalization of the wheat market offered further incentives to mechanization. After 1875, steamships made imported US wheat cheap and put Prussian landowners under pressure to increase their productivity by employing machinery. Agricultural machines were manufactured in series production and mostly imported from Britain. Still on the eve of the First World War, British manufacturers dominated the German market.[37]

Agricultural associations played an important role as mediators between the manufacturers of agricultural machinery on the one side and rural users on the other. In the late nineteenth century, the German Agricultural Society established itself on the model of its English counterpart. Its main goal was to foster the mechanization of German agriculture and diffuse the latest innovations through journals and exhibitions.[38] In addition, the scientification of agriculture began in the 1880s, inspired by research at the technical universities. In 1902, the Agricultural University of Berlin established the first professorship for agricultural engineering. For the most part, though, agricultural research still neglected technological aspects. Moreover, only from the mid-twentieth century onwards did farmers make widespread use of the novel knowledge.[39] The dissemination of knowledge was largely due to agricultural schools established in the early twentieth century and beyond. By that time, most German farmers conceived of themselves as living at the mercy of nature. There was little interest in scientific approaches towards agriculture. Young farmers learned about science and technology only little by little.[40]

Somewhat paradoxically, fertilizers mark the exception of early and comprehensive adoption of science in agriculture. There was no need for farmers to gain scientific knowledge about fertilizers. For a long time, they just followed the maxim 'a lot helps a lot'.[41] Already by the turn of the nineteenth century, there were forty-five natural fertilizers on offer to German farmers. After 1840, Peruvian guano boomed in Europe and had nearly a monopoly for the next two decades until the deposits were exhausted.[42] The use of external fertilizers in addition to the manure of one's own livestock began only after transport costs decreased in the wake of canal building and the establishment of tributary light railways. Comprehensive fertilizing paved the way for industrial agriculture, once pre-industrial agriculture had reached its maximum possible output. Now, fertilizers were transported to the rural areas in one direction while potatoes and pork were carried in the other to the industrial cities. As a result of these trade relations, areas like Oldenburg came to specialize in pig feeding around 1900.[43] Furthermore, new types of seeds were introduced after the turn of the twentieth century. This innovation had synergetic effects with the new commercial fertilizers: the new seeds had only high yields in combination with new fertilizers.[44]

The First World War brought a breakthrough with artificial fertilizers, which rapidly replaced the imports of nitrates from Chile. The chemist Fritz Haber, who later received the Nobel Prize, invented a nitrogen-fixation technique on the eve of war. For the German government and industry alike, nitrogen was particularly interesting because of its dual use: it was the basis for both fertilizers and explosives. Without this innovation, Germany would probably have 'run out of munitions as well as food by the spring of 1915'.[45] Even this innovation could not entirely alleviate the situation. Malnutrition and starvation caused between 700,000 and 800,000 casualties in Germany during the war. In particular, the Turnip Winter of 1916–17 was a national trauma.[46] After the war, Germany lost much of its cropland and food imports became increasingly important.[47] On the other hand, the land shortage offered incentives for increasing agricultural productivity. In this context, a novel collaboration of interest between the state and agriculture occurred during the Weimar period. In fear of the next famine, government agencies propagated artificial fertilizers. Contemporary observers praised artificial fertilizers as the 'big engine of agriculture'. Although fertilizer prices steadily declined, 10 to 12 per cent of total operating expenses in the late 1920s were in fertilizers.[48] As a result of the politics of fertilizer promotion, German farmers used far more artificial fertilizers than their counterparts in France or the United States.[49] In addition to the new artificial fertilizers, the old technology of natural fertilizers persisted. As late as 1929, roughly thirty-seven tons of bone meal fertilized agrarian soils in Germany. As extensive as that sounds, this was only 2 per cent of the total phosphates spread on croplands.[50]

Parallel to the changes in cultivation, the industrialization of milk production was established. In the latter half of the nineteenth century, the demand for meat and milk increased further, which gave incentives for better feeding and keeping as well as selective breeding of high-yielding breeds. In this context, the specialization of farmers in either arable farming or livestock breeding gradually began. Again, the transport revolution was crucial for agricultural development, with the railways facilitating the marketing of milk products. As a consequence, the cow transformed from a multipurpose animal to an entity whose single purpose was the production of milk. The process was gradual, and even in the mid-twentieth century small farmers occasionally used cows as draught animals.[51] In the last quarter of the nineteenth century, a certain 'bureaucracy of breeding' was established, operating with pedigrees, standardizing the animals and professionalizing the expertise. Herd books of dairy cattle helped to differentiate between breeds. Agricultural colleges established professorships for breeding sciences. Yet only the introduction of practical measures at the farms, namely milk records, provided the means for a great advance. A transfer from Danish farming to northern Germany at the turn of the twentieth century introduced regular testing by an inspector. With this oversight, farmers knew the exact relation between costs and yield. Due to improved feeding and targeted breeding, milk production per cow increased by 50 per cent from 1922 to 1932. The Nazis made the milk book and regular tests obligatory. However, these attempts to spark a further increase failed, as these methods had already reached their peak. It was only after the Second World War that milk yields increased to previously unknown heights.[52]

Before 1945, there was nearly no success in milking automation. Although machines were invented, milking remained manual labour in practice. In 1910, the German Agricultural Society presented a milking machine at a Hamburg exhibition.[53] The first British experiments on milking machines date even to 1836, but only in the 1920s was the technology ripe for practical application. Widespread usage in Germany, however, did not begin until after the Second World War.[54] In the first half of the twentieth century, milking machines diffused rather slowly due to high costs and susceptibility to failure.[55] In 1933, only 0.4 per cent of German farms used milking machines. Accordingly, milk-cooling systems were still very rare.[56]

In general, rural electrification was rather slow. Thus, when the electric threshing machine was introduced in 1893, only a small minority of farmers made use of it. Still in 1913, only 25 per cent of the rural regions in Germany were connected to the electrical grid. After the First World War, however, Germany and other European governments promoted rural electrification and greatly surpassed US rural electrification rates. However, the devil was in the detail: while some western German regions were heavily electrified by the 1920s, electrification was delayed in the underpopulated eastern rural regions. Agricultural structure and scale hindered the diffusion of electric farming even in those western regions where small farms that could not afford expensive machines dominated.[57] Thus, the promotion of rural electrification did not entirely succeed at that time. Nonetheless, farmers started to learn about the new options, which probably facilitated the appropriation of new technologies after 1945. For instance, a popular Dortmund exhibition presented an 'electric farm' in the 1920s.[58]

Agricultural motorization was also slow in Germany. In 1895, only 16.4 per cent of German farmers used motorized machines while the overwhelming majority relied on manual labour or animal traction. The agricultural workforce was still massive. In 1882, 43.6 per cent of Germans worked in agriculture. The number decreased over the next few decades, but even on the eve of the First World War, rural workers made up a third of the total labour force. Steam ploughs signified the beginnings of agricultural motorization; however, they were economically viable only under specific conditions, one of which was melioration. From the turn of the twentieth century onwards, steam ploughs were used to cultivate moors, which were then transformed into heath landscapes. In addition, drainage and fertilizers transformed the former moors into grassland.[59]

For agricultural applications in a narrow sense, it was almost exclusively the large farms in eastern Germany that purchased those expensive steam ploughs. They had the capital and the workforce (at least five operators were needed for ploughing). In addition, the soil in that region fit quite perfectly to the technology of steam ploughs, which provided deep ploughing. Ploughing was crucial for increasing agricultural yields and the new technology promised significant enhancement. While around 1800, the plough went only ten to twelve centimetres into the soil, the introduction of the steam plough increased the depth to thirty centimetres or even more.[60] Nevertheless, only 3,000 steam ploughs operated on less than 1 per cent of the total arable land in 1907.[61] In the early 1920s, this technology had an unlikely revival in a modified version. A regional manufacturer adapted the steam plough to the needs of Ruhr farmers by replacing the engine with the old technology of draught animals. He downsized the plough so that two horses were able to pull it.[62]

In the other direction of mechanization, motorization in the narrow sense of internal combustion engines had its first beginnings with the motor plough before the First World War. Initially, motorized ploughing was simple but energy-intensive. Some contemporaries also raised the objection that the motor plough missed the sensitivity of horses or oxen.[63] Only after the First World War did motorization gain momentum. A national committee of agricultural technology was established in 1920. In the following years, the German state promoted the motorization and mechanization of agriculture through a loan programme. In addition, enthusiasm for technology inspired by the American model reached German agriculture in the 1920s. While the early tractors had many flaws, the machine spread more comprehensively after 1924 when the import of the US Fordson tractor began. Initially, high custom duties prevented the widespread success of the Fordson tractor. However, the German manufacturer Lanz appropriated Ford's winning formula and introduced assembly-line production for its Bulldog tractor. Due to these developments, tractor prices fell. At the end of the Weimar period, half of the large-scale farms were motorized. Smaller tractors that mid-size farmers could also afford were introduced from 1930 onwards.[64]

Compared with the United States, where agriculture was industrialized during the 1920s, German farming transformed only gradually. In a certain sense, the 1920s were a decade of mere efficiency talk. As has been shown for German industry (see Chapter 1), debates surrounding technological innovation in agriculture greatly exceeded any actual changes. Nevertheless, machinery became more important on German farms after 1918. Before the First World War, only 30 per cent of the capital invested for livestock was spent on machinery. By 1928, the ratio doubled. By that time, the purchase of a tractor was not a rational decision for most mid-size farmers. Sometimes, younger farmers just decided that they wanted to be modern. In 1925, according to differing estimates, there were 7,000 to 12,000 tractors in German agriculture. After the Great Depression hampered further motorization, the number was still below 16,000 in the early 1930s. What is more, the number of motor ploughs deceased in these years. Before that time, motor ploughs had been particularly successful in Germany with 12,000 units in use by 1925. However, there were many disappointments with both the motor ploughs and the early tractors. In particular, the general difficulties with the tractor were increased by the fact that this was a new kind of machine, namely the first multipurpose machine on the farm. Furthermore, the average tractor of the interwar period had a lifespan of only five years. The main problem with the introduction of the tractor was that the farmers had little technological experience: for example, they often had no shed for the tractor. They knew little about the necessity of regular maintenance and greasing. In addition, as a rule, large farms employed specialized tractor drivers. Usually, the farmers did not accept that their employees needed so much time for the machine's maintenance. As a result, careless maintenance and ignorant use could easily destroy the expensive machine.[65]

In the long run, however, these small troubles laid the foundation for the post-1945 success of the tractor. First of all, those farmers who were early purchasers of a tractor gained some modest expertise by tinkering with the technology. In hindsight, the Lanz Bulldog of the 1920s was romanticized for its peculiarities. In particular, the procedure of starting the Bulldog tractor by blowlamp became the subject of myth in the German countryside. On the one hand, the Bulldog's advantage was that it was easy to maintain and repair. On the other hand, it was not easy to use. If the farmer wanted to drive from the field to the road, he first had to change the field wheels to the pneumatic wheels. These early tractors were also very slow and hindered traffic. Furthermore, the Bulldog did not even have a reverse gear. Therefore, the driver had to employ a high degree of machine sensibility to change the motor's direction of rotation. In general, sensory skills were needed for the correct handling of early tractors. As a consequence, drivers needed some time to learn about the machines. But by the early 1930s, some observers praised tractor drivers for recognizing any mechanical problems simply by listening to the engine's sound.[66]

Most German manufacturers were less sophisticated than Lanz. As shown in Chapter 1, mass production was still in its infancy in Germany. As late as 1935, the

Fendt company, which would become a leading manufacturer in the post-war era, produced a mere thirty tractors in the whole year. In general, the engineers were not familiar with the farmers' demands on the tractor: it had to drive slowly on the field and fast on the road; it had to be both not too heavy and not too lightweight; and most of all not too expensive. Moreover, the service infrastructure was missing: the blacksmiths usually lacked the skills for repairing tractors, and more professional repair shops were rare. On the other hand, this lack of infrastructure was an advantage for technological development in the medium term. Manufacturers such as Hanomag maintained a network of technicians, who drove to the farms and repaired Hanomag tractors. As a result, by visiting the rural workers the technicians learned about the actual problems of technology in use. As a contemporary observer stated, the rural workers became collaborators for further technological development and design.[67]

The Nazis tried hard to accelerate rural motorization. In contrast to their propaganda image as guardians of traditional rural life, the Nazis promoted agricultural technology such as machines, tractors and fertilizers by providing government subsidies. For the Nazis, modern organization and novel technologies promised to pave the way towards an alternative modernity.[68] Analogue to the 'people's car', or *Volkswagen*, the Nazis promoted the 'peasants' tractor', or *Bauernschlepper*. Unlike the *Volkswagen*, which was not marketable before 1945, the low-cost tractor manufactured by Deutz was at least a minor success with roughly 10,000 units sold until production was paused in 1942. What is more, Deutz resumed production of the slightly modified tractor in 1946.[69]

In another field, Nazi politics paved the way towards factory farming more comprehensively. They continued the efforts of pig breeding in the 1920s and combined it with the specific Nazi politics of autarky. From the 1920s onwards, agronomists successfully experimented with breeding a new breed of pigs that produced more fat and protein while needing less fodder. They considered German soil peculiarities in their breeding experiments, and pigs became scientific objects. The agronomists crossed traditional German breeds with English breeds for the purpose of replacing the traditional fat German pigs with fast-growing, leaner pigs. The new breed promised to supply the growing urban demands for lower-fat pork. However, within the 1930s, the rather fat combination of both breeds became the most popular variant. This marks a crucial difference between German pig breeding on the one side and its British and US counterparts on the other side. As Germany was a large net importer of fat, the pigs' fat content was more important. The issue became even more relevant in the context of the Nazi policies of autarky and war. By then, Germany urgently needed to increase its animal fat production. In a certain sense, livestock farming had both become scientific and ideological. Leaving aside the Nazi rhetoric of 'keepers of tradition', farming was modernized, and livestock breeds were transformed into 'technoscientific organisms'.[70]

In this way, pig farming underwent important changes before 1945. The transformations that followed in both German states were, to some degree, a continuation of the Nazi approach to scientific pig farming. In particular, the concept of technological fix still prevailed. Contemporaries were convinced that any issues caused by the new technologies of industrialized pig farming could be solved by technology. Farmers and politicians in both socialist East and capitalist West Germany shared these convictions. By and large, the same technological transformations of pig farming occurred in both states.[71] Concerning the animal itself, increased demand for low-fat and high-protein pigs resulted in lean hybrid breeds after 1970.[72] The new basic technologies – gestation crates, automatic chucking machines and slatted floors – provided the means for increased efficiency, saving labour power. Nevertheless, complete automation remained out of reach, with human labour still essential for taking care of the crucial functions of control.[73] Only from the 1980s onwards did computer-controlled pig feeding gradually break through.[74] Furthermore, the complex task of finding operations that both fit the needs of the animals and the farmer was tedious. As such, innovations like manure removal were largely due to a lengthy process of trial and error.[75]

Only after the Second World War was West German agriculture comprehensively mechanized. Now, the economic, institutional and infrastructural preconditions existed for that mechanization to occur. Until the mid-twentieth century, the areas of vocational training, consultancy and service lagged behind the possibilities that research and new technologies offered. For instance, agricultural schools now taught more than half of their classes in technology-related topics. Furthermore, only the adequate capitalization of farms after 1945 could pave the way for structural change in West German rural life. For the first post-war decades, mid-size farms embracing the new technologies prevailed while small farmers gradually disappeared. Thus, the former issue of land fragmentation vanished, and the agricultural areas became increasingly tractor-friendly.[76] Mechanization corresponded with a loss of labour and in the decade after 1950, the number of farm employees declined from 5.1 to 3.1 million in West Germany. This decline continued into the 1960s. By then, specialized and industrialized farms had become the norm.[77]

Agricultural technology had also become far more reliable. While the machines of the 1920s were prone to failure, the more reliable technology of the post-war period paved the way for the large technological systems of factory farming. By then, rural users had already had their first experience with agricultural technology and were more open to innovations. Only after this initial familiarity with new technology had been acquired during the interwar period could the comprehensive mechanization of agriculture be built upon a widespread acceptance of technology. The new generation of farmers much more readily embraced modern technology, particularly the tractor.[78] Between 1949 and 1959, the number of tractors in West Germany increased tenfold. By the mid-1960s, any mid-size or large farm had purchased at least one tractor and even half of the small farms had one. This was not always a rational choice. Often the tractor was largely a status symbol that symbolized optimism.[79]

Although the German state facilitated collective machine ownership for economic reasons, most young farmers were stubborn and wanted to have a tractor of their own. Therefore, most small- and medium-sized farms acquired a tractor in the 1950s and 1960s. In a certain sense, this was the inefficient outcome of technology euphoria.[80] There were, however, countertrends. In particular, the older generation of farmers born in the early twentieth century were often reluctant to purchase new machines. For example, the retrospective report of a northern German farmer's daughter states that her father had little interest in technology, because he enjoyed the manual labour on the farmland together with his four children. Only after the patriarch's death in

1962 did his heir acquire a tractor.[81] This account confirms the argument of historian Frank Uekötter about the traditional 'peasant ideal'. According to Uekötter, from 1800 to 1945, a patriarchal and holistic ideal of the 'peasant' dominated German rural life. Even in the 1950s, the ideal of the organic farm prevailed. This ideal explains why the motorization of German agriculture was rather slow. Even in the late 1950s, most farms kept horses in addition to tractors.[82] Draught animals were a rational choice for the small- and mid-size farmers who dominated German agriculture until the late twentieth century. In 1950, every third cow was a draught animal. This changed drastically within fifteen years. In 1965, only 2.8 per cent of livestock in West Germany were used as draught animals. As late as 1954, the entire horse-power of tractors exceeded the number of horses in West Germany. Yet even in the 1960s, small- and medium-sized rural landowners often held oxen in addition to tractors.[83]

However, in comparison with household technology, the tractor spread rather rapidly in the immediate post-war period. In the mid-1950s, West German farmers were more likely to own a tractor than a washing machine. While a new washing machine cost 1,800 deutschmarks, the price of a second-hand tractor was roughly 3,000 to 4,000 deutschmarks (while a new one cost 6,000). It became essential for a farmer to have basic technological skills, including a driver's licence and knowledge about the main agricultural machines.[84] In general, the younger generation of rural landowners lacked the reluctance regarding novel technologies that was typical of their parents and grandparents. By then, most farmers were proud to be in control of technology and to tinker with their machines. Interestingly, this increased the gender differentiation of farming tasks as men assumed responsibility for mechanical tasks on the field, while women's tasks were evermore reduced to housewifery and working in the barns.[85]

As mentioned before, the West German government welcomed this technology enthusiasm, although collective ownership was recommended as economically more rational than private ownership. With this in mind, the West German state supported machinery rings and cooperatives.[86] In socialist East Germany, of course, collective ownership became obligatory. Initially, machine stations loaned agricultural machinery to small farmers. In 1952, they were renamed 'machine tractor stations'. These establishments were important steps towards the collectivization of East German agriculture. When the process was completed in 1961, the machinery was transferred to the Agricultural Production Cooperatives. By and large, motorization was slower than in West Germany. Nevertheless, the number of tractors doubled within the 1960s, and the number of harvesters even tripled. With very different political motives, the East German administration proclaimed the same goal that their West German capitalist counterpart had a few years earlier: the transition to industrial farming.[87] In applying pesticides, East Germany even exceeded the quantities used in West Germany. Or more precisely, the East German farmers used roughly as many – 30,000 tons of pesticides in 1988 – as their West German counterparts, even though the GDR had much less cropland. The East German agrochemical centres were even spraying plant protection by helicopters and aeroplanes by the late 1960s.[88]

The West German government's motivation for a similar path towards agricultural modernization differed completely from the East German vision of a new society. West

German agriculture never reached the dimensions of the large-scale Eastern production cooperatives before 1990, but the similar mechanization approaches of both in the 1950s were pursuing different ends. For the conservative West German government of the 1950s, agricultural technologies served as a means of preserving farm and village life as well as the conservative values this kind of living implied. Most of all, politicians hoped that technology would facilitate the work of countrywomen.[89] In a certain sense, this was a version of an alternative modernity: novel technologies were not meant to establish progressive ideals of society but to preserve older values and slow down social change, namely urbanization. The milking machine was one technology central to this vision of both helping countrywomen and preserving rural gender roles. Milking was a basic skill in the post-war era. A survey of 1952 found that more West Germans were capable of manual milking than driving or typewriting.[90] Milking was also seen as women's work. While men overwhelmingly fulfilled most tasks beyond household chores, women did 53 per cent of milking work in 1953.[91] Milking machines were therefore seen as essential to the vision of easing female labour.[92] At the same time, the female agricultural labour force of East Germany was also trained how to use the new machines (see Figure 6.1).

However, the milking machines – whether in the East or West – of the 1950s and the early 1960s were not very efficient. They were single machines combined with buckets. Given their inefficiency, while milking machines widely prevailed on large farms, manual milking persisted at most small farms with less than ten hectares of land in the 1950s. Only the systematic approach of milking machines directly connected to

Figure 6.1 The agricultural teacher Josef Annich explains to two apprentices of the East German Barby agricultural college how to use Soviet milking machines, 27 November 1952. Bundesarchiv.

milk pipelines that led to cooling tanks in the milk room provided the desired effect of increasing efficiency by 1970.[93] The West German state promoted these milking and cooling systems, which all but the smallest farmers soon installed.[94] With the enhanced possibilities of pasteurization, cooling and transport, the marketing options for milk increased significantly. Yet, there were still clear limits to productivity growth. Although there were sharp increases in milking efficiency, the milking machine did not provide automation. Milkers still had to apply the machine before milking and to remove it afterwards. In addition, milkers had to control the cows' physical conditions. Despite the ongoing mechanization of milking, it still was the one subtask in husbandry that turned out to be a performance bottleneck. While milking was gradually mechanized, the parallel increase in feeding and cleaning efficiency was much more impressive. Working with a living being like the cow limited productivity growth to a certain degree. Only from the late twentieth century onwards would rotary milking parlours and robot arms reduce human labour comprehensively. Gradually, being a merchant became the farmer's foremost task.[95] Looking at the twentieth century as a whole, the increase in productivity was nonetheless impressive. While one worker in 1948 milked ten cows per hour, one farmer milked seventy-seven cows per hour in 1994.[96] Furthermore, the animal itself was transformed into a 'turbo-cow': in 2005 the average cow produced 3.5 times more milk than in 1900.[97]

In parallel with these developments in the barns, the work on the field was further mechanized. Unlike the tractor, small- or mid-size farmers rarely owned the larger and more-expensive harvester. Before the Second World War, harvesters were rare and only low-tech variants were available. In 1932, only fifteen harvesters driven by horses were used on large estates in eastern Germany. From the late 1930s onwards, tractor-driven harvesters became more common.[98] It was only in the 1950s, however, that modern harvesters gradually broke through in Germany. A new type of contractor who no longer worked together with the locals but sat alone on his harvester drove them. Contractors prospered in northern Germany, particularly where intensive farming prevailed.[99] In most regions of Germany, mechanization of farming had replaced manual tasks by 1970. The 1970s and 1980s were a period of purchasing more machines, enhanced engine horsepower and the shift to the usage of large tractors. German farmers had a certain 'horsepower fetishism', which saw the entire horsepower of West German agricultural tractors increased by 67 per cent. By then, machine-adapted processes and economies of scale dominated in West German agriculture.[100] A new type of farmer who – in contrast to his predecessors – was a technophile realized this comprehensive transformation of German agriculture. Without this necessary acceptance of technological change, the transformation most probably would have been a failure. Even repairing the machinery became ever more popular. In 1990, the distributors sold half of the total spare parts directly to the farmers who did the repairs on their own.[101]

The previously unimaginable rise in agricultural productivity between 1950 and 1980 paved the way for factory farming. As has been shown, the productivity increases mainly relied on a new generation of technologically euphoric farmers and the comprehensive infrastructures of service stations and contractors. Equally importantly, agricultural chemistry had enhanced crop yields sharply. In that entire period, farmers perceived mechanization and the widespread use of artificial fertilizers and pesticides as a practical constraint.[102] In northern Germany especially, the usage of artificial fertilizers multiplied between 1950 and the mid-1980s. Prices were very low, and there was no incentive to use fertilizers sparingly. At the same time, international competition was the argument that made factory farming seem unavoidable. The transformation to factory farming was the most rapid in poultry keeping. From the 1950s onwards, some regions in northwest Germany specialized in poultry and introduced a system of cage-rearing, automatic feeding and automated egg collection. Those regions could rely on their historical practice in pig farming, which facilitated the introduction of US-style large-scale poultry keeping. This was an example of factory farming at its extreme.[103] In the long run, the local differences between the various agrarian regions in Germany vanished by and large: intensive agriculture was the ultimate leveller.[104]

Besides artificial fertilizers, a new variant of natural fertilizers became more and more important: liquid manure. Unfortunately, its environmental effects were disastrous. The history of liquid manure demonstrates how a regional technology became widespread. Originally, liquid manure was popular in the Alpine regions that lacked straw. The regional experience showed that liquid manure was an efficient fertilizer for pastureland: milk production increased twofold within a few years. After the Second World War, liquid manure broke through for intensive farming in most parts of West Germany. Complemented by modern disposal facilities, it was a very efficient kind of animal keeping. In the 1990s, it was used for 30 per cent of cattle breeding and even 70 per cent of pig farming. Corn was the ideal crop for liquid manure, as masses of fertilizers could not harm it. Just as importantly, it was the ideal fodder for cattle and pig fattening. Leaving aside the environmental hazards, it seemed to be a perfect cycle. Factory farming with masses of livestock produced unknown masses of excrements. The technology of liquid manure provided the means for efficiently fertilizing the field where crops grew. These crops were perfect for feeding livestock in factory farms. It all appeared to be a matter of technological efficiency.[105]

The Danish Pattern: A rural path to modern industry

During the period of burgeoning factory farming, food-processing companies entered poultry and pig keeping on a large scale.[106] This was one outcome of a long history of interconnectedness between agriculture and industry. As farms became more company-like, some industrial companies broadened their scope by entering the farming business. Historians Karin Zachmann and Per Øsby have asserted that food production gradually turned into a large technological system during the twentieth century, when the boundaries between agriculture and industry blurred.[107] However, particularly in the northwest, German rural regions have a long history of close relationships between agriculture and industry ranging back into the nineteenth century. In these regions, industrialization largely occurred on a rather small scale, an agricultural variant described by economic historians as the 'Danish pattern'. As enhanced railway infrastructure facilitated the marketing for farmers, they could easily sell their foods at the towns' weekly markets and deliver beets to sugar refineries. In the end, farming was modernized and new rural industries such as sugar factories, distilleries and food-processing factories emerged. In particular, the northern Prussian province of Schleswig-Holstein – a direct neighbour and former part of Denmark – specialized in cabbage, sausages and flourmills. Farmers also established tinned food factories for sauerkraut and sausages.[108]

Similar developments occurred in Westphalia. The farmers in this region had problems selling their livestock to the Ruhr district, although the urban food demand was rising. The high costs of transporting live cattle by rail hampered their marketing. Therefore, some traditional farmers became foodstuff producers who sold processed foods such as dry sausages. In this way, transport costs decreased significantly. Modern food companies also emerged that specialized in products for the consumer society such as confectionaries. In addition, former artisans turned their shops into industrial wood-processing businesses. The Westphalian furniture industry evolved out of these modest beginnings. In the mid-nineteenth century, saw mills began to sell the pit prop to the mines of the Ruhr district. Moreover, enhanced inland steamship navigation and the ever-growing connection of rural areas to railways gave rise to cigar factories in the Westphalian hinterlands. Due to this modern means of transport, traditional manual labour persisted. While labourers were abundant and cheap there was little incentive to mechanize the work process. An additional factor in favour of the evolving rural industries was the proximity of resources.[109]

Some of the evolving rural industries were especially modern inasmuch as they applied science and technology to their business models. Before the famous German chemical industry started to dominate the global markets, the beet-sugar industry had already established itself as the 'first major science-based industry in Germany'. The basic process of beet-sugar production had been developed by the mid-eighteenth century in Berlin, but it failed commercially due to its high production costs.[110] Only in the mid-nineteenth century did the central German rural areas become a centre of the emerging beet sugar industry. The recipe for success was that farmers established sugar factories on their own, beginning on the side-business level and growing larger by the 1860s. This combination of agriculture and industry provided a steady supply of beets, which non-agrarian businessmen missed. Among the beet farmers were several noblemen who made the transition to becoming successful capitalists. By the 1840s, steam engines were widely employed for sugar production, while chemical experts entered the factories on a large scale from the 1860s onwards. The prospering industry had manifold synergistic effects on its surrounding areas. In particular, machine construction profited from nearby demand.[111] By and large, the beet-sugar industry neatly encapsulates the effects of the interconnection between agriculture and industry. Once an importer of sugar, Germany became one of the world's most important sugar producers with 43 per cent of the global share of beet sugar in 1884.[112]

Nutrition patterns changed drastically from the nineteenth century onwards. First of all, meat consumption increased sharply and processed foods entered the market. German meat consumption rose from 16.4 kilogrammes per capita annually in 1816 to 51.1 kilogrammes in 1907. The main increase occurred in the latter part of the nineteenth century in a period of high industrialization and urbanization.[113] Afterwards, the rate grew steadily, reaching a peak in 2018 with 90.1 kilogrammes per capita.[114] Also in the last third of the nineteenth century, processed foods broke through. Initially, these new foods succeeded in niche sectors such as military provision or children's food. The standardized industrial mass production of stock cubes, meat extract and seasoning began at a time of rapid urbanization accompanied by new eating habits. A rising standard of living enabled many levels of the population to consume new foodstuffs. Instead of the traditional foods like gruel, porridge or bread and milk soup, common people could now afford to eat meat broth; Liebig's meat extract was the first industrial mass product on the food market, but soon serious competitors arose.[115] In general, food chemistry paved the way for a new understanding of food and taste. In this context, the realm of technology grew as a product's bad taste was now perceived as a technological issue.[116]

Also in the late nineteenth century, refrigeration technology enabled the transport of fresh meat and bananas from South America to Europe. In Germany, bananas were particularly popular; the first German banana steamboat operated beginning in 1912.[117] Even more importantly, the steamboats paved the way for the transnational entanglement of rural industries with commercial food manufacturing, the most prominent example being Liebig's meat extract. The philosopher and mathematician Gottfried Wilhelm Leibniz had had the idea of developing preserved food and meat extract for soldiers in 1714. Roughly half a century later, Prussian soldiers tested meat meal during the Seven Years' War. At the same time, the Saxon army also introduced a similar kind of food, a certain 'meat grit'. However, both experiments had no follow-ups and were soon forgotten. Thus, there was a long pre-history of meat extract before chemist Justus von Liebig explored the chemical composition of meat in the mid-nineteenth century. His disciple Max von Pettenkofer produced meat extract albeit on a small scale.[118]

The industrial production of meat extract began with the German engineer Georg Christian Giebert, who had constructed roads and railways in Brazil during the 1850s and contacted Liebig and Pettenkofer in 1861. Liebig allowed Giebert to employ his name for the marketing of the novel product and Giebert established the 'Liebig's Extract of Meat Company Ltd.' with a plant in Uruguay in 1865. Now, for the first time, meat extract was produced on an industrial scale. In South America, cattle breeding took place on a very large scale and beef was much cheaper than in Europe. As such, Uruguay provided perfect conditions for large-scale production of a relatively cheap product. From the beginning, this was a multinational enterprise: two Belgian merchants were co-founders and the firm's headquarters was in London. Already by the 1870s, the rationalization of industrial meat processing peaked with the Uruguayan enterprise. The company expanded its product range to canned corned beef and canned bouillon. The efficiency-minded business even used the beasts' foot oil as machine oil. In the 1880s, the multinational dimensions of the company increased further with land purchases in Uruguay, Paraguay and Argentina. Still, the machinery for meat extracting, such as meat flakers, huge pans and filters, was delivered from Germany.[119] In sum, the food industry was by no means less advanced than other industries.

During the First World War, the issue of food security motivated the German state and food industry to be particularly innovative in producing substitute foods or foods with a long shelf life. After 1933, Nazi autarky politics also gave rise to state funding for agricultural and nutrition research. These fields received more than 30 per cent of the state's research expenditure between 1937 and 1943. Substitute foods, especially fats, became more important than ever due to import blockades during the war. In contrast to these politics of transformation, the Nazis still celebrated traditional farming as an essential element of Germanness.[120] After the Second World War, the interconnectedness between agriculture and industry increased in many areas. For instance, modern plant breeding affected industrial production. Only standardized vegetables made mechanized cleaning economically viable. Before standardization, the multitude of vegetable variants in length and width led to unwanted losses, which made manual labour by unskilled women more economical than automated processes.[121]

However, while there were important innovations in rural areas that historical researchers often miss, there was also a significant persistence of old technologies. As late as 1935, only 35 per cent of German farms had a central water supply.[122] The business of wood processing is another good example. Even if iron and steel replaced timber in many agricultural devices from the mid-nineteenth century onwards, there was still good business for cartwrights: first, machinery, for instance that of threshing machines, had become so complicated that farmers were no longer able to manufacture the appliances themselves. Second, farming had become more intensive and relied on machines. Although a growing share of agricultural machinery was made of steel, the absolute number of wooden machines did not decline at first. Thus, even in the age of steel, timber works persisted until after the Second World War. The spread of rubber tyres in agricultural trailers eventually ended the profession of cartwrights in the 1960s. The transition to rubber tyres had been delayed in many places until then due to high prices and the long lifespan of cartwright trailers. In the agriculturally dominated province of Lower Saxony, there were even more horses in 1957 than in 1870; the number reached its peak in 1950 due to a lack of fuel during the Korean War.[123] At that time, modern road building had just begun in West German rural areas: former dirt roads were tarred or cobbled with interlocking paving stones.[124]

Conclusion

Germany's strong heritage of feudalism and a large number of small farmers were important factors in slowing down the mechanization of agriculture in the nineteenth century. Furthermore, there were sharp disparities between rural areas in different parts of Germany. Large estates dominated the east while small holdings prevailed in the south and the west. By contrast, industrial agriculture and rural industries evolved in the northwest at a relatively early stage. These disparities shaped German history for a long period of time. The process of further mechanization of farming, however, made these regional disparities gradually disappear. Of course, the history of technology cannot explain this process as a whole. The loss of the agrarian eastern regions in 1945 changed the conditions for the restart of German agriculture entirely. What was more, the antagonism between cities and rural areas lost much of its strength in the post-war era. There are still different forms of voting behaviour and daily routines, but urban technologies of supply and disposal diffused to the countryside by the 1960s and made those differences less sharp. So too, technological standards of livings were levelled to a large degree. But even today, the latest technologies such as fast internet connections take longer to reach rural regions.

The state's role was that of a leveller of disparate regional standards of living. As such, political history is of utmost importance for understanding the rural dimension of the history of technology in Germany. Since the nineteenth century, the state had promoted agricultural technology and enhanced fertilizing. The period of intense mechanization in the twentieth century was also largely dependent upon state subsidies and loans. A certain comingling of interests evolved between state politics and agricultural business. It became contested only in the context of growing awareness of the environmental burden of factory farming (see Chapter 8). The state had also played the role of arbiter between differing interests. Small landowners, for example, had interests that were different from large farmers. While the West German politics of enhanced mechanization after 1945 at first helped mid-size landowners, in the long run, the economic urge to invest and grow made both small- and mid-size landowners disappear by the end of the twentieth century. Only factory farms survived. The support for factory farming occurred on both sides of the Iron Curtain during the Cold War, albeit in different variants. After reunification in 1990, these developments even sped up. The economic pressure of globalization seemed to be an incontrovertible argument for ever-increasing factory farming.

In some cases, the state was also an arbiter between industrial and agricultural interest groups. As a rule, large construction projects such as dam building mostly favoured industry. Yet in many fields, a partial blending of interests between industry and agriculture evolved. The state also functioned as a mediator by promoting agricultural chemistry and modern farming technology. The particular strength of the German chemical industry was an important factor in the high rate of artificial fertilizing. In general, the close relationships between agriculture and industry did not only evolve with factory farming. On the contrary, there had been close links between both sectors since the nineteenth century.

7

Everyday technologies

Historian Joachim Radkau has pointed out that the *Sputnik* satellite was, at its time, less important to contemporary history than the ready-to-serve pizza.[1] Since then, however, satellite technology has become essential to daily tasks of mobile communication, media entertainment and navigation. Nevertheless, Radkau's attention to small technologies offers a useful starting point for this chapter. If history's basic interest is in the transformation of ordinary people's lives, those technologies that affect everyday life are crucial to the historian of technology. So too, a history of technology beyond design and innovation is required, given that they are only a small part of the lifecycle of technological artefacts, whereas the manifold forms of usage often last for many decades. Usage, in many ways, defines what a technology means – both to individual people and to society as a whole. Equally important is how users' creative appropriation of technology often causes designers to modify it. Of course, these appropriations too are always embedded in cultural concepts of technology and progress (see Chapter 3).

The users' creativity, however, should not be mistaken for absolute freedom. It is a problematic feature of some histories of technology that they overrate personal freedom. In this context, technological innovations such as the automobile, the internet or the mobile phone star in whiggish narratives of ever-increasing freedom through consumption. Following Mikael Hård, it 'ought to be the historian's task to deconstruct this hype' about technology and freedom.[2] Accordingly, it is of utmost importance to consider 'the social and cultural constraints on user-technology relations', while still remembering that, as Nelly Oudshoorn and Trevor Pinch have demonstrated, technology users do matter.[3] This chapter will explore the users' agency between the script of technological objects and the self-reliance and ingenuity of the users. As will be shown, technologies are often designed with the intention to limit the range of possible usage. This does not mean that the object always determines the users' action. Nonetheless, for the forms of usage to expand outside the box, sometimes a lot of creativity – and perhaps tinkering – is required.

Following these lines, this chapter discusses the cultural appropriation of technology in German everyday life. It examines the relevance of small technologies and the technological transformation of daily routines, which started in the nineteenth century and accelerated in the age of mass consumption that began in post-war Germany. The focus will be first on household technologies and second on the various types of tinkering with technology, both bearing class and gender dimensions. Machines entering private homes did not transform everyday life for the whole family evenly.

The ever-increasing use of household technology challenged established gender roles; or so it seemed. Partially, this chapter is about social roles *not* changing, although the introduction of new machines laid the foundations for a rearrangement of gendered power relations. This persistence of both social roles and cultural concepts within a period of comprehensive technological innovation will also be analysed in this context.

The second section of this chapter deals with more openly creative appropriations of technology: the manifold forms of tinkering, ranging from refurbishing the house to handicrafts, from car tuning to the legal and illegal activities of computer hobbyists. Bit by bit, computers have entered private life. Furthermore, many privately used technologies were digitalized – often without users recognizing this transformation. Tinkering, both analogue and digital, often oscillated between the poles of subversion and conformity. Tinkering could belong to a certain subculture, or even counterculture, but sometimes these subversive practices turned into mass phenomena over time and lost any subversive potential. In several contexts, this chapter takes a close look at the complex relationship between intended and actual forms of technology usage.

Household technologies

During the nineteenth century, domestic work changed considerably. In particular, the cholera epidemic of the 1830s and germ anxiety beginning in the 1880s helped to establish new standards of hygiene (see Chapter 2). In the context of early hygiene awareness, laundresses had already opened small businesses at the outskirts of most German cities in the 1860s. In addition, home-based regular hot washing and ironing became more common in the following decades. At the turn of the last century, the first tenements were equipped with laundries. After the First World War, laundry increasingly turned into a task of the housewife while professional laundry declined. By then, a monthly 'big laundry' day had become a common feature both of middle- and working-class families. This took place either at the tenement or in commercial self-service laundry facilities with simple washing machines.[4]

Electricity was another crucial innovation which would transform everyday life and lay the foundation for the next laundry revolution in the second half of the twentieth century, having entered middle-class homes from the late nineteenth century onwards. The retrospective account of the writer Edwin Redslob gives a good example of the excitement that the introduction of electricity sparked in families in the 1890s. Although Redslob's Weimar family was among the early adopters of this new technology and thus obviously innovation friendly, they still struggled with the conversion. The usage of electric light was, probably not only in this family, not as intended. They were proud of being innovative and well-off, but the light was felt as 'unpleasantly bright' due to unfamiliarity with the novel brightness and maladjusted usage. Electricity was still very expensive and, thus, 'one did not think of dimming the bulbs or hiding them behind lamp-shades', which would have been adequate. Instead, old practices persisted in this electrified home. Redslob's father hung on to his kerosene study lamp, which both gave a pleasant light and was a good substitute for heating when the stove was turned off in the evenings.[5] This example demonstrates that it was often a lengthy process of adapting novel technologies to private routines. In this case, as often, the initial awkward usage of a new technology, which the designers did not intend, was not the result of an arbitrary individual act. Instead, individual usage was embedded in cultural patterns that were hard to overcome.

Sometimes, designers took up users' habits to ease the transition to a new technology. For instance, revolving switches that copied the operation of gas taps initially controlled electric light, even if it was not necessary to gradually turn the light on or off. This signifies the dual marketing strategy of the electric companies: on the one hand, electric light symbolized technological modernity, and its promoters emphasized the difference to gas lighting; on the other hand, they knew that users were not ready for a too-rapid break with common habits. The establishment of gas lighting itself had not been easy: in the 1860s, most German middle-class dwellers used gas lighting only in kitchens and corridors because the traditional lighting of candles and oil lamps seemed more sophisticated. Gas lighting conquered the living rooms only in the late nineteenth century. Historian Wolfgang Schivelbusch argues that the middle class initially rejected gas and electric lighting because it had origins in factory and street lighting. Through the new technology of lighting and connections to the new infrastructure, the public sphere and the industrial world entered middle-class homes.[6] In the medium term, it was crucial for business to win the acceptance of the next generation. Stefan Poser has demonstrated the importance of 'playful celebrations of technology', be it at the fairground, at sports or at home.[7] Such playful appropriations of technology paved the way for the future distribution of innovations. For example, soon after the advent of the telephone in the late nineteenth century, telephone toys became popular. Children were fascinated, even if the 'telephone' communication that they attempted by means of string and cans worked only over a very short distance. This toy had no practical advantage over everyday, plain conversation, but it had the aura of technological progress.[8]

Issues of progress and efficiency were central to the debates about household technology. For a short period of time, two different paths to household technology were proposed: collective modes or individual consumption. By the turn of the twentieth century, socialists had a vision of collective household technologies. The supporters of that vision sought tenement houses with one collective kitchen. However, these concepts were not a top priority on the socialist agenda. Only a few reform houses were realized, and they were rather expensive.[9] Soon, individual technology consumption remained the sole path to household modernity without any serious alternative. The first half of the twentieth century was characterized by a slow, but steady, diffusion of household technology. Yet, in many cases, there were obstacles and in particular users' objections against specific technologies. As will be shown, the issue of cautious attitudes towards or even open rejection of new technologies was often complex. What appeared like a self-reliant attitude at first glance was often based upon the material necessity of budgeting.

The reform residential estates of the Weimar period were the most comprehensive attempt to establish electricity and modern technology in the households of the lower middle and working classes. Some historians argue that the residential estates had been a large technological system, which contained a certain script urging the inhabitants to follow the hygienic ideals of the reformers.[10] While the reformers in fact intended to educate the inhabitants, they did not succeed easily. First, there is some evidence that although the apartments were standardized, the dwellers did not live standardized lives. Many expressed their self-reliance by rearranging the home furnishings to counteract the implied usage patterns.[11] The first 'fully electrified' (1929) housing estate Römerstadt in Frankfurt (see Chapter 2) is a prominent example of the complicated relationship between intended use and reality. Initially, there were significant protests against the dependence on electricity. A critical interest group formed, which had an impressive membership of 600 persons, considering that the settlement consisted of 1,220 accommodation units. The protest was both for practical reasons and due to high costs. Not only was electricity expensive, the residents also faced switchover costs. The electric kitchen range, for example, was not compatible with some of the old pots and pans. Moreover, the new technology did not function as well in the beginnings. Residents complained about prolonged cooking times. Some even used their own camping equipment and made coffee on the gas cooker.[12] As such, residents did not reject the new technology entirely but rather were creative in finding transitional solutions. By 1931, most residents had familiarized themselves with electric household appliances. Falling electricity prices surely helped. By then, more than 70 per cent of the households used these appliances comprehensively, whereas only 5 per cent did not use them at all.[13] However, Römerstadt had not become the reform housing programme for the working class as initially intended. Due to high rents, white-collar workers of the lower middle class mostly inhabited the tenements.[14]

The famous Frankfurt kitchen, which was part of the Römerstadt reform settlement (see Chapter 2), also demonstrates the limits to users' creative modifications of household technology. In general, the kitchen was a contested area: users' expectations challenged the introduction of electricity. Most technology users still associated heating with visibility. Therefore, municipal offices initially discussed if it would be appropriate to design open hotplates for making the glowing wire visible and thus make heating seem visible. These plans, however, came to nothing.[15] Or, more precisely, they came into existence only some decades later with ceramic hobs. Furthermore, some gender sensitivity was needed in marketing the new electrical appliances. Women usually rejected men's explanations about household work. Thus, housewives' associations played an important role in the diffusion and acceptance of new household technologies.[16] In the case of the Frankfurt kitchen, the users did not entirely conform to the expectations of the architect. In 1979, inhabitants of the housing estate that were interviewed stated that while in hindsight they had enjoyed the enhancement of living standards in the new settlement, they had rejected the original concept of a functional working kitchen. Instead, they had transformed the kitchen, in contrast to the architect's intention, into a traditional eat-in kitchen.[17] By and large, however, the Frankfurt kitchen offered only limited possibilities for such tinkering. The floorplan of a modern functional kitchen was hard to modify, not least due to its limited space. Besides the initial difficulties in gaining general acceptance for modernist reform, the architects' expectations came true: in the long run, users adopted 'the new lifestyle that was inscribed in the kitchen'; not entirely, but mostly. This took quite a long time, however, due to the Nazis' rejection of the modernist kitchen. Finally, in 1968, 30 to 40 per cent of West German apartments were equipped with a built-in kitchen.[18]

It is widely known that the Cold War was a high-tech conflict with nuclear weapons as its most impressive expression.[19] A lesser-known aspect of the Cold War was the East–West competition about everyday technologies in private homes.[20] In the reconstruction period, showcase apartments in East Berlin were not yet modelled after modernist architects' visions as would become the East German standard in later years (see Chapter 2). Instead, the initial post-war housing projects followed German traditions with wooden furniture and stuffed sofas. At that time, the communist administration wanted to demonstrate that the GDR 'was the more legitimate Germany'.[21] In the following decades, however, competition was focused on modernity and which system offered the most home convenience to its citizens. Besides heating and water supply, any important household technology was based upon electricity consumption. Beginning in the 1950s, West German households widely adopted new attitudes and behaviours regarding hygiene as well as food storage and cooking. In that decade, West Germany made the 'transition towards a Western high-energy society' with East Germany following up a decade later.[22] Once again the laundry and kitchen were at the forefront of change.

Before private washing machines made a breakthrough in West Germany from the 1960s onwards, US companies introduced modern washing techniques in the post-war years: self-service laundrettes with automatic washing machines were a huge success. German customers quickly became familiar with the revolutionary chemical and technological processes which put an end to the tedious traditions of the monthly laundry day.[23] Then, in the 1960s and 1970s, the rapid proliferation of household technology again transformed the everyday life of German housewives. Before, the 'big laundry' had filled a whole day once a month. Now, the laundry was done more regularly and on the side. The second single most important artefact of the household was the refrigerator, which obviated the daily food shopping routine. However, alongside the breakthrough of new technology, hygiene standards and expectations rose: clothes were changed more often. Therefore, as historian Ruth Schwartz Cowan has demonstrated, new household technology in fact meant 'more work for mother' for the time being.[24]

These developments took place in both German states, albeit with a certain delay in the Eastern part. In both cases, electrical kitchen equipment symbolized affluence. Yet, there were some features of the kitchen that turned it into a controversial object of Cold War politics.[25] By and large, in West Germany, the new household technology facilitated the persistence of traditional gender values, whereas the case was more complicated in the East. In the West, conservatives supported the electrical remodelling of the kitchen as the only realistic way to preserve the traditional family model.[26] This combination of new household technologies and traditional gender stereotypes had already been promoted by manufacturers' advertising in the interwar era. As a result, the concept of the housewife did not vanish. Instead, it was modernized and paved the way for women continually doing the household chores even with increasing female (part-time) employment. In the interwar period, household technology had not had an effect on female employment rates. Due to high prices, it was mostly upper- and middle-class families that purchased the new artefacts. This changed incrementally with the diffusion of new appliances in the 1950s and afterwards. In the long run, the widespread use of modern technology in the home laid the foundation for an unintended, but important, change in gender roles. It had become common for women to cope with technology. They proved that they had no problems dealing with it.[27]

The situation in East Germany was different. On the one hand, there was the same persistence of gender roles that gave East German women the 'double burden' of being housewives besides their full-wage labour. As late as 1970, an East German survey demonstrated that women took over thirty-seven hours of the average forty-seven total weekly hours of domestic work.[28] Thus, at home the situation was similar to West Germany. On the other hand, school lunches were far more important in East Germany, which freed mothers from the need to prepare lunch daily for the children. The difference in this respect was enormous. While West Germany rejected school lunch programmes, the GDR's school lunch was a showcase project in worldwide comparison. In the early 1960s, four out of five West German mothers prepared lunch daily and two out of five even prepared two hot homemade meals daily. By contrast, the socialist GDR increased school lunch funding in the 1970s. This programme was part and parcel of politics of establishing a socialist consumer society.[29] In addition, cafeterias became more popular in both East and West after the war. From the late 1950s onwards, in both states, a majority of employees ate lunch at cafeterias.[30] Therefore, even if domestic work still was obligatory for women, some tasks were gradually outsourced.

New kitchen technology contributed to the transformation of traditional cultural habits of purchasing and preparing food. In the 1950s, the West German government actively sought to establish a consumer society. The economic ministry put the proliferation of the electric refrigerator in the centre of these politics. Within the decade, the rate of households with a fridge increased from 4 per cent to over 40 per cent, which was very high compared to other European countries.[31] Together with the automobile, which became more common in the 1960s (see Chapter 2), the refrigerator enabled the storage of vast amounts of food. Accordingly, consumer habits changed during the 1960s. The weekly bulk purchase with the car replaced daily shopping.[32] These two products signified the coming of the consumer society. However, users had to integrate the technologies into their daily routines. Moreover, social acceptance had to be won. An oral history interview with an early purchaser of these two consumer goods gives a good impression of the ambivalent situation in West Germany. The interviewee admitted that she was privileged due to the fact that she and her husband were a middle-class double-income couple of two teachers. As such, they could afford a refrigerator and a Volkswagen car before they became common. While they proudly presented the fridge to any visitor in 1951, it was different with the automobile three years later: the couple was quite ashamed to own a car because they feared the envy of their neighbours. Consequently, they never parked in front of their house, but always around the corner.[33]

Electric stoves also became a huge success during the 1950s. As in the late 1920s, they still represented modernity and rationalization. But now, housewives welcomed the modern stoves with open arms. This was due not least to the promotional campaigns for modern technology in cookbooks and women's journals.[34] Among others, the very popular cookbooks of the food company Dr. Oetker promoted electric cooking in the 1950s. The introduction of one of these volumes proclaimed that only the electric household was a rational household. Electricity was vital for household efficiency, but a comprehensive approach was needed. As the kitchen was the 'housewife's workshop', a small built-in kitchen with perfectly fitted appliances and furniture was proposed as most appropriate.[35] Electrical appliance manufacturers such as AEG or Linde also published cookbooks to support the marketing of their stoves, fridges and freezers; later followed by microwaves. In the late 1950s, the AEG Company combined the manual for its electric stove with recipes. It praised the modernity of its stove, which would support vitamin-friendly cooking. The manual emphasized that although this was a very modern appliance, AEG had seventy-five years of experience with electrical appliances already. The manual also praised the stove as a particularly unpretentious 'friend'.[36] In addition, the stove advertising promised pleasure instead of hard work: electric cooking offered 'fully automated food preparation', by which cooking was transformed in mere 'button-pushing and dial-turning'.[37]

The combination of novel cooking and cooling technology with a new supply of convenience food transformed West German housewives' food preparation entirely.[38] In general, preserved foods were nothing new in post-war Germany. The German canning industry began its success story in the last third of the nineteenth century. Initially, the canning companies targeted not the masses but rather wealthy people. By 1872, twenty-nine companies in the rather small duchy of Brunswick were canning food, mostly asparagus but also peas and beans.[39] Additionally, there were important innovations in DIY canning taking place in the 1890s. The German entrepreneur Johann Weck invented a simple system of glass and a rubber ring. Due to 'aggressive propaganda and marketing methods', Weck's system soon replaced traditional preservation techniques such as drying and salting and became a huge success in a number of European countries.

After the Nazi takeover, home canning turned into something like a national duty. Private DIY techniques perfectly matched with the Nazi politics of autarky. During the war, the Germans urged the occupied countries to adopt these home-canning methods.[40] Mass consumption of canned food, however, started after the Second World War; between the mid-1950s and the mid-1960s the consumption of canned vegetables tripled in West Germany. This challenged the traditional household technology of homemade preserving; while DIY canning was dependent on harvest seasons, commercial tinned food was available all year. A survey from the mid-1950s showed that rather than time saving, the increase in consumer choice was the main reason for most housewives to switch from homemade preserving to canned foods.[41]

The proliferation of freezer technology broadened the range of convenience foods. By the mid-1960s, frozen food sales had increased sharply, with nearly half of West German households using these foods from time to time and 25 percent regular consumers.[42] Even with the breakthrough of the new techniques and the widening of commercial supply, the traditional practice of preserving did not vanish. This was mainly due to the fact that the image of a competent housewife depended on such practices. In addition, many people preferred the flavour of homemade dishes. Thus, homemade cooking did not disappear with the advent of processed foods and ready meals. The new technologies of convenient foods offered timesaving, but they did not displace the social and cultural meaning of cooking. Furthermore, in preparing homemade food, there were certain limits to kitchen technology given the fact that users were convinced that a certain sensual expertise or tacit knowledge was necessary for cooking. Although the food processor was capable of producing dough, only while kneading the dough with her own hands could the experienced housewife know exactly when the dough was ready. The senses of hands and fingers were not easily to replace. Automation surely had its limits, in the household as well as in the factory.[43] Besides any practical matters, it was of utmost importance that customers had the impression that they were in control of the operation. Thus, around 1980 washing machines with only one control button vanished from the West German market. Devices with more buttons, which symbolized more control and agency for housewives, replaced them.[44]

The freezer diffused more slowly through Germany than the refrigerator. In 1995, when almost any German household was equipped with a fridge (most of them including small freezing compartments) but still only 45 per cent owned a freezer, a cookbook published by a subsidiary of the appliance manufacturer Linde advertised the advantages of modern freezing. This technology would increase, it claimed, the meal planning opportunities for the housewife and at the same time save her a lot of time. Furthermore, the book presented this modern lifestyle as closely linked to ancient times – freezing in crevasses – and to the pre-modern but ingenious practices of the Inuit. Still at the end of the twentieth century, these cookbooks promoted modern household technology on the basis of traditional gender roles: This book from 1995 still argued that the 'gourmet housewife' stimulated by the wish to please her husband's culinary demands needed a freezer: most of all, this appliance saved her time to decorate the plates more attractively.[45] Along with the freezer, the microwave also prevailed in Germany from the 1980s onwards. Again, the Linde Company presented a cookbook with its appliance, which was addressed as 'the microwave wonder' that promised ever-more timesaving opportunities. Again, gender roles were very traditional in its promotion: the cookbook's target group was housewives who had to prepare lunch and dinner for their husbands and children 'on time'. By contrast to former cookbooks, now the 'special case' of preparing readymade meals under time constraints played an important role.[46]

By and large, the household technology of the post-war era changed social patterns of everyday living in West Germany. Fridges, freezers and microwaves gradually dissolved the tradition of family meals and individualized food consumption.[47] In the long run, patchwork approaches to preparing food prevailed: ready-made meals became more widespread, but they replaced homemade cooking only in some families. Occasional usage was still most common. For its part, East Germany lagged behind in household technology diffusion, although the socialist state had the promotion of processed food high on its agenda. In 1970 when 85 per cent of West German households owned a refrigerator, the rate was only 50 per cent in the East. Even worse, infrastructural failure hampered the distribution of frozen foods in East Germany: freezing capacity was insufficient and unbroken cold chains were rare.[48]

One common feature of both German states was the success of self-service stores in the 1960s and 1970s. While at the end of the 1950s, only 10 per cent of retailers had switched to self-service, in the early 1970s, 90 per cent of West German retail revenues were accounted to self-service stores.[49] Following the American example, the first German self-service grocery store opened in Osnabrück in 1938. However, it flopped and closed in 1940: customers did not like the concept. In any case, with food rationing during the war, failure was unavoidable.[50]

More comprehensively, self-service increased rapidly in West German food retailing from the mid-1950s forward. Initially, customers experienced self-service as a shock and felt insecure in the presence of unknown choices. In particular, they were afraid of buying more than needed. But soon, they became familiar with the new situation. While there were only thirty-nine self-service shops in 1951 and a mere 203 in 1955, by 1960 the figure had risen to 17,132.[51] In general, the concept of self-service was adapted from the United States. However, technological and economic know-how transfer had to be adapted to German cultural assumptions of shopping, selling and consumption. The triumph of self-service, supermarkets and discounters in the 1950s and 1960s was not particularly German; it took place in most European countries, West and East. The combination of different technologies – of household, distribution and mobility – engendered a seismic change in consumption patterns. From the 1960s, inner-city retailers suffered from the new competition of supermarkets on the outskirts, now accessible due to motorization. With the shift to a weekly bulk purchase with the car instead of daily shopping, supermarkets needed parking space, which was rare and expensive in the inner city.[52] In socialist East Germany too, self-service retail gained ground from the late 1950s onwards. The state party established a programme with the slogan 'modern people shop modern', seeking an increase in efficiency. Self-service, ready meals and frozen food symbolized modernity in both German states.[53] These general beliefs were shared by the capitalist West as well as by the socialist East. Thus, both German states faced a new problem resulting from their common food technology politics by the 1970s: obesity.[54]

While this problem represents some of the worst consequences of the consumer society, there were aspects to it that increased individual agency, in particular for women. From the mid-1950s onwards, a new generation of young women became technology users at home in a very different sense to their mothers. While the previous generation had used irons, vacuum cleaners and had just begun to try out electric stoves and washing machines, their adolescent daughters had everyday contact to leisure technology. This started with the automatic record player, which was a low-threshold offering and very comfortable: technological skills were not needed. Most record players were sold to young people, in particular to young women. Usage of these new artefacts introduced young women to public life. They used technology at home, but unlike the usual household technology, this was not about cleaning or cooking. Instead, it gave rise to the emerging youth culture based upon the consumption of popular music.[55]

Novel radio technology supported this trend. In contrast to the TV, which had become common by the mid-1970s and which was at that time usually used by the whole family, the innovation of the transistor radio provided adolescents with a technology that laid the foundation for new cultural distinctions. Transistor technology had both minimized and cheapened the radio. This process had already occurred in the United States, but its transfer to West Germany was delayed for economic reasons. Only the upswing in the 1950s made it a huge success in West Germany. For the first time, many young people had their own radio and the chance to listen to their kind of music whenever they liked. Before, huge radio devices had enjoyed a monopoly. Now, adolescents purchased portable radios.[56] Consumer electronics were crucial to the rise of youth culture, with adolescents having a higher affinity to everyday technologies. As importantly, the devices – most of all portable radios, record players and tape recorders – became affordable from the late 1950s onwards. In particular, the tape recorder offered opportunities for users' creativity. However, the still relatively high prices and the rather challenging operation limited the number of users. It was mainly middle-class young men who owned a tape recorder in the 1960s. Only from the late 1960s onwards did cassette recorders become more popular, due to lower prices and high user-friendliness. Initially, during the 1960s, young consumers did not care much about sound quality. To them, low prices and a modern design were more important aspects. Many young people did not even mind interfering noises on the radio so long as the device guaranteed the reception of many international stations. The interference noises even represented a certain global-mindedness.[57]

Devices for listening to music implied very different forms of user agency, which was dependent upon one's generation. In contrast to young people, most adults preferred the traditional (and huge) radiogram. This combination of radio and record player in a large piece of furniture, which prevailed until 1960, could not offer users much opportunity for tinkering or experimenting. This appeared only with the success of the modern stereo system. The stereo contained many more controllers, demanded cable laying and, most of all, gave the user the option to choose the components for the system he or she needed (or those that were affordable for the moment). Its later extension, for instance, with a tape deck, was simple. This is a good example of the profound change in consumers' mentality. From the 1960s onwards, mass consumption afforded a sense of individual consumption. In this context, practices of individual distinction went very well together with mass-produced uniformity. Importantly for the history of technology, the stereo's design emphasized its nature as a technological artefact whereas its predecessor, the radiogram, concealed the technology under the disguise of a piece of furniture. Due to high prices, the stereo nonetheless remained a status symbol of the middle class until the 1970s.[58]

By and large, leisure technology offered the opportunity for enhanced user agency. As the example of the tape recorder demonstrates, the boundaries between mere consumption and tinkering blurred: a minority of enthusiasts did not only record music from the radio or from vinyl, but also went further and experimented with making their own sound recordings or even audio dramas.[59] In particular, leisure technologies opened up a space for the creative appropriation of technology at home.

Tinkering: Analogue and digital

Since the second half of the twentieth century, different types of tinkering with technology have been omnipresent in the lives of millions of Germans. But even long before DIY became a widespread phenomenon, a certain DIY culture had already begun with the arrival of the automobile at the turn of the twentieth century. At that time, motorists needed technological skills. Lacking any comprehensive infrastructure of mechanical workshops, they had to know how to maintain and repair their own automobile; or rather, in this early period of luxury driving, their chauffeur had to have the necessary know-how. Early cars were also very prone to technical glitches. For the first time, non-specialists had to cope with technological issues in their leisure time. To some extent, they became experts on their own. This tradition of tinkering remained important to drivers for a long time: as late as the 1950s, warnings were common that one needed basic technological skills if one did not want to become beholden to one's car.[60]

Automobile tinkering had different meanings in the two German states after the Second World War. While in East Germany, tinkering was a necessity due to the lack of service facilities, West Germans' automobile tinkering was primarily a hobby in many cases. Thus, the GDR experience was a reminder of the early period of automobiles when technology was unreliable and the users had to help themselves. As will be shown herein, this particular kind of DIY was central to the ideology of the socialist consumer society. In this context, consumption was a means to activate consumers. From the very beginning, a new car needed some mechanical care from the purchaser. Before driving, the buyer had to mount the wipers of a brand-new Trabant car on his or her own. To some degree, the consumer was part of the production process.[61] Even the Trabant manual was not simply the usual user guide Westerners would have expected. Rather, it was 'a basic course in automotive engineering'.[62] As a result of the lack of garages, most drivers learned to do basic repairs on their own. This was a crucial advantage of the Trabant: its two-stroke engine was easy to repair.[63] The same was true for the second GDR marque, Wartburg. It was more expensive and a kind of luxury alternative to the Trabant, but still one had to know how to maintain and repair one's own car (see Figure 7.1).

What at first glance looks like a shared hobby in Cold War times had in fact a very different meaning in East and West. Only car enthusiasts – although they numbered hundred thousands of men – in West Germany tinkered with their automobiles and thus established a masculine subculture. By contrast, *any* East German car owner had to tinker periodically with his or her car as a matter of necessity. Both East and West Germany had become consumer societies, but consumption was more demanding for technology users in the GDR.[64] After the fall of the Berlin Wall, it became obvious that there was a specifically East German car culture beyond this realm of necessity. In 1990, many Trabants and Wartburgs that had been aesthetically modified were spotted on West German roads. The cars had, among others, special wing mirrors or colour schemes in mimicry of Western brands.[65]

The very term 'do-it-yourself' arrived in West Germany after the Second World War. US middle-class men had enjoyed a DIY subculture since the first half of the twentieth century, and after the war American influence helped to establish it in West Germany.

Figure 7.1 Repairing a Wartburg 311 car at the East German Meyersgrund camping site 1964, photo by Köhne, Bundesarchiv.

There, DIY really broke through in 1957, when the first DIY journal *Selbst ist der Mann* became popular. By that time, the term 'do-it-yourself' became commonly used as an Anglicism. There was, however, a certain prehistory to DIY around 1900. Many German educators were impressed by the average level of technical skills in the United States and supported handicrafts for boys. This movement was motivated by a 'fear of lost skills', in particular in international comparison: educators were afraid of Germany losing ground to the United States and other competitors. Yet only in the mid-1950s did the establishment of the five-day week pave the way for the breakthrough of DIY and hardware stores, when the United States was a role model, both for products and for the stores themselves.[66] Indeed, the entrepreneur Otmar Hornbach established the first building supplies supermarkets in West Germany after having visited the United States in order to study those stores in the mid-1960s.[67]

The social and cultural context of West Germany in the 1950s and 1960s also explains the rapid success of DIY. According to historian Jonathan Voges, DIY was conceived of as an antidote to the contemporary cultural pessimism about automation and the widespread fear of technology taking command. DIY represented the possibility of designing one's own environment,[68] serving as a remedy both for stressed managers and for workers challenged by automation.[69] This new discourse in some ways appealed to the tradition of 'joy in work', which has been strong in the Weimar and the Nazi period.[70] Moreover, do-it-yourself projects held the allure of saving money. In the end, however, the acquisition costs for tools were quite high, meaning DIY often turned out to be a costly hobby. Nevertheless, the boom continued in a period of prosperity.[71]

Furthermore, the first DIY wave of the late 1950s occurred during a severe crisis of masculinity. After the end of Nazism, the image of the heroic soldier ceased to be a practical model of masculinity. DIY offered new images of manliness that ranged from 'rough manhood' to 'respectable' models of masculinity based upon 'togetherness' with the children. The father became the 'repairman' and earned respect; most do-it-yourselfers represented a 'gentle fatherhood'. They cared for the home and educated their sons – more rarely also their daughters – in practical tasks. Yet, this was not a linear trend from rough to softer images of manliness. Tinkering with machines gave the DIY men a certain symbolic power and strength. This became more important again in the 1960s and 1970s, at a time when male physical strength increasingly disappeared from the workplace. DIY partially compensated for that loss. For many, drilling machines, which most do-it-yourselfers owned, symbolized male strength.[72]

At the same time, a counter-culture added new social groups to the DIY movement as the environmental movement and the alternative scene adopted traditional practices of handicraft and gave them a novel meaning.[73] More and more women also discovered DIY from the 1970s onwards.[74] At that time, the numbers of West German women do-it-yourselfers were significantly lower than in Western European countries. Therefore, hardware stores started a marketing offensive that was very successful: already in the mid-1980s, a third of the roughly thirteen million West German do-it-yourselfers were women. By then, DIY had become a mass phenomenon. More than 50 per cent of West Germans of all classes, including an increasing share of women, refurbished and repaired their homes. During the DIY boom of the 1980s, more professional single-purpose renovation devices gradually replaced multipurpose machines, while the professional equipment had become affordable. In general, DIY technology was now easy to handle, small and rather cheap.[75]

In socialist East Germany, a distinctive DIY culture established itself. While do-it-yourselfers in East Germany used similar machines and read journals comparable to those of their Western counterparts, the ideological content was different. The East German self-help literature commonly used the term 'join-in movement', or *Mach-mit-Bewegung*, instead of do-it-yourself. This appealed to the handymen's sense of collectivity. Accordingly, the individual practices of refurbishing the house were integrated in the role model of a socialist consumer who was helping to build socialism.[76] As the GDR lacked the infrastructure of hardware stores, materials were hard to obtain. Most machines were available only in special stores for foreign currencies. Consequently, after reunification, many West German chains opened new hardware stores in the East and DIY boomed again.[77]

In both German states, the DIY spirit entered into a new field of activity from the 1970s onwards with the arrival of computers. Of course, there was an important prehistory of computer usage before computer hobbyists took over. By 1973, the impressively high number of 15,000 computers were in use in West Germany, mostly in public administration, banking and universities. Only the United States had a higher number of computers at that time.[78] In contrast to the United States, the military is not central to the West German history of computing. Instead, state administration, universities and businesses were the most important players in the proliferation of computers from the 1960s onwards.[79] Microelectronics also gradually invaded the home, sometimes unnoticed, during the 1980s as digital technology was integrated into many household technologies such as the washing machine, the dishwasher or the vacuum cleaner. Not least, the automobile was also incrementally digitalized.[80] This process began early. Volkswagen was the first car manufacturer worldwide to integrate digital technology into automobiles in 1967. In the following year, Volkswagen proudly advertised this innovation in the United States: 'Now. A car with a computer in it'. Volkswagen also introduced computer diagnosis to garages in 1971 in order to increase their productivity. This was a reaction to ever-increasing sales, which had the consequence that garages could not cope with the high quantity of consumer-demanded maintenance by the established methods. The early computer diagnosis system was a disaster due to regular malfunctions, but in the long run, this became the future model of car repair.[81] By and large, technology users got accustomed to the diffusion of computer technology in everyday life.

The invention of the microchip in the mid-1970s was essential in many respects. Often overlooked, a shift of innovative power was engendered by this fundamental technological change. No longer were network operators the most important actors of innovation. Now, the device manufacturers or even the users had become the decisive forces of innovation. Tinkering was essential to the early period of the private use of computers. Initially, there had been no fixed usage patterns: any user had to find his or her own way to use computers be it for communication or any other application. Often, this meant the early users had to tinker around with the devices or write programmes on their own. Only if one knew how to programme could computers offer a multitude of applications. Therefore, the early users – the hobbyists-turned-experts – were crucial for the further development of home computers. Many of these tinkerers had already belonged to familiar scenes before the advent of the home computer; most often radio amateurs or 'phone freaks'. This combination of interests and skills gave birth to the Computerized Bulletin Board System in the United States in 1978. A few years later in West Germany, this communication innovation became established, albeit on a rather low level, where it was known as 'mailboxes'.[82]

From the beginning, there were strong incentives for West German computer hobbyists to break the law. The official procedure of renting a modem from the Federal Post Office, which was the only legal option until 1986, was not very attractive to computer hobbyists due to the high costs involved and the rather slow connection it provided. Therefore, many home computer users illegally imported modems or used DIY modems.[83] As a consequence, new subcultural communities emerged that were prone to half-legal activities: hackers, gamers, crackers and bulletin board system users. By and large, the triumph of the home computer and the personal computer in the 1980s and early 1990s established new cultural technologies such as gaming and programming.[84] Most users of home computers were male adolescents who received the computer as a present. The gender gap was enormous: in the late 1980s, more than a third of boys and young men in West Germany owned a computer, whereas the rate of female adolescent owners was less than 7 per cent.[85]

The first subcultural computer group to become well known to the German public were hackers. In particular, the 1981-founded Chaos Computer Club (CCC), until today the largest association of hackers in Europe, earned national fame in 1984 when a group of CCC members succeeded in hacking into the online videotext system of the German postal service and got the Hamburg Savings Bank to transfer more than 100,000 deutschmarks. They retransferred the money the next day because they only wanted to inform the public about security issues. However, there are serious doubts about the veracity of the CCC members' self-perception as invincible hackers. Contemporary computer experts suspected that the famous case was not a hack in the proper sense. Instead, there is some evidence that it was a mere matter of spying out the password. Nevertheless, it helped the CCC to put data security issues on the political agenda and establish itself as a 'watchgroup' for data security.[86] Although the hacker subculture was crucial for the early development of personal computers, the more official usage in research and development was at least as important.[87] At West German universities, computer science was established in the 1970s, a decade later than in the United States. In addition, there was a vivid discussion about obligatory computer education at school in West Germany during the mid-1980s. However, this never became obligatory.[88] Nevertheless, computing was established as a political topic relevant for business, education and everyday life.

By contrast, most contemporaries missed the perhaps most important factor for the success of private computer usage: gaming. The common visions of the coming computer age that were most prevalent in the 1970s entirely underrated the impact of gaming. Yet, arcade games and video games paved the way for the proliferation of computer game culture. Even before the advent of the home computer, there were roughly 100,000 arcade machines in West Germany. In the end, computer games were of utmost importance for the proliferation of home and personal computers from the 1980s onward. Furthermore, games were the reason for an ever-increasing demand for more processing power. As early as the end of the 1980s, more than 80 per cent of adolescents had played video or computer games, with many of them daily users.[89] Beyond their presence in children's rooms, the new subculture became visible by occupying public space. In the early to mid-1980s, the electrical departments of the big stores in West Germany became the showplace for computer kids. Those kids had rare skills and used the department stores to meet other members of the emerging subculture. These computer kids transformed the store into a contact zone for kindred spirits.[90]

The advent of home computers and the emerging scene of tinkering youths assisted in changing public perceptions of computers. While many intellectuals and leftist critics were initially sceptical about computer technology, which they associated with state surveillance and corporate power, this changed with the advent of the new subculture. Hackers, crackers and the bulletin board scene demonstrated that there were actually ways to creatively appropriate the computer. Most fascinatingly, the majority of the new subculture protagonists were young.[91] Most of them were not overtly political, but mainly interested in gaming. This was particularly true for the crackers. In the 1980s, all over the Western World the triumph of home computers gave rise to a male-dominated, juvenile subculture of security hackers who specialized in breaking the copy protection of gaming software. These 'cracker groups' were neither interested in commercial software piracy, nor did they focus on the political agenda of hacker groups such as the famous CCC.[92] Basically, crackers did not share the open-source values of the hackers. In contrast to hackers, crackers concealed their programming tricks. Their goal was not to give free software to the public, but to sign the cracked software with their self-made intros. They did not sell the cracked games: their currency was public attention and fame within the scene.[93]

The crackers' subculture was all about competition between different groups. In a certain sense, the crackers were both subversive and conformist: on the one hand, they broke copyright law (or at least they assumed that they would); on the other hand, the computer kids appropriated neo-liberal values. They wanted to have success and to beat their competitors. Furthermore, high efficiency was their upmost goal, in their organization as well as in distribution channels. Thus, data transfer through bulletin board systems soon displaced postal dispatch. As a result, the scene became transnational, making the crackers 'global players'.[94] The cracker scene was particularly vivid in West Germany, where the *Commodore 64* was a huge success in the mid-1980s. Strictly speaking, they did not even break copyright law in Germany initially, because there was no specific law for software copyright until 1985.[95] Nevertheless, the crackers enjoyed their outlaw image. On a European scale, there was little danger of criminal persecution before the European standardization of digital copyright laws in 1993. Before that, the worst that happened to West German crackers were house searches and small fines.[96]

Notwithstanding the carefully cultivated outlaw image, the crackers did not perceive themselves as antagonists of the game developers. By contrast, the transitions between professional programmers and crackers were fluid. After breaking the copy protection, the crackers introduced a new intro to the game, by which they sought to make their own group famous. The animations and the music of these intros became ever more advanced, finally introducing a novel subculture of computer hobbyists: the demoscene. The software industry recruited some of the most talented members of both the crackers and the demoscene. The German game developer *Magic Bytes* even cooperated with the cracker group *Radwar*. The young crackers helped with developing new games, while the company gave preliminary versions of new games to the cracker group in return. The deal was that the crackers were allowed to distribute the game only to their intimate friends within the scene while all of them promised to respect the copyright for the time being. Thus, the most potent crackers were out of the game and the company had at least a few weeks to sell original copies untroubled.[97]

By contrast to the vivid West German scene, computer hobbyists in the socialist GDR led a shadowy existence. This was a result of the low output of the East German computer manufacturers (see Chapter 1). Domestic industry only produced 30,000 low-performance home computers, and most of those were for export. Nevertheless, roughly 200,000 East Germans owned a home computer. These machines were either tolerated presents from West German relatives or official purchases in *Intershops* where rare Western products were offered for foreign currencies. In principle, hardware imports were legal; only software imports were banned.[98] Therefore, a certain scene of computer enthusiasts existed in the GDR. Even before the two states formally reunited, the computer hobbyists of East and West Germany unified in East Berlin in February 1990. The unification was quite easy because they shared similar values and practices, even if they had established themselves in two very different political systems. Both shared a do-it-yourself mentality. Yet, they came from different directions. While East Germans in general had to do-it-yourself because of a lack of consumer goods, the West German hackers' DIY signified a voluntary opposition towards the doctrines of the consumer society. There was countercultural response to the mainstream computer user mentality. All hackers, however, shared the belief that their hobby was about understanding and controlling the machines.[99]

Conclusion

Histories of technology usage tend to focus on spectacular cases of amateurish expertise and subversive practices. As has been shown, these cases did exist and sometimes they were crucial for further technological development. However, one should beware of false impressions regarding the quantitative impact of these forms of usage. In fact, most consumers were not creative, but rather merely responsive. Predominantly, technology users simply followed the given use and consumption patterns. Nevertheless, technology usage was always a contested area where the designers' intentions were never entirely identical with the appropriations of everyday life. Often there was a certain persistence of cultural patterns that hindered novel technologies from transforming everyday life immediately. In particular, as the case of household technologies demonstrates, traditional gender values often halted modernist reformers' intentions.

However, consumer society proved to be fortified against any subversive strategies. In general, capitalist consumer society demonstrated its integrative power. In many cases, subversive or illegal subcultures were integrated into corporate structures. Even semi-criminal computer kids often made their ways into IT industries. Yet, technology users always mattered. Their agency should not, however, be mistaken for resistance. Particularly in periods of transition, the creativeness of users was vital for the further development of specific technologies. This was most obvious in the case of computing. Less spectacularly, it was crucial for novel everyday technologies, such as household technologies, that users adopted them, accepting any teething troubles and overcoming traditional habits.

With respect to the issue of possible German peculiarities in the history of technology, the history of consumption gives fewer examples of these than the history of production. There seems to be a certain process of convergence in consumption patterns towards a common Western model. To a certain degree, however, cultural traditions and material necessities of class resulted in a persistence of some German peculiarities. For example, economic state dominance was crucial for particular developments, such as the West German state monopoly on communication giving rise to a strong semi-legal subculture of computer hobbyists who had been looking for alternatives to the official supply. During the Cold War, political differences between East and West were of more importance than any broad German tradition. East Germans were more likely to share their DIY spirit in tinkering with technology with the citizens of other socialist countries. In this case, shared transnational political and economic constraints of the Cold War era outranked any national historical traditions.

8

Apprehensions of uncertainty

The economic crises of the 1970s saw an increase in the number of critical interpretations of the idea of technological progress. After the Club of Rome published its report *The Limits to Growth* in 1972, some of the basic assumptions of modernity were challenged. The concept of modernity as a (twisted) road to progress, which had been the dominant narrative thus far (see Chapter 4), began to totter. Furthermore, there was a growing awareness that technological innovations could easily be used for politically totalitarian ends or result in the accidental destruction of the environment. However, the master narrative of progress was not replaced with visions of regress or decline, but rather a sense of uncertainty.

The report's high impact was due to globally growing environmental awareness, and the report became a bestseller only with the nascent boom in environmentalism. Crucially, the report did not reject modern technology as such – in fact, it was even based upon big data collected by mainframe computers at the Massachusetts Institute of Technology.[1] Technological innovations had inspired people to envision both sides: humanizing technologies on the one hand as well as exploitation and total computerized surveillance on the other.

Yet the story is even more complicated. Although historian David Edgerton rightly pointed to the persistence of old technologies in global history, linear accounts of 'history as progress' still dominate for the most part.[2] This chapter will demonstrate that there is not only a 'shock of the old' but also a 'shock of the surprisingly early' that should be considered in the history of technology. For instance, environmental criticism of novel technologies was already an issue in the nineteenth century. As such, the depiction of the modern history of technology as being dominated by the unquestioned notion of progress until the 1960s is not accurate. On the other hand, neither did the critical debate of the 1970s mean that the very idea of progress had vanished. Rather, visions of progress survived to a certain degree, albeit integrated into the new visions of uncertainty.

The first section of this chapter deals with social protest against novel technologies. Most prominently, machine breakers signified the trouble with industrialization. Furthermore, from the advent of the railway forward, traffic was a contested field. In particular, the car-friendly city caused much criticism from the latter half of the twentieth-century onwards. Indeed, from the outset early automobiles had been contentious. The confrontation, however, was not clear-cut. Around 1900, technological euphoria and technophobia often went hand in hand. On the one hand, rapid transformations raised fears about the changes that modern technology was about to bring. On the other hand, even sceptics – from socialists to nationalists – were convinced that they needed these very technologies for their political visions to come true. As such, it was not merely opposition between supporters and opponents of modern technology but most often the emotional conflict that took place inside singular observers that heightened ambivalent attitudes towards technological innovations.[3]

In general, emotions played an important role in the history of technology's acceptance. Nonetheless, the assertion that opponents were simply driven by emotions also deserves critical inspection. Historians must take care not to affirm the prejudicial propositions of their sources. This is essential for the second section of this chapter, which addresses environmental issues. For instance, the rejection of certain technologies such as nuclear fission does not imply that the opponents were afraid of technology. Rather, their rejection was often predicated on forms of counter-knowledge and the political rejection of technocratic rule. Thus, it was not mere irrational fears but rather rational concerns about the environmental, social and political outcomes of a large technological system. In this context, it will be critically discussed if the notion of 'German Angst' gives an accurate representation of a particularly strong German environmental movement. The relationship between this movement and technology was ambivalent, leading West Germans to debate an ostensibly new technophobia and its dangers in the 1970s and 1980s. Accusing political opponents of being machine breakers came into fashion again, especially concerning nuclear power or computer technology.

Social protest against new technologies

Often overlooked is the fact that not every reformer of the early nineteenth century was an advocate of industrialization. For instance, two eminent Prussian reformers of education – Wilhelm Süvern and Johannes Schulze – hindered the establishment of vocational schools with industrial training during the 1820s. The study of classical languages seemed more important to them than any technical education.[4] More widely known is the resistance of workers towards machinery during the initial period of industrialization. The Luddites or machine breakers were, however, often misunderstood. As the eminent historian Eric Hobsbawm has shown, the British Luddites were not hostile to machines or technological progress as such. More precisely, they were concerned 'with the practical twin problems' of avoiding unemployment and maintaining their standard of living.[5]

Historians have demonstrated similar findings for nineteenth-century machine breakers in Germany. Most of all, the machine breakers directed their actions against a specific new technology which threatened their jobs. Their demolitions were even disciplined. Generally speaking, they did not plunder and usually deliberately destroyed only the newest machine of a certain type in their region. Their goal was always to crush technology in the introductory phase. Most of the machine breakers remembered the guilds' restrictions against new technologies and expected the authorities to control the introduction of new technologies. Machine breakers were seldom romantics who longed for the 'good old pre-industrial times', but rather saw themselves as defending their rights to survive as artisans. Thus, they were neither anti-progress nor anti-technology. Rather than a fight against innovation, theirs was a struggle for socially adequate technologies in defence of their livelihood. A closer look at the structure of German machine breaker groups shows that their protests had little to do with atavistic hostility against technology or reactionary politics. Instead, an anti-capitalist element is discernible. Most prominent among these groups were those whose occupations were moving in the direction of capitalistic wage labour. Artisans such as calico printers, cloth shearers and book printers had worked for merchant putter-outs or later in factories. They had experienced the conflict between capital and wage labour at an early stage. It was not the poorest workers such as spinners but rather skilled and well-paid craftsmen who protested against new machines: strength was needed for such resistance. Belonging to the labour aristocracy, these workers had long experience in representing their interests that dated back to the eighteenth century.[6]

By and large, machine breakers were rather rare in the early nineteenth century. They were mostly found in regions with pre-industrial traditions. The artisans of Solingen are good examples of a late machine-breaking action. In March 1848, they destroyed a cast steel plant because this specific technology endangered their craft. Before the destruction, the artisans had petitioned the authorities without success to regulate wages. These petitions show a conception of technology that significantly differed from the labour movement, which established itself several years later. The Solingen activists did not perceive technology as a necessary step towards a better future.[7] This crucial difference explains why the modern German labour movement that strongly believed in the opportunities of technological progress distinguished itself from the machine breakers of the early period of industrialization.[8] The Social Democrats criticized the machine breakers as undisciplined and reactionary because they had not grasped the Marxian difference between 'the machine as such' and the capitalistic use of machinery (see Chapter 4). Through to the present, the German labour movement has always emphasized that they were no machine breakers.[9] The fear of machine breakers and social uprisings, however, was the reason for some of the old elites, who were afraid of upsetting the traditional social order, to oppose industrialization as such. These rather small groups combined hostility towards technology with reactionary politics.[10]

While the quantity of machine breaker actions was rather low, broader civilizationary criticism began at the end of the nineteenth century.[11] The crucial aspect to the acceptance of technology, whether in industry, energy production or traffic, was the issue of technological risks. As will be shown in this chapter, critical debates about technological risks were not anti-modern as such. Instead, risk awareness and risk acceptance helped modern technology to gain crucial enhancements and to find public acceptance.[12] In many cases, critics helped novel technologies to prevail because they demanded modifications. They articulated their criticism and opted for socially adequate technologies. By contrast, technology enthusiasts dismissed any criticism of new technologies and claimed that their opponents were rejecting progress as such.[13]

In the interwar period, characterized by widespread debates on Taylorism and Fordism (see Chapter 1), there was surprisingly little attention to the option of machine breaking. This was largely due to the politics of the Social Democrats and the unions, who unequivocally admired technology. Only the Communists openly opposed efficiency measures. Even the ambivalent outcome of rationalization did not shake the Social Democrats' 'belief in the necessity of rationalization'. For instance, in 1928, a miners' union newspaper denounced any fantasies of machine breaking as 'nonsense, a waste of power'.[14] Yet the rank and file did not so unequivocally welcome new technologies. In particular, the older generation of miners was 'hostile to machines' because they feared unemployment. However, miners' refusal vanished when they experienced the benefits of mechanized work, which reduced their exhaustion. Furthermore, most skilled factory workers were not usually affected by the rather unpopular new technology of assembly-line production. Non-skilled, often female, workers were employed with these tasks as a rule. In general, workers did not reject technical change as such. Most of all, their problem with rationalization was symbolized by icons of capitalist exploitation such as the stopwatch, which accompanied the efficiency measures.[15] After the takeover of power, the Nazis made use of the widespread aversion against this specific technology of time-and-motion studies and did away with the iconic symbols of exploitation, without changing industrial work routines as such. A Cologne newspaper propagated in 1939 the idea that the feared 'man with the stopwatch' of the past was gone by now (see Figure 8.1). However, efficiency measures did not disappear. As the article insisted, workers had to adapt themselves to the machine, now that the machine no longer threatened their jobs under the Nazi politics of full employment.[16]

Figure 8.1 'The man with the stopwatch, feared in past times' at the Humboldt-Deutz plant, Cologne. Kölnische Zeitung, 12 February 1939. Rheinisch-Westfälisches Wirtschaftsarchiv, file 107-19-1, image 10.

Only the computerization wave of the 1970s and 1980s brought a revival of the machine breaker debate. In general, the West German trade unions did not reject automation technology. Unionists consistently considered both sides during this debate: on the one hand, there was the fear of unemployment, and, on the other, there were the hopes associated with the social and political effects of technological progress that the labour movement had harboured ever since its inception.[17] Consequently, the automation debates of the 1960s and 1970s largely resembled the comparatively tame Weimar discussions about Fordism. Only the advent of the microchip, beginning to transform working life in the mid-1970s, brought established union politics into question. The printing industry was one of the first sectors to be transformed by computerization (see Chapter 1), and the printing labour disputes of the late 1970s demonstrated how the traditionally technology-friendly attitude of workers was challenged by novel computer technology.

In March of 1978, the printing union prevented the production of many daily newspapers, while the union itself published so-called strike or emergency papers during the dispute. On 6 March 1978, one of these 'emergency papers' wrote about an incident that took place on the street during a strike rally in Wuppertal. Any potentially fictitious editorial amendments to this account aside, the passage certainly testifies to the future expectations of the striking workers. It described a bewildered worker from a different industry, a patternmaker, asking the picketer from the printing union why on earth the striking workers were opposed to this new technology. The picketer, a shop steward with the union, replied to the sceptical joiner:

> Imagine there is an order for a new workpiece, and your employer also has these cheeky computerized systems. Then all you do is take the wood, the glue, and the drawing of the model and enter that into the computer, which then produces the finished piece. Also, you won't be needed any longer to press the button for the machine to start, that'll be done by someone else. How do you like that?

According to the report, the joiner was quite shocked, responding, 'Oh, my God, that can't be true', and donated five deutschmarks to the strike fund.[18]

At that time, most labourers in printing industries still had a strong belief in technological progress and their own irreplaceability (see Chapter 4). It was not until they were confronted with the 'cheeky computerized systems' that some workers in early-computerized print shops felt compelled to reject the new technology or even developed Luddite fantasies. One worker, who attended the annual print trade fair with several of his colleagues to inform visitors about the anticipated dreadful consequences of computerization, pointed out that he would 'gladly like to grab and smash up' these machines 'because they are taking away my job'.[19] However, there were no cases of machine breaking in German print shops, even though typesetters were keenly aware of the threat to their jobs. If they had any aversion to new technology or daydreamed of Luddism, the workers discarded those ideas themselves. A phototypesetter told an interviewer about his bad feelings about the 'silly automats' and his insight that the end of typesetting was near. The option of throwing those 'crappy things' out of the window would make no difference anyway, because typesetters would 'be right out of it' soon.[20]

Figure 8.2 'Small causes – big effects'. Poster of the German trade union federation (DGB), 1979. Archive of Social Democracy/Friedrich Ebert Foundation.

Printing was only the harbinger of general change. In this context of the broad challenge of computerization, the German trade union federation published a poster in 1979 showing a microprocessor with the slogan 'small causes – big effects' (see Figure 8.2). The poster's text warned of the effects of computerization. Microprocessors would turn into 'job killers'. At the same time, the text on the poster made clear that unionists were 'no machine breakers'. Instead, the union federation advocated for socially just technological progress.[21] There is some evidence that the mood was different among the rank and file, although no incidents occurred. According to historian David Noble – an activist himself – many workers opposed technological change in the 1970s and 1980s, or they at least wanted to slow it down. Only the trade union leaders were afraid of the accusation of being Luddites and resigned themselves to the idea of progress. Critical European workers met in Hamburg in 1982 to discuss the challenge of computerization. At this meeting, a Hamburg dockworker asserted that his colleagues knew exactly that new technologies were never in favour of their own cause. An employee representative of a print shop agreed. Workers were interested in the here and now, not in the outcome of progress in a distant future. Another printer stated that the labour base was ready to resist technological change whereas the employees' representatives were more reluctant.[22] In the end, no rebellion occurred. In all likelihood, the congress represented only the minority of worker-activists.

The computer expert Ulrich Briefs, one of the sharpest critics of computerization, was possibly the only prominent unionist in Germany who advised workers on Luddite actions. He gave the stealthy clue that liquids could harm computers without mentioning the possibility of sabotage.[23] But even Briefs stated in 1984 that the destruction of jobs was but one side of the coin. On the other side, work time was won. Therefore, a transformation of society was thinkable: spending time for new social tasks or shorter working weeks appeared as attractive opportunities.[24] Accusing one's political opponent of being hostile to technology per se was, however, far more frequent than calls for machinebreaking. This grand accusation was an effective means of delegitimizing any criticism of a specific technology. Usually, people had an issue with very specific technologies, often due to environmental risks or the fear of job losses. In public debate, these critics were accused of being afraid of change. In summer 1979, even West German chancellor Helmut Schmidt employed similar rhetoric to warn of the danger that the fear of technology could turn into hostility towards technology.[25]

The advent of computers also engendered concerns beyond the workplace. In particular, concerns about data protection were widespread in Germany. While the debates about data protection in the United States and West Germany had many overarching similarities, German data protectionists successfully exerted pressure early on that led to the first worldwide legislative response to the issue. By 1970, the federal state of Hessen had introduced a data protection act. A national act followed in 1978. By that time, however, data protection had become a political issue in the wider Western World. The United States and Sweden, for instance, also adopted laws in the early 1970s.[26] The fear of the computerization of state administration concentrated on two aspects: the danger of an Orwellian Big Brother state of total surveillance and the threat of data embezzlement. From the mid-1970s onwards, the general public began to show interest in data protection issues. In particular, the computerized manhunt for terrorists in the late 1970s generated fears about the emergence of a surveillance state that reached wider political groups. Even the liberal press covered the topic. The national census of 1983 offered a focus to the critique of electronic data collection. Many action groups expressed their fears of computer dominance and total surveillance. Now, the range of protagonists widened further: computer criticism was no longer the sole business of the New Left. Middle-class Germans, intellectuals and politicians of different parties joined in the critique of the computerized census.[27]

However, political excitement disappeared soon after this peak as a result of the rapid spread of computers. Many people worked with computers, and kids played games on home computers (see Chapter 7). With the ongoing diffusion of computers into workplaces and everyday life, there was no realistic chance of avoiding the new technology. People had to cope with computers. Accordingly, they adapted themselves and familiarized themselves with the new technology. The formerly dominant scepticism regarding computers ebbed away. By the mid-1980s, the former hefty public discussion had abated and the image of computers had changed drastically. If until the early 1980s, computers had symbolized an uncontrollable technology of surveillance, by the mid-1980s, computers had become icons of modernity and entertainment.[28]

There seem to be a general pattern to the decline of social protest against new technologies. After the technologies had gained momentum, total rejection became impossible, leading people to feel that they had to adapt to the new technological reality. Nevertheless, a degree of agency in coping with the new technology and using it for their own purposes was preserved. This was obviously true for computer users (see Chapter 7). Similar developments occurred in the history of modern traffic after the advent of the railway.

The railway was the first contact with modern technology for the rural population that did not work in factories. Many country people deeply mistrusted the changes the railway symbolized. The railway was a harbinger of an industrial revolution that was yet to reach their homes. As a key technology of industrialization, the railway both symbolized and accelerated transformation.[29] The new form of transport also brought with it concrete dangers: during the early days of the railway, fears of accidents were omnipresent. Nonetheless, by the mid-nineteenth century, passengers had culturally and psychologically adjusted to the railway journey.[30] As has been demonstrated, this is very typical for the public reaction towards novel technologies. People needed both the time and opportunity to adopt the technology in their everyday lives.

The emergence of the automobile caused similar reactions in the countryside as the railway had a few decades earlier. To some degree, the rural protests against the early automobiles were protests against urban modernity entering the countryside. Criticism of urban automobile traffic had already begun around 1900, in accordance with the previously depicted patterns. Early in the era of the automobile, the new technology lacked widespread public acceptance. In this transitional period, cars did not fit into the traditional patterns of everyday life. A German peculiarity among the Western countries was the endurance of anti-automobile protest. Even in the 1920s, protests against automobiles were stronger in Germany than in the United States, France or Britain. This protest disappeared with the Nazi takeover of power because the new regime aggressively promoted automobiles.[31] Nonetheless, in the early 1930s when car traffic was still relatively modest, future problems were already visible. In 1931, the right-wing cultural pessimist Oswald Spengler asserted that the car was already too widespread and thus had lost its desired effect of accelerating transport in crowded cities. The automobile was no real help in many German cities. Instead, walking was faster.[32]

After the end of the Second World War, the mounting trouble with urban traffic became increasingly acute. Increasing awareness of its seriousness was betrayed in the language used to describe it. While 'traffic difficulties' was the main term that was established in the interwar years, in 1954, when most West Germans still did not own a car, a book had the popular title lamenting the 'cities' traffic misery', or *Die Verkehrsnot der Städte*. Since the beginning of the 1960s, these serious traffic problems have dominated debates about urban transport. At that time, urban criticism of automobiles was a common feature of German politics because it was the municipal authorities that had to bear the burden of the 'car-friendly city'. Munich mayor Hans-Jochen Vogel even complained about mass motorization 'paralyzing' the city. Finally, from the 1990s onwards, the dramatic term 'traffic collapse' prevailed.[33] More or less, any Western country had similar issues. Automobile emissions too came into view as a health risk. Cancer sensitivity on a large scale reached Germany later than the United States but from the late 1970s onwards, carcinogenic pollution played a role in changing the image of cars. This change accelerated in the late 1980s, when the degree of an automobile's eco-friendliness was carefully assessed in line with new concerns about its effect on climate change.[34]

Considering the large number of traffic fatalities, which was particularly high in West Germany until the 1970s, there was surprisingly little protest against this danger. Despite all its dangers and problems, the car was a central icon of the promises of modern life. Obviously, people were willing to accept the fatal downside of individual mobility. Only the peak of fatal accidents in the 1960s caused some public awareness of the issue, resulting in some measures to increase traffic safety mostly by technological means.[35] By contrast, the introduction of obligatory measures to decrease car accidents was highly controversial. A significant group of drivers and lobby groups opposed the increasingly cautious traffic politics; these were relatively large and especially loud groups that tried to dominate the public debate. Initially any means of state restriction of traffic caused public protests, including setting a blood-alcohol limit, the use of radar surveillance, the introduction of speed limits and the obligatory use of seat belts.[36]

As late as the end of the twentieth century, the eighteenth-century Luddites still played a role in public debates on new technologies. Protesters against the projected magnetic levitation train Transrapid, which later proved to be a failed innovation, were labelled machine breakers even in 1997.[37] In general, social protest usually declined as people increasingly familiarized themselves with novel technologies. However, the case was different with environmental protests against technology, which grew steadily. While it was sometimes easy for users to adapt themselves to changing social circumstances, environmental hazards were hard facts, which did not leave any place for individual agency in the worst cases of health risks or the seemingly looming threat of total destruction.

Environmental protest and 'German Angst'

There was no broad public debate on environmental issues in Germany until the late nineteenth century. However, the main topics in that context, water and air pollution, had been of local interest in many cases from the mid-nineteenth century onwards. In particular, earlier attempts to tackle the issues of urban hygiene and public health (see Chapter 2) signify to a certain degree a prehistory of later environmentalism.[38] At that time, there was hardly any environmental awareness in the proper sense. Instead, commercial competition for resources dominated the disputes. Clean water was a particularly important resource for industries such as sugar, paper and textiles. Usually, the respective companies simply discharged their sewage into rivers – and complained about the sewage of the other industries.[39] Air pollution issues also arose in the mid-nineteenth century, even before coal smoke became a huge problem, with copper smelters the cause of early political criticism.[40]

Over the course of the nineteenth century, the increasing popularity of coal as an energy resource caused twin problems. First, the emissions were more problematic than smoke from wood burning. Second, factories had become independent of water power and were increasingly established in densely populated areas, thus irritating more people in their vicinity. Some characteristics of modern environmental protest had already evolved in the nineteenth century. Scientific and technological experts had a say in lawsuits, and citizens' initiatives were formed on many occasions. Furthermore, the distinction between private and public interests was established. Aspects of health, vegetation and property were central to any disputes about industrial air pollution from the beginning. In general, the aversion to coal smoke was increasingly widespread.[41] Until 1900, a good part of the criticism regarding industrial smoke or water pollution was based upon aesthetic discomfort, not on environmental protection.[42] Already in the early nineteenth century, however, the industrial usage of coal had caused intensive medical debates about the harmfulness of coal smoke.[43]

Thus, some characteristic features of twentieth-century debates emerged before 1900. Attempts to assess the impact of technology had a prehistory before its institutionalization in the 1970s. Observers were, for example, already complaining in the early nineteenth century about the future damage that could be caused by potash sewage.[44] By the 1840s, technological solutions to environmental problems were beginning to replace traditional social compensations. In the industrial age, environmental damage was identified as a technological problem for which technological solutions had to be found.[45] Until the mid-nineteenth century, the environmental burden was still rather low. That changed rapidly during the following decades as industrial production increased sharply and new processes were introduced. In consequence, emissions skyrocketed. Now, organized resistance and hefty conflicts proliferated. After 1860, the increased usage of coal caused more and more lawsuits against firms that produced air pollution. Engineers began looking for a technological solution within the next decades but with little success. This failure was only partly due to technological reasons. One primary difficulty was that the authorities showed little interest in the issue. In particular, the Prussian Ministry of Commerce blocked any advances. From the ministry's point of view, air pollution was a mere nuisance, from which no health risks flowed.[46]

By following the vague idea that a technological fix could alleviate ongoing problems, the authorities nevertheless ordered steel mills to install mechanical technologies for reducing emissions in the early 1870s. Those measures were not very effective for reducing environmental harm; mostly they served to appease opponents. The same was true for chemical-technological measures, which prevailed soon afterwards. In spite of this progress, those new technologies acted mainly as a smokescreen and simplified approval procedures for factories.[47] In general, the factory owners and their engineers claimed to be the only relevant experts on technological questions. From the early nineteenth century onwards, the defence in lawsuits argued that the complaints were 'technologically ignorant'. This argument was at the core of the defence strategy for the whole century. Factory owners repeatedly argued that earlier smoke issues had been solved by new technologies. Indeed, there was some success but far less than promised. Diffusive emission was a problem that was hard to solve and pipeline leaks were very common.[48] The same belief in a technological fix regarding water pollution issue replaced any steps towards prevention in the medium term.[49] In a certain sense, these early answers to pollution criticism structurally consolidated the very problems they were about to solve.

Vested interests had the best chance to succeed with environmental protest. Steglitz, a residential villa area near Berlin, forbade the establishment of factories in 1890 due to 'harmful smoke' and noise. Although the Berlin government later cancelled the act, the high property prices of Steglitz prevented any industrial activity in the town.[50] Steglitz is a good example of the motivations underlying environmental critique in the nineteenth century. Some people feared financial disadvantages caused by industrial environmental damage.[51] From the turn of the twentieth century onwards, however, pollution control was more broadly discussed. Contrary to the relatively strong interest in environmental issues in the United States at that time, most German municipalities were rather reluctant to act on environmental measures. While Hamburg was an exception with the establishment of a 'Society for Fuel Economy and Smoke Abatement', most officials just demanded that factory owners raise their smokestacks.[52] The First World War abruptly halted the few promising approaches to air pollution protests. This continued during the crises of the Weimar Republic and the rapid rearming efforts of the Third Reich.[53] Of course, the problem did not cease and some observers continued to criticize it harshly. For instance, the famous journalist Egon Erwin Kisch criticized the chemical corporation IG Farben in 1927 for polluting the air of Leverkusen, discolouring residents' skin and causing serious diseases.[54]

Also beginning in the late nineteenth century, the issues of water shortages and water pollution became the subject of growing public concern. In the first decades of the twentieth century, parliamentarians worried about the state of different German rivers. In particular, the coal and chemical industries were heavily polluting the rivers, which were consequently given unflattering nicknames. The Emscher was dubbed the 'river of hell', while the Wupper was known as the 'river of ink'. As in the case of air pollution, few political sanctions followed. The first concerns regarding biodiversity were also expressed. Protest groups agitated against dam construction in the Rhine town of Laufenberg due to the endangered salmon runs. The reaction was comparable to the rejection of machine breakers. The protesters were ridiculed as reactionary romantics.[55]

At first sight, the initial environmental awareness of the early twentieth century was only modest. In hindsight, however, there is some evidence that this period was formative in the history of environmentalism. The vivid life reform movement in Germany was the beginning of critical consumption. In addition, cancer debates cast renewed attention on pollutants. Although the participants of that time were rather conservative, the later leftist environmentalists of the 1970s were ready to modernize these existing traditions.[56] Some observers even imagined climate catastrophe. In 1931, the far-right writer Oswald Spengler expected climate change with disastrous consequences as the outcome of technological developments in industrial modernity. Forests would cease to exist and many animal species would become extinct within the next decades, as would many 'human races'.[57] Actual politicians, however, were still reluctant to act. Although the Nazis employed some 'green' rhetoric, the way of dealing with 'air pollution after 1933 was a continuation of previous procedures'. Nonetheless, the Reich Food Estate, the farmers' organization in the Third Reich, emboldened farmers to demand 'compensation for crop damage caused by industrial pollution'.[58]

After the end of the Second World War, environmental policy in the proper sense gradually established itself. There is also an often-missed history of environmental protests in the 1950s. Mostly, these were protests against local problems and protesters had no broader agenda of social or political change. At this time, environmental protests took the concrete form of opposition against canal-building projects or industrial projects that endangered natural sites.[59] On the political level, the West German parliament demanded in 1955 that the government provide information about air pollution and suggestions for effective solutions to the problem. In the following years, the beginnings of air pollution policy incrementally evolved. It was by no means a leftist affair. Instead, the political mainstream was convinced that it was possible to combine economic growth and pollution control. At that time, Germany was still far behind the more comprehensive American efforts in environmental protection politics. Beginning in the 1960s, an environmental consciousness gradually evolved among the public. The modern environmental movement took root in this context and there were some initial successes. From the mid-1960s until the end of the twentieth century, the emission of sulphur dioxide in the Ruhr area sharply decreased, although this was mostly due to the decline of coal, iron and steel industries.[60] Yet, German environmental politics gained momentum at that time. Now, a previous disadvantage of the early twentieth century turned into an advantage. The strong German bureaucracy and autonomous engineers who 'had hindered the smoke debate of the early 1900s' now formed a 'counterweight against industrial interests'. While US industry has dominated air pollution policy from the 1960s onwards, industrial influence was significantly less in Germany due to these peculiar structural features established in the long history of the relationship between engineering and the state.[61]

A shift in the direction of environmental policy occurred only slowly. In the initial post-war period, commodity and energy production were the sole targets of environmental criticism, whereas consumption was widely ignored.[62] This ignorance faded away due to the obvious significance of increasing environmental problems. The emerging consumer society fostered individual lifestyles of technology use and resource consumption. Standards of industrial production and living changed in the 1950s in a way that was characterized by immense amounts of energy and resource spending. Today's environmental problems have their foundations in this period of consumption growth. The relative price decline of fossil fuels caused this pattern of wasteful energy and resource consumption. Environmental harms originated in this period.[63]

As the historians Ruth Oldenziel and Mikael Hård have shown, the post-war environmental movement that gradually established itself in most Western countries finally succeeded in overcoming an inherent contradiction. Although the traditions of reuse and thrift were common to most Europeans, the emerging consumer society offered the very attractive option of consumption without caring about the concomitant waste of resources. The DIY culture imported from the United States partially solved this problem, as tinkering, repair and recycling became integral parts of the modern consumer society. There seemed to be an alternative path for modern consumption with less harmful effects on the environment.[64] With regard to some aspects of environmental protest, however, the arguments of the early industrial era have persisted until the present. Water pollution control was denounced as early as 1876 as a means to harm industrial interests and destroy jobs. In 1994, the chairman of the Federation of German Industries still applied the same reasoning: environmental policy should not challenge industry's and workers' interests.[65]

A certain technology provoked the formation of especially strong protest groups: nuclear power. There are particular reasons for why West German anti-nuclear activists established what probably was the 'most powerful environmental movement in the Western World'.[66] First of all, the fear of nuclear weapons was particularly strong in Germany because the country would obviously have been ground zero and turned 'into a nuclear wasteland' if the Cold War had escalated.[67] It is important to emphasize that scientific experts shared these fears of nuclear war. Indeed, Germany's most renowned physicists initiated the protest against incipient plans of the West German government to install nuclear arms in 1957. Nobel laureate Werner Heisenberg (see Chapter 3) and seventeen more nuclear experts formed the famous 'Göttingen Eighteen'. Their protest manifesto received wide public attention.[68] For instance three years later, activists protested against the presentation of US Army short-range nuclear missiles on the occasion of the premiere of the Wernher von Braun biopic *I Aim at the Stars* (Figure 8.3). The very combination of a celebration for a Nazi rocket scientist with the proud presentation of Cold War weapons caused fear and anger.[69]

Figure 8.3 Demonstration against the premiere of the film *I Aim at the Stars*, Munich, 19 August 1960. Bundesarchiv.

Often-missed opposition against civil nuclear power also arose in the mid-1950s. The historian Dolores Augustine is generally correct in stating that anti-nuclear protests were rather small at that time.[70] However, there is some evidence that there was a widespread awareness of the risks of radioactive radiation. The fear of cancer was most probably the hidden driving force of the emerging environmental awareness in general. In particular, it fuelled the nascent anti-nuclear movement.[71] Even before the large anti-nuclear protests of the 1970s, activists were quite successful on the local level. For example, they hindered the establishment of a projected nuclear research centre near Cologne. The government had to find a new site outside the metropolitan area. Obviously, public trust in the new technology was from the outset far lower than politicians and nuclear experts had expected.[72]

Some of the most prominent nuclear experts joined in the criticism. Heisenberg stated in 1959 that the scientific-technological complex was no longer socially controllable.[73] The physician Bodo Manstein also warned of the 'dangers of nuclear energy' in his book *In the Chokehold of Progress*, or *Im Würgegriff des Fortschritts* in 1961.[74] Yet most West German scientists employed with nuclear plant projects were significantly less critical than their US counterparts. While the American experts criticized reactor safety from within, it was left to West German counter-experts to point out safety risks. Obviously, anti-nuclear activists did not resemble the cliché of blind Luddites. From the very beginning, knowledge of science and technology was part and parcel of the protests.[75] At the same time, West German culture had changed significantly with regard to fear. As the historian Frank Biess had shown, fear was no longer pathologized in general. Instead, a new perception of fear as a 'healthy reaction to external dangers' broke through.[76]

Thus, the relationship between anti-nuclear activists and modern technology was rather complicated: fear of nuclear risks was embedded in knowledge about nuclear plants and safety issues. Furthermore, the activists derived from very diverse groups. This was clear with the first large protest against the projected nuclear plant in the south-western German village of Wyhl. The long protests began right after the public announcement of the project in summer 1973. It culminated when tens of thousands of protesters demonstrated in 1975. After two decades, the project was finally abandoned in 1994, which was an initially unexpected success for activists and the world's first anti-nuclear protest that had successfully hindered a nuclear plant.[77] The first large demonstration in 1975 was thus arguably the 'birthplace of one of the most powerful and influential anti-nuclear power movements in the world'.[78]

As has been shown, this was the outcome of scientific debates and local protests for two decades. One key feature of the movement's strength was its diversity: locals demonstrated side-by-side with outside activists, partly from nearby France. At first glance, their interests seemed incompatible. The locals had concrete concerns without any real interest in expressing broader opposition to the political system, whereas most external activists rejected the capitalist state as such. Even the respective object of fear was different: the local winegrowers' first concern was the risk of local climate change resulting in a lower quality of their wine, not the risks of nuclear power in general. However, the different groups of Wyhl protesters shared a critical conception of modernity: they did not reject modernity as such but were sceptical about some outcomes of industrial modernity. While the conservative locals sought to defend their traditional rural lifestyles, most external protesters shared the post-materialist values of the New Left.[79] The protesters were at odds with a specific conception of modernity that subordinated social and environmental concerns to large technological projects. This certain conception of progress had lost its appeal to very different groups in West German society. Yet in West German society as a whole, they were still a minority in the 1970s. Moreover, even the environmentalists shared the planning euphoria. While they rejected the notion of permanent growth, many environmentalists put their trust in systematic and scientific politics, albeit of a different kind.[80]

In the following period of large protests against different nuclear sites in the 1970s and 1980s, the trend of diverse protest groups continued. Most protests were not organized by outsiders but by locals who felt compelled to resist the projected nuclear plant in their neighbourhood.[81] However, any protest triggered harsh reactions from the supporters of nuclear energy. Conservative defenders of the contested technology appealed to public fears with prophecies that 'the lights would go out' without nuclear power plants.[82] To a certain degree, the rigid adherence to nuclear power was due to vested interests. Public utility companies partly owned by the West German federal states operated the nuclear plants. As such, the political establishment was prone to defend nuclear power investment decisions. The confrontation between supporters and opponents of nuclear power became stronger after the protests against the building of the Brokdorf site near Hamburg. Although the demonstration groups were diverse again, the media coverage concentrated on militant activists. Protests were increasing, but still a majority of 53 per cent of Germans opted for the expansion of nuclear power in an opinion poll in 1977.[83]

In international comparison, the West German protesters were among the strongest. This strength of the West German anti-nuclear movement was probably partly due to the fact that West Germany was not a nuclear power in a military sense. Although the US and French movements were also strong, there was significantly more acceptance for nuclear-fission technology in those nuclear-power countries than in West Germany.[84] Nonetheless, there were many transnational similarities, transfers and entanglements. By and large, the anti-nuclear movements in Western countries 'closely resembled one another'.[85] As already mentioned, French activists participated in the Wyhl protests. Moreover, the Smiling Sun badge ('Nuclear Power? No Thanks'), which became very popular in West Germany, had been introduced in Denmark in 1975 and became a transnational icon of the anti-nuclear movement across national languages.[86] In the early 1980s, West German anti-nuclear activists invited groups of Native Americans to join their protests. Their participation at several demonstrations made clear that parts of the anti-nuclear movement endorsed visions of 'natural life' that opposed the idea of progress at any cost. At the same time, the political supporters of nuclear power, namely the conservative government, accused the protesters of risking a social throwback to a 'primitive level'.[87]

Yet, the anti-nuclear activists of the 1970s and 1980s were not only complaining about nuclear power stations. In addition, some activists tinkered with alternative technologies of energy production such as low-tech wind energy systems following the lead of Danish activists. The West German environmental movement distributed around forty construction manuals for such systems between 1975 and 1990. DIY was central to their approach, as was a 'return to the human dimension' as an alternative to large technological systems that humans could no longer effectively control as Heisenberg warned in 1959. Thus, the activists were by no means foes of technology. Instead, they tinkered with technologies that aligned with their hopes for the future. The opposition was not between support and rejection of technology but rather between small low tech and large high tech.[88] Of course, the quantity of alternative energy produced at that time was negligible. However, the activists made their point. They had demonstrated their desire for controllable and humanized technology and their opposition to an uncontrollable, large technological system. In this sense, they successfully revolted 'against technocratic thinking'.[89] In the end, the anti-nuclear protests even gave birth to the state's interest in industrial-scale wind power.[90]

Even after no more new nuclear plants were planned, the Asse and Gorleben storage sites and the transportation of nuclear waste were the sites of many protests during the 1980s and 1990s.[91] The most important incidents for the later decision for nuclear phaseout were, however, nuclear disasters; in the Soviet Union in 1986 – Chernobyl – and Japan in 2011 – Fukushima. In the context of an already powerful anti-nuclear movement, the Chernobyl catastrophe provoked particularly strong public reactions. The growing rejection of nuclear power can only partly be attributed to the relative geographical proximity of Germany to the Ukraine. In neighbouring countries such as France, Switzerland or the Netherlands, discomfort with the danger of a nuclear disaster also became more widespread, but the street protests were not as strong as in West Germany. Around that time, the notion of 'German Angst' was established outside of Germany to describe the allegedly heightened role of fear as a feature of German political culture. As has been demonstrated earlier, the specific fear of nuclear power was not irrational or anti-rational but rather the outcome of long years of debate about the scientific evaluation of nuclear risks.[92]

From the mid-1980s onwards, it was absolutely clear even to conservative politicians in favour of nuclear energy that a majority of Germans rejected any new nuclear building projects. In the early twenty-first century, though, the nuclear phaseout was highly controversial in Germany. The coalition government of Social Democrats and Greens had opted out of the nuclear energy programme in 2000 with the final goal of leaving it completely in 2020, but then the newly elected centre-right government cancelled it in 2010. Instead, a lifetime extension for nuclear plants was granted.[93] Only one year later, the Fukushima disaster changed German nuclear politics again. The conservative chancellor Angela Merkel, who had just extended the lifetime of nuclear plants, decided two and a half months after the catastrophe to withdraw from the nuclear energy programme by 2022. This change of mind occurred, however, only after the Green Party – the strongest political opponent to nuclear power – had won the elections in the state of Baden-Württemberg. Thus, in the aftermath of the nuclear disaster, a Green politician became head of a German state for the first time. The establishment and electoral good fortune of the Green Party marks one decisive feature of the German environmental movement, which partly helps to explain its singular success. Together with counter-expertise and a particularly strong association of nuclear energy with nuclear war, this political dimension was among the peculiarities of the German story of anti-nuclear activism. Thus, the nuclear phaseout was not a mere reaction to the catastrophe of Fukushima but rather the result of decades of critical debates about a contested technology.[94]

Conclusion

In 2015, four years after the nuclear phaseout decision and seven years before the projected shutting down of the last nuclear power plants in Germany, two artists created a series of nineteen porcelain plates depicting all former (or still existing) nuclear plants in Germany: 'atom plates', or *Atomteller*. For the artists, the nuclear sites were 'monuments of error'. This work of art mocked both the tradition of hanging wall plates and the failure of the national nuclear energy project. The artists employed the romantic or even kitschy tradition of wall plates for the cause to ironically remind viewers of what soon will be the nuclear past. Usually such porcelain plates depict romantic natural sites. The atom plates, by contrast, depicted moribund high-tech artefacts, which were sites of a long history of acrimonious political dispute. Obviously, the artists were keen to celebrate the end of nuclear power in Germany. Moreover, the atom plates reflected the long history of protest against this technology that has been explored in this chapter. In a certain way, these pieces of art can be seen as a comment on the way any criticism of certain technologies was treated within the last two centuries. Usually, the protesters were denounced as anti-modern or romantic foes of technological progress. Now, however, the former icons of high technology, nuclear power stations, had themselves become sites of the past to be remembered in a mock-nostalgic way. The plate depicting the nuclear power plant Brokdorf even combines the traditional motifs of porcelain plates with the artistic project: sheep are grazing in the meadows in front of the nuclear power plant (Figure 8.4).

As has been demonstrated, the history of protests against technology would not be complete without exploring the reactions of both the authorities and the public against the protesters. In broad terms, the accusations against the protesters have not changed much since the early nineteenth century. According to the respective defenders of the contested technologies, those protests were driven by some mixture of irrational factors: fears, romanticism, unworldliness or anti-modernism. Usually, the supporters of the technologies claimed to be the sole spokespersons of rationality and science, whereas the protesters were mere Luddites motivated by an atavistic hostility towards technology or even progress itself. A closer look at the protests shows, however, that as a rule, the protesters did not reject new technologies per se. Usually, the protesters had good cause to feel threatened by technological change. They ran the danger either of losing social status and income or being exposed to health risks. In the course of the twentieth century, the protesters' most important demand was for human or social control of technology. On many occasions, protesters fought against large technological systems that were outside the realm of democratic control.

The historian Bernhard Rieger has asserted that the 'public belief in technological modernity' had significantly decreased after the end of the Second World War. According to Rieger, the widespread fears of nuclear war and environmental disaster

Figure 8.4 Atomteller by Mia Grau and Andree Weissert. A set of nineteen porcelain plates, 2015 (www.atomteller.de). This 'atom plate' depicts the nuclear power plant Brokdorf. Photograph by the artists.

give evidence of this change of mentality.[95] At first glance, Rieger gives an accurate account of an important change. However, as this chapter has demonstrated, there is good evidence that the belief in technological progress and in technological fixes specifically did not decrease after 1945. In a certain sense, the Club of Rome's warning of 'the limits to growth' failed. Although the report was a bestseller and many politicians quoted it, a fundamental change of mind did not occur. Almost never was a reduction of economic growth the result of the environmental harms that modern technology had generated. Instead, new technologies have been conceived of as a panacea. Even now, governments do not seriously discuss a sharp reduction of energy consumption as a means of leaving behind the problematic technologies of coal and nuclear power. Instead, they place their hope in technological innovations such as solar and wind power. Similarly, electric car technology is projected to replace internal combustion engines, whereas only radical environmentalists envision a reduction of private transport. Furthermore, some contemporaries even suggest a return to nuclear power as an answer to global warming. In general, visions of progress are still part and parcel of technology debates. Yet it seems uncertain to what kind of future society such novel innovations will lead.

Concluding remarks

This short conclusion first reflects upon some of the general topics in the history of technology. Thereafter, I will reconsider the peculiarities of technology in modern German history. The two sections of the book – Chapters 1 to 4 and Chapters 5 to 8 – are interconnected in many ways. The impulses given in the latter section seek to enrich, and not to replace the more common topics of the history of technology – the history of the body and history from below – are important aspects of a new history of industrialization or urbanization. On the other hand, ongoing industrialization processes have changed the perception of the body. Likewise, visions of progress are always linked to more sceptical views. In addition, the old and the common remain important to high tech history. The traditional view of the urban as the foremost pacesetter of technological development needs modifying: rural areas too played an important role in the history of industrialization. Above all, the technology user is decisive to the history of technology and its social and cultural applications.

Nonetheless, a central theme of this book has been the limits of users' agency. The persistence of long-lasting structures that were hard to overcome is not to be ignored. In general, social and economic as well as mental structures limited the agency of technology users. One example is that the focus on individual consumption patterns misses the crucial point that the evolving high-energy society of the latter part of the twentieth century was largely the result of pre-existing technological infrastructure and the momentum of oil and gas heating. The long-lasting effects of cultural assumptions about technology are another example. In many regards, these assumptions fostered gender inequalities: once the male body had been associated with technology, certain skilled tasks at work had become the sole realm of men. In this context, it is important to state that technological innovations did not automatically correlate with social progress. Instead, in several cases, new technologies helped traditional values and social structures to persist. For instance, the persistence of the gendered division of labour after 1945 was dependent upon technological innovation in kitchens and in farms.

Historical actors asked themselves on many occasions if new technologies enhanced their agency. At first glance, novel technologies enriched the freedom to act for many people in everyday life, but at the same time, technology users in a certain sense became dependent upon these very technologies and their reliability. Sometimes, the impression was overwhelming that these new machines almost took command over the technology users' bodies and minds. The historian of technology should consider the effects of these historical emotions. Often, first-time technology users felt subjected to the new objects – and, over time, got used to this feeling of subjection. Passengers on train rides are a good example. Thus, there is some ambivalence between the widespread understanding of technology as the will to dominate nature, most often shared by inventors, engineers and politicians alike on the one side, and on the other later users who at least sometimes were accustomed to being dominated by machines or large technological systems. Yet, we must also analyse the concrete outcome of such developments beyond contemporaries' emotional impressions.

Users' agency usually did not lead to acts of resistance. First and foremost, users followed given patterns. Yet, there were some particular cases of creativity that influenced technology development. In particular, users' tinkering was central to the acceptance, improvement and diffusion of novel technologies. Especially at the workplace, the other side of users' agency became important: the reluctance or even resistance of workers towards some novel technologies that were potential threats to their skilled positions or even to their very jobs. In general, the expansion of modern technology also meant the expansion of contested political areas. It created more room for conflicts that centred on the issue of who profited from a specific form of technological innovation and its concrete usage. Furthermore, certain technologies were door openers for interconnected developments in faraway spheres of society. Most of all, the transport revolution of the nineteenth century was of utmost importance for innovation in different fields such as industry, agriculture and individual mobility. Above all, the new technologies of mobility had a strong impact on technology acceptance in general.

The sometimes rather difficult first experiences of users proved in many cases to be in favour of turning partly interested, partly sceptical users into technology lovers. In particular, the time-consuming practices of tinkering and repairing established emotional bonds for those who finally succeeded with their efforts to control the stubborn technological artefact. Combined with cultural concepts such as technological progress and modernity, these individual, but widespread, practices were essential for the success and acceptance of novel technologies. On the one hand, technological artefacts symbolized modernity. On the other hand, even severe problems of first-time technology users in dealing with these innovative, but complicated, artefacts increased the aura of a sometimes mysterious modernity. To comprehensively grasp these developments, the historian of technology needs to examine closely the history of emotions and expectations as well the cultural appropriation of technology. Consequently, both cultural systems of making sense of emotions and social practices of making use of technologies combined give us the framework for the history of technology.

Pre-existing structures of social inequalities also had a huge impact on the implementation of novel technologies. Conversely, the introduction of technologies affected social structures. Contrary to the still widespread image of technological progress as a door opener for social progress, novel technologies often increased social inequalities with regard to class and gender (not to mention its role in colonialism). The planning and implementation of large technological systems was, particularly in the context of urban development, largely dependent upon middle-class pressure groups that followed their own agenda and neglected working-class interests. These intentions, however, did not always correlate with the outcome. This is especially true for the area of industrial labour. Of course, the managerial interest in productive bodies was crucial for the organization of technified workplaces. Yet, the relationship between efficiency and the humanization of work was multidimensional and opened up spaces for workers' self-reliance in dealing with machines at the factories. At the same time, workers' self-reliance was the basis for even more effective ways of making use of worker's skills and hidden potentials.

It is not an easy task to identify particular national features in the history of technology in modern Germany. For example, the concept of the 'technological fix' prevailed in different eras of German history, most obviously in issues of urban development or environment. Yet, this was a common trend in modern Western history and not a German peculiarity. Moreover, it is crucial not to neglect regional differences within German history. At first glance, there seem to be many contradictions in the history of technology in modern Germany. Since the late twentieth century, it has established one of the world's largest and most powerful environmental movements, which is very critical of harmful technologies. Accordingly, it is the self-proclaimed world champion of recycling. On the other hand, German industry has a long history of producing harmful chemicals; German agriculture has used enormous amounts of fertilizers and pesticides on farmland since the beginning of the twentieth century.

As another example, nature, in particular forests and rivers, was central to a certain self-perception that many Germans had of their homeland. As has been shown, however, such socionatural sites had been integrated into large technological systems from the nineteenth century onwards, at least. There seem to be even more contradictions in many aspects of modern technological development in Germany, in particular during the Third Reich. The Nazis propagated traditional rural life and at the same time systematically transformed the countryside by technological means. The same occurred with industry, where Nazi propaganda rhetorically opposed 'Americanism' but continued efforts towards rationalization. Furthermore, high-tech projects like rocket developments during the Second World War made use of slave labour employed with rather primitive technological means. As has been shown, these phenomena are only contradictive if the history of technology is viewed as inherently linear. What is more, many historians still tend to equate social progress implicitly with technological progress. However, not only a 'progressive' society (whatever that might be) can make use of modern technology. So too can totalitarian regimes.

This book has also suggested the need to reconsider Gerschenkron's classic concept of economic backwardness and a need for a renewed close look at the historical sources. While Germany was not in fact discernibly backwards, contemporaries' fear of being backwards in the nineteenth century remains crucial for an understanding of the history of technology in Germany. This fear helps to explain why so many actors in different strata of the state bureaucracy voted for heavy state support both for high tech and for the proliferation of everyday technology. In general, German state bureaucracies made a strong impact on industrialization. A structural peculiarity of Germany was thus in favour of the strength of state bureaucracies in this regard, fostering a close relationship between engineering and the state. Moreover, the German states in the nineteenth century and the unified German state since 1871 were effective arbiters between different interest groups. This certainly helps to explain the long history of interconnectedness in Germany between agriculture and industry.

The crucial strength of the German state is nonetheless not foremost found in supporting technological development in a narrow sense. Instead, the sometimes-overlooked institutional frameworks of technological change were established at relatively early times, whereas the innovations as such often broke through only gradually. It was, however, a particular strength of the German state to participate in the establishment of such frameworks. The state functioned as an effective mediator of technological development. Associations and state institutions disseminated knowledge and became mediators between the manufacturers and the users. Furthermore, the German state was very effective in its role as a promoter of mechanization, electrification and motorization. The main focus on state technology promotion was, however, on production, not on consumption, at least until the 1960s. This was true not only for politicians but also for technological experts such as architects, whose concepts of rational housing clearly differed from US approaches.

Around the turn of the twentieth century, technology was the one aspect of modernity that most Germans embraced. The rapid changes of the previous decades, which had occurred faster than in most industrializing countries, was taken by many as proof that technological improvements would bring further improvements to social life. At the same time, the rise of the nation was perceived as being a result of German cultural supremacy leading to technological supremacy. Confident assertions of feasibility were omnipresent. After the defeat in the First World War, these optimistic conceptualizations of technology and nationalism turned into more aggressive dreams of national rebirth and revenge. It was at this point that Germany became 'a nation of troubling modernity'.[1] It was, of course, not an entirely singular path of Germany, and German developments bore many resemblances to the American movement of technocracy. However, they had a different social and political context rooted in the very developments of German history of the late nineteenth and early twentieth centuries.

Germany's technological rise in the nineteenth and twentieth centuries helps to explain its political ambitions. In a certain sense, widespread technological euphoria gave birth to modern nationalism in Germany. The grassroots enthusiasm for airships and other spectacular artefacts produced a modern kind of nationalism. Complex emotions were critical to these processes. Tragic accidents were perceived as heroic disasters and many Germans perceived technological change as the outcome of superior German culture and the Germans' capability to overcome initial failures. Already around 1900, modern technology was accepted as an integral part of Germanness. Still, a certain emotional ambivalence was prevalent. Many Germans associated both fears of cultural pessimism and visions of omnipotence with technological innovation. After 1918, the combination of high-tech chauvinism and self-victimization that was widespread among middle-class Germans helped National Socialist resentments break through, a drastic example that shows how the spheres of emotions, technology and politics were interwoven.

With respect to the more common aspects of modern technology, most notably industrial labour, there were also national peculiarities. There is some evidence to suggest that the human factors of production were particularly important in Germany. To some degree, the early interest in human capital and factory workers' skills were key to Germany's industrial success. In later periods, this feature partly explains the flexibility of shifting between mass production and diversified quality production. From the perspective of a history from below, there are also certain distinctive features in the German working class's self-perception. Many male skilled workers from the late nineteenth century onwards were proud of their technological affinity, their embodied knowledge and the ostensibly national tradition of German quality work.

Modern technology held many promises for different wings of the political spectrum. While progressives hoped that new technologies would speed up unstoppable social progress, German conservatives sought for ways to slow down social transformation. For the latter group, specific technological applications offered the chance for old-fashioned, rural ways of work and life to compete with the modern challenges of urbanization and industrialization. The Nazis too perceived technological innovation as a promising path towards an alternative modernity of national autarky. The notion of progress served manifold ends. We still tend to conceive of the combination of progress and rationality as the natural outcome. Yet, German history teaches us that there were competing notions of modernity and very different versions of technological progress.

Technological development after 1945, not least including nuclear power, the space age and the digital revolution, ended any remaining dreams of German autarkic high-tech politics. The Cold War urged Germany, along with the other European countries, to collaborate on European high-tech projects if they wanted to have any chance of competing with the global superpowers. No insight into the failures of German history or moral re-education was needed; it was just a matter of fact that the old aspirations of power through technology were unrealistic in the face of the dominant superpowers of the Cold War. In this context, German visions of technological progress, by and large, conformed to the broader Western model.

Notes

Introduction

1 Peter Fritzsche, *A Nation of Fliers: German Aviation and the Popular Imagination* (Cambridge: Harvard University Press, 1992).
2 Mary Nolan, *Visions of Modernity: American Business and the Modernization of Germany* (New York: Oxford University Press, 1994); Karsten Uhl, 'Giving Scientific Management a "Human" Face: The Engine Factory Deutz and a "German" Path to Efficiency, 1910–1945', *Labor History* 52, no. 4 (2011): 511–33.
3 Jeffrey Herf, *Reactionary Modernism: Technology, Culture, and Politics in Weimar and the Third Reich*, repr. edn (Cambridge: Cambridge University Press, 2003).
4 Raymond G. Stokes, Roman Köster and Stephen C. Sambrook, *The Business of Waste: Great Britain and Germany, 1945 to the Present* (New York: Cambridge University Press, 2013), 32–3; Arndt Sorge and Wolfgang Streeck, 'Diversified Quality Production Revisited: Its Contribution to German Socio-Economic Performance over Time', *Socio-Economic Review* 16, no. 3 (2018): 598–9.
5 Joachim Radkau, *Technik in Deutschland: Vom 18. Jahrhundert bis heute* (Frankfurt: Campus, 2008), 29.
6 Nelly Oudshoorn and Trevor Pinch, 'Introduction: How Users and Non-Users Matter', in *How Users Matter: The Co-Construction of Users and Technologies*, ed. Nelly Oudshoorn and Trevor Pinch (Cambridge: MIT Press, 2003), 5, 8.
7 Alf Lüdtke, *Eigen-Sinn: Fabrikalltag, Arbeitererfahrungen und Politik vom Kaiserreich bis zum Faschismus* (Hamburg: Ergebnisse, 1993), 253–4; see also Joan Campbell, *Joy in Work - German Work: The National Debate, 1800–1945* (Princeton: Princeton University Press, 1989).
8 Ulrich Wengenroth, *Technik der Moderne: Ein Vorschlag zu ihrem Verständnis*, Version 1.0. (Munich, 2015), 52–3. Available online: https://www.edu.tum.de/fileadmin/tuedz01/fggt/Wengenroth-offen/TdM-gesamt-1.0.pdf
9 Reinhold Bauer, *Gescheiterte Innovationen: Fehlschläge und technologischer Wandel* (Frankfurt: Campus, 2006).
10 David Edgerton, *The Shock of the Old: Technology and Global History since 1900* (London: Profile Books, 2006).
11 Radkau, *Technik in Deutschland*, 37, 422.
12 Wengenroth, *Technik der Moderne*, 33.
13 Ruth Oldenziel and Mikael Hård, *Consumers, Tinkerers, Rebels: The People Who Shaped Europe* (Basingstoke: Palgrave Macmillan, 2013).
14 Mikael Hård and Andrew Jamison, *Hubris and Hybrids: A Cultural History of Technology and Science* (New York and London: Taylor & Francis, 2005), 303.
15 Langdon Winner, 'Do Artifacts Have Politics?', *Daedalus* 109, no. 1 (1980): 121–36.
16 Thomas P. Hughes, 'The Evolution of Large Technological Systems', in *The Social Construction of Technological Systems: New Directions in the Sociology and History of Technology*, ed. Wiebe E. Bijker, Thomas P. Hughes and Trevor Pinch (Cambridge: MIT Press, 1987), 51–82.

17 Donald Mackenzie and Judy Wajcman, 'Introductory Essay: The Social Shaping of Technology', in *The Social Shaping of Technology*, 2nd edn, ed. Donald Mackenzie and Judy Wajcman (Buckingham: Open University Press, 1999), 4.
18 See in particular the *Making Europe* series: Johan Schot and Phil Scraton, ed. *Making Europe: Technology and Transformations, 1800–2000*. Six volumes (Basingstoke: Palgrave Macmillan, 2013–19).
19 David Arnold, 'Europe, Technology, and Colonialism in the 20th Century', *History and Technology* 21, no. 1 (2005): 85–106.
20 Jürgen Habermas, *Technik und Wissenschaft als "Ideologie"*, 4th edn (Frankfurt: Suhrkamp, 1970), 93.
21 David Blackbourn and Geoff Eley, *The Peculiarities of German History: Bourgeois Society and Politics in Nineteenth-Century Germany* (Oxford: Oxford University Press, 1984).
22 Jürgen Kocka, 'Looking Back on the *Sonderweg*', *Central European History* 51, no. 1 (2018): 137–42.
23 Edward Ross Dickinson, 'Biopolitics, Fascism, Democracy: Some Reflections on Our Discourse About Modernity', *Central European History* 37, no. 1 (2004): 5.
24 Heinrich August Winkler, *Germany: The Long Road West*, 2 vols. (Oxford: Oxford University Press, 2006–7).
25 Herf, *Reactionary Modernism*, 189–90, 214, 224–5; Jeffrey Herf, 'Comment by Jeffrey Herf', *History and Technology* 26, no. 1 (2010): 36.
26 Ulrich Wengenroth, 'Die Flucht in den Käfig: Wissenschafts- und Innovationskultur in Deutschland 1900–1960', in *Wissenschaft und Wissenschaftspolitik. Bestandsaufnahmen zu Formationen, Brüchen und Kontinuitäten im Deutschland des 20. Jahrhundert*, ed. Rüdiger vom Bruch and Brigitte Kaderas (Stuttgart: Steiner, 2002): 52–9.
27 Radkau, *Technik in Deutschland*, 92.
28 Wengenroth, *Technik der Moderne*, 34, 50.

Chapter 1

1 Sheilagh C. Ogilvie, 'Proto-Industrialization in Germany', in *European Proto-Industrialization*, ed. Sheilagh C. Ogilvie and Markus Cerman (Cambridge: Cambridge University Press, 1996), 121.
2 Gerschenkron's writings inspired this approach since the 1950s; see Alexander Gerschenkron, *Economic Backwardness in Historical Perspective: A Book of Essays* (Cambridge: Belknap Press, 1962).
3 For a critical retrospect see Kocka, 'Looking Back on the Sonderweg'.
4 Harold James, *Krupp: A History of the Legendary German Firm* (Princeton: Princeton University Press, 2012), 2.
5 Sidney Pollard, 'The Industrial Revolution – an Overview', in *The Industrial Revolution in National Context: Europe and the USA*, ed. Mikuláš Teich and Roy Porter (Cambridge: Cambridge University Press, 1996), 386; David S. Landes, *The Unbound Prometheus. Technological Change and Industrial Development in Western Europe from 1750 to the Present*, 2nd edn (Cambridge: Cambridge University Press, 2003), 105.
6 Robert C. Allen, 'Why the Industrial Revolution was British: Commerce, Induced Invention and the Scientific Revolution', *Economic History Review* 64, no. 2 (2011): 380.

7 Kenneth Pomeranz, 'Political Economy and Ecology on the Eve of Industrialization: Europe, China, and the Global Conjuncture', *American Historical Review* 107, no. 2 (2002): 437.
8 Ibid., 444–5.
9 Robert C. Allen, *The British Industrial Revolution in Global Perspective* (Cambridge: Cambridge University Press, 2009), 104.
10 Allen, 'Why the Industrial Revolution Was British', 366.
11 Günter Vogler, *Europas Aufbruch in die Neuzeit, 1500–1650* (Stuttgart: Ulmer, 2003), 274–5.
12 Heinz Duchhardt, *Europa am Vorabend der Moderne, 1650–1800* (Stuttgart: Ulmer, 2003), 121.
13 Toni Pierenkemper, *Gewerbe und Industrie im 19. und 20. Jahrhundert*, 2nd edn (Munich: Oldenbourg, 2007), 9, 14.
14 Maxine Berg, *The Age of Manufactures: Industry, Innovation and Work in Britain, 1770–1820* (Oxford: Blackwell, 1985), 220.
15 Jan de Vries, 'The Industrial Revolution and the Industrious Revolution', *The Journal of Economic History* 54, no. 2 (1994): 256.
16 Thomas J. Misa, *Leonardo to the Internet: Technology and Culture from the Renaissance to the Present*, 2nd edn (Baltimore: Johns Hopkins University Press, 2011), 92.
17 Pollard, 'The Industrial Revolution', 373.
18 Joyce Appleby, *The Relentless Revolution: A History of Capitalism* (New York/London: Norton, 2010), 165.
19 Fernand Braudel, *Civilization and Capitalism, Vol. 3: The Perspective of the World* (London: Collins, 1984), 548.
20 Günter Bayerl and Ulrich Troitzsch, 'Mechanisierung vor der Mechanisierung? Zur Technologie des Manufakturwesens', in *Technik und industrielle Revolution. Vom Ende eines sozialwissenschaftlichen Paradigmas*, ed. Theo Pirker, Hans-Peter Müller and Rainer Winkelmann (Opladen: Westdeutscher Verlag, 1987), 124.
21 Sheilagh Ogilvie, 'Consumption, Social Capital, and the "Industrious Revolution" in Early Modern Germany', *Journal of Economic History* 70, no. 2 (2010): 320–1.
22 Karin Zachmann, 'Die Kraft traditioneller Strukturen: Sächsische Textilregionen im Industrialisierungsprozess', in *Landesgeschichte als Herausforderung und Programm*, ed. Uwe John and Josef Matzerath (Stuttgart: Steiner, 1997), 518.
23 Hans-Jürgen Teuteberg, 'Britische Frühindustrialisierung und kurhannoversches Reformbewußtsein im späten 18. Jahrhundert', *Vierteljahrschrift für Sozial- und Wirtschaftsgeschichte* 86, no. 2 (1999): 173.
24 Georg Christoph Lichtenberg, *Lichtenberg's Visits to England as Described in his Letters and Diaries*, trans. Margaret L. Mare and W. H. Quarrell (Oxford: Clarendon Press, 1938), 47.
25 Ibid., 97.
26 Helga Schultz, *Handwerker, Kaufleute, Bankiers: Wirtschaftsgeschichte Europas 1500–1800*, 2nd edn (Frankfurt: Fischer, 2002), 81–2.
27 Ibid., 81.
28 Rudolf Muhs, 'Englische Einflüsse auf die Frühphase der Industrialisierung in Deutschland', in *The Race for Modernization: Britain and Germany since the Industrial Revolution*, ed. Adolf M. Birke and Lothar Kettenacker (Munich: Saur, 1988), 39.
29 Hans-Joachim Braun, 'Technologietransfer im Maschinenbau seit dem 18. Jahrhundert: Wandel und Kontinuität', in *Wissenschafts- und Technologietransfer*

zwischen industrieller und wissenschaftlich-technischer Revolution, ed. Klaus-Peter Meinicke (Stuttgart: GNT-Verlag, 1992), 86–7.
30 Muhs, 'Englische Einflüsse', 38–9.
31 Teuteberg, 'Britische Frühindustrialisierung', 175–7.
32 Otto Keck, 'The National System for Technical Innovation in Germany', in *National Innovation Systems: A Comparative Analysis*, ed. Richard R. Nelson (New York: Oxford University Press, 1993), 117.
33 Dieter Ziegler, *Die industrielle Revolution*, 3rd edn (Darmstadt: WBG, 2012), 37.
34 Muhs, 'Englische Einflüsse', 38.
35 Landes, *The Unbound Prometheus*, 141–2.
36 Pierenkemper, *Gewerbe und Industrie im 19. und 20. Jahrhundert*, 50.
37 Landes, *The Unbound Prometheus*, 141–2.
38 Wengenroth, *Technik der Moderne*, 112.
39 Landes, *The Unbound Prometheus*, 142.
40 Toni Pierenkemper, 'Vorrang der Kohle: Wirtschafts-, Unternehmens- und Sozialgeschichte des Bergbaus 1850 bis 1914', in *Geschichte des deutschen Bergbaus, Vol. 3. Motor der Industrialisierung. Deutsche Bergbaugeschichte im 19. und frühen 20. Jahrhundert*, ed. Klaus Tenfelde and Toni Pierenkemper (Münster: Aschendorff, 2016), 68.
41 Hans Otto Gericke, 'Die erste Dampfmaschine Preußens in der Braunkohlengrube Altenwedding (1779–1828)', *Technikgeschichte* 65, no. 2 (1998): 98, 100–1, 115–16.
42 Pierenkemper, *Gewerbe und Industrie im 19. und 20. Jahrhundert*, 50.
43 Karsten Uhl, 'Work Spaces: From the Early-Modern Workshop to the Modern Factory', in *Europäische Geschichte Online (EGO)*, ed. Leibniz-Institut für Europäische Geschichte (IEG), Mainz 2 May 2016. Available online: http://www.ieg-ego.eu/uhlk-2015-en URN: urn:nbn:de:0159-2016042200 (Accessed 9 May 2019), 18.
44 Pierenkemper, *Gewerbe und Industrie im 19. und 20. Jahrhundert*, 50.
45 Muhs, 'Englische Einflüsse', 43.
46 Ibid., 37.
47 Ursula Klein, *Humboldts Preußen: Wissenschaft und Technik im Aufbruch* (Darmstadt: WBG, 2015), 300.
48 Teuteberg, 'Britische Frühindustrialisierung', 155–6, 163–4, 170.
49 Klein, *Humboldts Preußen*, 7, 301.
50 Michael Schäfer, *Eine andere Industrialisierung. Die Transformation der sächsischen Textilexportgewerbe 1790–1890* (Stuttgart: Franz Steiner, 2016), 63–4.
51 Adolf M. Birke and Lothar Kettenacker, 'Introduction', in *The Race for Modernization: Britain and Germany since the Industrial Revolution*, ed. Adolf M. Birke and Lothar Kettenacker (Munich et al.: Saur, 1988), 7.
52 Gary Herrigel, *Industrial Constructions: The Sources of German Industrial Power* (Cambridge: Cambridge University Press, 1996), 72.
53 Ibid., 18, 20, 43, 71.
54 Gerschenkron, *Economic Backwardness*, 26.
55 Hal Hansen, 'Rethinking the Role of Artisans in Modern German Development', *Central European History* 42, no. 1 (2009): 40.
56 Caroline Fohlin, *Finance Capitalism and Germany's Rise to Industrial Power* (New York: Cambridge University Press, 2007), 40, 331.
57 Schäfer, *Eine andere Industrialisierung*, 230, 283–4.
58 Ziegler, *Die industrielle Revolution*, 39.

59 Zachmann, 'Die Kraft traditioneller Strukturen', 525.
60 Schäfer, *Eine andere Industrialisierung*, 300.
61 Hubert Kiesewetter, *Industrielle Revolution in Deutschland: Regionen als Wachstumsmotor* (Stuttgart: Franz Steiner, 2004), 174.
62 Charles Sabel and Jonathan Zeitlin, 'Historical Alternatives to Mass Production: Politics, Markets and Technology in Nineteenth Century Industrialization', in *The Textile Industries, Volume 1: General Concepts*, ed. Stanley David Chapman (London: Tauris, 1997), 63, 68; Frank Dittmann, 'Geschichte der elektrischen Antriebstechnik', in *Alles bewegt sich. Beiträge zur Geschichte elektrischer Antriebe*, ed. Kurt Jäger (Berlin and Offenbach: VDE-Verlag, 1998), 49.
63 Lars Magnusson, *The Contest for Control: Metal Industries in Sheffield, Solingen, Remscheid and Eskilstuna during Industrialization* (Oxford: Berg, 1994), 118; cf. 100, 108–9, 113.
64 Ogilvie, 'Proto-Industrialization in Germany', 136.
65 Pierenkemper, *Gewerbe und Industrie im 19. und 20. Jahrhundert*, 57.
66 Klein, *Humboldts Preußen*, 10, 295, 303.
67 Keck, 'The National System for Technical Innovation in Germany', 120.
68 Hansen, 'Rethinking the Role of Artisans in Modern German Development', 36, 43.
69 Ibid., 47–8, 50–1, 62.
70 Kiesewetter, *Industrielle Revolution in Deutschland*, 289.
71 Pollard, 'Industrial Revolution', 382.
72 Richard Tilly, 'German Industrialization', in *The Industrial Revolution in National Context. Europe and the* USA, ed. Mikuláš Teich and Roy Porter (Cambridge: Cambridge University Press, 1996), 112.
73 Pierenkemper, *Gewerbe und Industrie im 19. und 20. Jahrhundert*, 7; Tilly, 'German Industrialization', 101–2.
74 Keck, 'The National System for Technical Innovation in Germany', 116.
75 Kiesewetter, *Industrielle Revolution in Deutschland*, 60–2.
76 Ziegler, *Die Industrielle Revolution*, 16.
77 Magnusson, *The Contest for Control*, 113.
78 Pierenkemper, *Gewerbe und Industrie im 19. und 20. Jahrhundert*, 14.
79 Pierenkemper, 'Vorrang der Kohle', 99.
80 Eric Dorn Brose, *The Politics of Technological Change in Prussia. Out of the Shadow of Antiquity, 1809–1848* (Princeton: Princeton University Press, 1993), 136; Pierenkemper, 'Vorrang der Kohle', 100.
81 Brose, *Politics of Technological Change*, 135.
82 Pierenkemper, 'Vorrang der Kohle', 100.
83 Franz-Josef Brüggemeier, *Grubengold: Das Zeitalter der Kohle von 1750 bis heute* (Munich: Beck, 2018), 20.
84 Pierenkemper, 'Vorrang der Kohle', 45, 65.
85 Brüggemeier, *Grubengold*, 43.
86 Pierenkemper, 'Vorrang der Kohle', 68.
87 Ziegler, *Die Industrielle Revolution*, 87.
88 Pierenkemper, 'Vorrang der Kohle', 46–7, 59.
89 Ibid., 59, 67–8; Brüggemeier, *Grubengold*, 9.
90 Kiesewetter, *Industrielle Revolution in Deutschland*, 219; Pierenkemper, 'Vorrang der Kohle', 65.
91 Brüggemeier, *Grubengold*, 9.
92 Pierenkemper, 'Vorrang der Kohle', 46–7, 52.

93 Dietmar Bleidick, 'Bergtechnik im 20. Jahrhundert: Mechanisierung in Abbau und Förderung', in *Geschichte des deutschen Bergbaus, Vol. 4: Rohstoffgewinnung im Strukturwandel. Der deutsche Bergbau im 20. Jahrhundert*, ed. Dieter Ziegler (Münster: Aschendorff, 2013), 365, 380.
94 Tobias A. Jopp, 'Did Closures Do Any Good? Labour Productivity, Mine Dynamics, and Rationalization in Interwar Ruhr Coal Mining', *Economic History Review* 70, no. 3 (2017): 944.
95 Brüggemeier, *Grubengold*, 105.
96 Helmuth Trischler, 'Partielle Modernisierung: Die betrieblichen Sozialbeziehungen im Ruhrbergbau zwischen Grubenmilitarismus und Human Relations', in *Politische Zäsuren und gesellschaftlicher Wandel im 20. Jahrhundert*, ed. Matthias Frese and Michael Prinz (Paderborn: Schöningh, 1996), 147–8.
97 Brüggemeier, *Grubengold*, 99.
98 Bleidick, 'Bergtechnik im 20. Jahrhundert', 377.
99 Pierenkemper, 'Vorrang der Kohle', 91–3.
100 Ibid., 51.
101 Dieter Ziegler, *Eisenbahn und Staat im Zeitalter der Industrialisierung: Die Eisenbahnpolitik der deutschen Staaten im Vergleich* (Stuttgart: Franz Steiner, 1996), 19.
102 Kiesewetter, *Industrielle Revolution in Deutschland*, 234.
103 Ziegler, *Eisenbahn und Staat im Zeitalter der Industrialisierung*, 16–17.
104 Kiesewetter, *Industrielle Revolution in Deutschland*, 205.
105 Ziegler, *Eisenbahn und Staat im Zeitalter der Industrialisierung*, 11–12, 24–5, 534.
106 Ibid., 290, 546.
107 Tilly, 'German Industrialization', 106.
108 Burkhard Beyer, *Vom Tiegelstahl zum Kruppstahl: Technik- und Unternehmensgeschichte der Gussstahlfabrik von Friedrich Krupp in der ersten Hälfte des 19. Jahrhunderts* (Essen: Klartext, 2007), 15.
109 Pierenkemper, 'Vorrang der Kohle', 46.
110 Beyer, *Vom Tiegelstahl zum Kruppstahl*, 12–13, 17.
111 James, *Krupp*, 28, 34.
112 Beyer, *Vom Tiegelstahl zum Kruppstahl*, 11, 598.
113 Pierenkemper, 'Vorrang der Kohle', 61, 73.
114 Landes, *The Unbound Prometheus*, 249.
115 Ulrich Wengenroth, *Enterprise and Technology: The German and British Steel Industries, 1865–1895* (Cambridge: Cambridge University Press, 1994), 1, 264–6, 272–3.
116 Pierenkemper, *Gewerbe und Industrie im 19. und 20. Jahrhundert*, 60–1; Tilly, 'German Industrialization', 103.
117 Tilly, 'German Industrialization', 112.
118 Pierenkemper, *Gewerbe und Industrie im 19. und 20. Jahrhundert*, 49.
119 Tilly, 'German Industrialization', 110–12.
120 Joachim Radkau, *Technik in Deutschland: Vom 18. Jahrhundert bis heute* (Frankfurt: Campus, 2008), 151.
121 Landes, *Unbound Prometheus*, 183.
122 Ibid., 149–50; Ziegler, *Die Industrielle Revolution*, 79.
123 Ralf Richter and Jochen Streb, 'Catching-Up and Falling Behind: Knowledge Spillover from American to German Machine Toolmakers', *Journal of Economic History* 71, no. 4 (2011): 1008.
124 Ibid., 1023.

125 Herrigel, *Industrial Constructions*, 100.
126 Schäfer, *Eine andere Industrialisierung*, 299.
127 Pierenkemper, *Gewerbe und Industrie im 19. und 20. Jahrhundert*, 18.
128 Hans-Werner Hahn, *Die industrielle Revolution in Deutschland*, 3rd edn (Munich: Oldenbourg, 2011), 128; Werner Abelshauser, 'From New Industry to the New Economy', in *German Industry and Global Enterprise: BASF – the History of a Company*, ed. Werner Abelshauser (Cambridge: Cambridge University Press, 2004), 1; Cornelius Torp, 'The "Coalition of Rye and Iron" under the Pressure of Globalization: A Reinterpretation of Germany's Political Economy before 1914', *Central European History* 43, no. 3 (2010): 403.
129 Hård and Jamison, *Hubris and Hybrids*, 53.
130 Wolfgang König, *Der Gelehrte und der Manager. Franz Reuleaux (1829–1905) und Alois Riedler (1850–1936) in Technik, Wissenschaft und Gesellschaft* (Stuttgart: Franz Steiner, 2014), 24, 250.
131 Werner von Siemens: Letter of March 13, 1872 (trans. Erwin Fink), *German History in Documents and Images, Volume 4. Forging an Empire: Bismarckian Germany, 1866–1890*, ed. James Retallack, www.ghdi-dc.org; see Jürgen Kocka, 'Modernisierung im multinationalen Familienunternehmen', *Themenportal Europäische Geschichte*, 2006, www.europa.clio-online.de/essay/id/fdae-1337.
132 Radkau, *Technik in Deutschland*, 191. On different waves of Americanization, see Harm G. Schröter, *Americanization of the European Economy: A Compact Survey of American Economic Influence in Europe Since the 1880s* (Dodrecht: Springer, 2005).
133 Radkau, *Technik in Deutschland*, 197.
134 Jeffrey Allan Johnson, *The Kaiser's Chemists: Science and Modernization in Imperial Germany* (Chapel Hill: University of North Carolina Press, 1990), 14.
135 Appleby, *The Relentless Revolution*, 253–4.
136 Peter Alter, 'Deutschland als Vorbild britischer Wirtschaftsplanung um die Jahrhundertwende', in *The Race for Modernization: Britain and Germany since the Industrial Revolution*, ed. Adolf M. Birke and Lothar Kettenacker (Munich et al.: Saur, 1988), 52–4.
137 Keck, 'The National System for Technical Innovation in Germany', 120–3.
138 Wolfgang König, 'Science-Based Industry or Industry-Based Science? Electrical Engineering in Germany Before World War I', *Technology and Culture* 37, no. 1 (1996): 76–7.
139 Johnson, *The Kaiser's Chemists*, 13.
140 König, 'Science-Based Industry or Industry-Based Science?', 77–8.
141 Keck, 'The National System for Technical Innovation in Germany', 115, 130.
142 Hård and Jamison, *Hubris and Hybrids*, 212.
143 Kiesewetter, *Industrielle Revolution in Deutschland*, 208.
144 König, 'Science-Based Industry or Industry-Based Science?', 74.
145 Johnson, *The Kaiser's Chemists*, 14.
146 König, 'Science-Based Industry or Industry-Based Science?', 100; Wolfgang von Hippel, 'Becoming a Global Corporation: BASF from 1865 to 1900', in *German Industry and Global Enterprise: BASF – the History of a Company*, ed. Werner Abelshauser (Cambridge: Cambridge University Press, 2004), 51.
147 Abelshauser, 'From New Industry to the New Economy', 2.
148 Johnson, *The Kaiser's Chemists*, 20.
149 Ibid., 21.
150 von Hippel, 'Becoming a Global Corporation', 102–3, 107.

151 Keck, 'The National System for Technical Innovation in Germany', 126.
152 von Hippel, 'Becoming a Global Corporation', 24, 50.
153 Ursula Klein, 'Technoscience avant la lettre', *Perspectives on Science* 13, no. 2 (2005): 227, 243.
154 Johnson, *The Kaiser's Chemists*, 10.
155 Radkau, *Technik in Deutschland*, 164.
156 Angelika Epple, 'Globale Machtverhältnisse, lokale Verflechtungen: Die Berliner Kongokonferenz, Solingen und das Hinterland des kolonialen Waffenhandels', in *Ränder der Moderne – Neue Perspektiven auf die Europäische Geschichte (1800–1930)*, ed. Christof Dejung and Martin Lengwiler (Cologne: Böhlau, 2015), 81–4, 89.
157 Zachmann, 'Die Kraft traditioneller Strukturen', 527, 530.
158 Michael Mende, 'Verschwundene Stellmacher, gewandelte Schmiede. Die Handwerker des ländlichen Fahrzeugbaus zwischen expandierender Landwirtschaft und Motorisierung 1850–1960', in *Prekäre Selbständigkeit. Zur Standortbestimmung von Handwerk, Hausindustrie und Kleingewerbe im Industrialisierungsprozess*, ed. Ulrich Wengenroth (Stuttgart: Franz Steiner, 1989), 101–2.
159 Anita Kugler, 'Von der Werkstatt zum Fließband. Etappen der frühen Automobilproduktion in Deutschland', *Geschichte und Gesellschaft* 13, no. 3 (1987): 315, 328.
160 Landes, *Unbound Prometheus*, 306, 317.
161 David Meskill, *Optimizing the German Workforce: Labor Administration from Bismarck to the Economic Miracle* (New York: Berghahn, 2010), 126–9.
162 Werner Abelshauser, 'The First Post-Liberal Nation: Stages in the Development of Modern Corporatism in Germany', *European History Quarterly* 14, no. 3 (1984): 288–9, 296.
163 Werner Abelshauser, 'Umbruch und Persistenz: Das deutsche Produktionsregime in historischer Perspektive', *Geschichte und Gesellschaft* 27, no. 4 (2001): 520.
164 Jürgen Bönig, *Die Einführung von Fließbandarbeit in Deutschland bis 1933: Zur Geschichte einer Sozialinnovation*, Vol. 2 (Münster: LIT, 1993), 689.
165 Matthew Jefferies, *Politics and Culture in Wilhelmine Germany: The Case of Industrial Architecture* (Oxford: Berg, 1995), 184–5, 229.
166 Frederick Winslow Taylor, *The Principles of Scientific Management* (New York: Harper & Brothers, 1911); Robert Kanigel, *The One Best Way: Frederick Taylor and the Enigma of Efficiency* (New York: Viking, 1997).
167 Christian Kleinschmidt, *Technik und Wirtschaft im 19. und 20. Jahrhundert* (Munich: Oldenbourg, 2006), 24.
168 Guido Buenstorf and Johann Peter Murmann, 'Ernst Abbe´s Scientific Management: Theoretical Insights from a Nineteenth-century Dynamic Capabilities Approach', *Industrial and Corporate Change* 14, no. 4 (2005): 573–4.
169 Christian Kleinschmidt, *Rationalisierung als Unternehmensstrategie: Die Eisen- und Stahlindustrie des Ruhrgebiets zwischen Jahrhundertwende und Weltwirtschaftskrise* (Essen: Klartext, 1993), 84.
170 Jefferies, *Politics and Culture in Wilhelmine Germany*, 186.
171 Anson Rabinbach, *The Human Motor: Energy, Fatigue, and the Origins of* Modernity (Berkeley: University of California Press, 1992), 259–62.
172 Mikael Hård, 'German Regulation: The Integration of Modern Technology into National Culture', in *The Intellectual Appropriation of Technology: Discourses of Modernity, 1900–1939*, ed. Mikael Hård and Andrew Jamison (Cambridge: MIT Press, 1998), 35, 51.

173 Karsten Uhl, 'Der Erste Weltkrieg als Ausgangspunkt der Humanisierung des Arbeitslebens im 20. Jahrhundert: Der Beginn des staatlichen und unternehmerischen Interesses am "menschlichen Faktor" in der Produktion', in *Humanisierung der Arbeit: Aufbrüche und Konflikte in der rationalisierten Arbeitswelt des 20. Jahrhunderts*, ed. Nina Kleinöder, Stefan Müller and Karsten Uhl (Bielefeld: transcript, 2019), 9–32; see Karsten Uhl, 'The Ideal of Lebensraum and the Spatial Order of Power at German Factories, 1900–1945', *European Review of History/Revue d'histoire européenne* 20, no. 2 (2013): 287–307.
174 Daniel T. Rodgers, *Atlantic Crossings. Social Politics in A Progressive Age* (Cambridge: Belknap, 1998), 375.
175 Thomas P. Hughes, *American Genesis: A Century of Invention and Technological Enthusiasm, 1870–1970* (New York: Viking, 1989), 284–6.
176 Hans-Liudger Dienel, '"Hier sauber und gründlich, dort husch-husch fertig": Deutsche Vorbehalte gegen amerikanische Produktionsmethoden 1870–1930', *Blätter für Technikgeschichte* 55 (1993): 19, 22.
177 Fritz Söllheim, *Taylor-System für Deutschland. Grenzen seiner Einführung in deutsche Betriebe* (Munich: Oldenbourg, 1922), 229; see Nolan, *Visions of Modernity*, 45.
178 Nolan, *Visions of Modernity*, 71.
179 Rüdiger Hachtmann, 'Industriearbeiterschaft und Rationalisierung 1900 bis 1945. Bemerkungen zum Forschungsstand', *Jahrbuch für Wirtschaftsgeschichte* 37, no. 1 (1996): 219.
180 Heidrun Homburg, 'The "Human Factor" and the Limits of Rationalization: Personnel-Management Strategies and the Rationalization Movement in German Industry between the Wars', in *The Power to Manage? Employers and Industrial Relations in Comparative-Historical Perspective*, ed. Steven Tolliday and Jonathan Zeitlin (London: Routledge 1991), 168.
181 Kleinschmidt, *Rationalisierung als Unternehmensstrategie*, 268; see Jennifer Karns Alexander, *The Mantra of Efficiency: From Waterwheel to Social Control* (Baltimore: Johns Hopkins University Press, 2008), 101.
182 Katja Patzel-Mattern, *Ökonomische Effizienz und gesellschaftlicher Ausgleich. Die industrielle Psychotechnik in der Weimarer Republik* (Stuttgart: Franz Steiner, 2010), 28, 40, 102, 270.
183 Andreas Killen, 'Weimar Psychotechnics Between Americanism and Fascism', *Osiris* 22, no. 1 (2007): 51.
184 Matthias Kipping, 'Consultancies, Institutions and the Diffusion of Taylorism in Britain, Germany and France 1920s to 1950s', *Business History Review* 39, no. 4 (1997): 70–1; see J. Ronald Shearer, 'The Reichskuratorium für Wirtschaftlichkeit: Fordism and Organized Capitalism in Germany, 1918–1945', *Business History Review* 71, no. 3 (1997): 567–602.
185 J. Ronald Shearer, 'Talking About Efficiency: Politics and the Industrial Rationalization Movement in the Weimar Republic', *Central European History* 28, no. 4 (1995): 485.
186 Bönig, *Die Einführung von Fließbandarbeit*, 699.
187 Rüdiger Hachtmann, *Industriearbeit im 'Dritten Reich': Untersuchungen zu Lohn- und Arbeitsbedingungen in Deutschland 1933–1945* (Göttingen: Vandenhoeck & Ruprecht, 1989), 68–9.
188 Michael Stahlmann, *Die erste Revolution in der Autoindustrie: Management und Arbeitspolitik von 1900–1940* (Frankfurt: Campus, 1993), 60, 62, 68–9, 72.

189 Kugler, 'Von der Werkstatt zum Fließband', 336–7.
190 Jopp, 'Did Closures Do Any Good?', 948; Brüggemeier, *Grubengold*, 243; Bleidick, 'Bergtechnik im 20. Jahrhundert', 385.
191 Marcel P. Timmer, Joost Veenstra and Pieter J. Woltjer, 'The Yankees of Europe? A New View on Technology and Productivity in German Manufacturing in the Early Twentieth Century', *Journal of Economic History* 76, no. 3 (2016): 881.
192 Ibid., 878, 888, 894, 897–9.
193 Cristiano Andrea Ristuccia and Adam Tooze, 'Machine Tools and Mass Production in the Armaments Boom: Germany and the United States, 1929–44', *Economic History Review* 66, no. 4 (2013): 957, 971–2.
194 Jonas Scherner, Jochen Streb and Stephanie Tilly, 'Supplier Networks in the German Aircraft Industry during World War II and Their Long-Term Effects on West Germany's Automobile Industry during the "Wirtschaftswunder"', *Business History* 56, no. 6 (2014): 997, 1014.
195 Pierenkemper, *Gewerbe und Industrie im 19. und 20. Jahrhundert*, 115.
196 Kleinschmidt, *Technik und Wirtschaft im 19. und 20. Jahrhundert*, 53; Werner Abelshauser, *Deutsche Wirtschaftsgeschichte seit 1945* (Munich: Beck, 2004), 67–74; Hermann Weber, *Die DDR 1945–1990*. 5 edn (Munich: Oldenbourg 2012), 32.
197 Jonathan Zeitlin, 'Introduction: Americanization and Its Limits: Reworking US Technology and Management in Post-War Europe and Japan', in *Americanization and Its Limits. Reworking US Technology and Management in Post-War Europe and Japan*, ed. Jonathan Zeitlin and Gary Herrigel (Oxford: Oxford University Press 2000), 35.
198 Volker Berghahn, *Industriegesellschaft und Kulturtransfer. Die deutsch-amerikanischen Beziehungen im 20. Jahrhundert* (Göttingen: Vandenhoeck & Ruprecht, 2010), 197, 199.
199 Susanne Hilger, *"Amerikanisierung deutscher Unternehmen". Wettbewerbsstrategien und Unternehmenspolitik bei Henkel, Siemens und Daimler-Benz (1945/49–1975)* (Stuttgart: Franz Steiner, 2004), 174, 179.
200 Mary Nolan, '"Varieties of Capitalism" und Versionen der Amerikanisierung', in *Gibt es einen deutschen Kapitalismus? Traditionen und globale Perspektiven der sozialen Marktwirtschaft*, ed. Volker R. Berghahn and Sigurt Vitols (Frankfurt: Campus, 2006), 104–5; see Rüdiger Hachtmann and Adelheid von Saldern, '"Gesellschaft am Fließband": Fordistische Produktion und Herrschaftspraxis in Deutschland', *Zeithistorische Forschungen/Studies in Contemporary History* 6, no. 2 (2009): 200.
201 Christian Kleinschmidt, *Der produktive Blick: Wahrnehmung amerikanischer und japanischer Management- und Produktionsmethoden durch deutsche Unternehmer 1950–1985* (Berlin: De Gruyter, 2002), 157.
202 Werner Abelshauser, *Kulturkampf: Der deutsche Weg in die neue Wirtschaft und die amerikanische Herausforderung* (Berlin: Kadmos, 2003), 130.
203 Volker Wellhöner, *"Wirtschaftswunder" – Weltmarkt – westdeutscher Fordismus: Der Fall Volkswagen* (Münster: Westfälisches Dampfboot, 1996), 109, 113, 123.
204 Ibid., 114, 126–7.
205 Abelshauser, *Deutsche Wirtschaftsgeschichte seit 1945*, 372–7; see Werner Abelshauser, 'Two Kinds of Fordism: On the Differing Roles of the Industry in the Development of the Two German States', in *Fordism Transformed: The Development of Production Methods in the Automobile Industry*, ed. Haruhito Shiomi and Kazuo Wada (Oxford: Oxford University Press, 1995), 269–96.
206 Peter Hübner, *Arbeiter und Technik in der DDR 1971 bis 1989: Zwischen Fordismus und digitaler Revolution* (Bonn: Dietz, 2014), 17, 433.

207 Jeffrey Kopstein, *The Politics of Economic Decline in East Germany, 1945–1989* (Chapel Hill: University of North Carolina Press, 1997), 167.
208 Abelshauser, *Deutsche Wirtschaftsgeschichte seit 1945*, 59.
209 Sorge and Streeck, 'Diversified Quality Production Revisited', 595–7, 599.
210 Hachtman and von Saldern, '"Gesellschaft am Fließband"', 189, 193–4.
211 Werner Plumpe, 'Das Ende des deutschen Kapitalismus', *WestEnd: Neue Zeitschrift für Sozialforschung* 2, no. 2 (2005): 12.
212 Wolfgang Hindrichs, Uwe Jürgenhake, Christian Kleinschmidt, Wilfried Kruse, Rainer Lichte and Helmut Martens, *Der lange Abschied vom Malocher: Sozialer Umbruch in der Stahlindustrie und die Rolle der Betriebsräte von 1960 bis in die neunziger Jahre* (Essen: Klartext, 2000), 7, 14, 30–1.
213 Bleidick, 'Bergtechnik im 20. Jahrhundert', 394, 399; Brüggemeier, *Grubengold*, 9, 360.
214 Stephan H. Lindner, *Den Faden verloren. Die westdeutsche und die französische Textilindustrie auf dem Rückzug (1930/45–1990)* (Munich: Beck, 2001), 9, 30,63, 69, 247, 250, 255.
215 Werner Abelshauser, 'BASF since Its Refounding in 1952', in *German Industry and Global Enterprise: BASF – the History of a Company*, ed. Werner Abelshauser (Cambridge: Cambridge University Press, 2004), 378.
216 Ibid., 451, 477–8, 487.
217 Karsten Uhl, 'Eine lange Geschichte der "menschenleeren Fabrik". Automatisierungsvisionen und technologischer Wandel im 20. Jahrhundert', in *Marx und die Roboter. Vernetzte Produktion, künstliche Intelligenz und lebendige Arbeit*, ed. Florian Butollo and Sabine Nuss (Berlin: Dietz, 2019), 74–90.
218 Karsten Uhl, *Humane Rationalisierung? Die Raumordnung der Fabrik im fordistischen Jahrhundert* (Bielefeld: transcript, 2014), 321–2; Thomas Bardelle, '150 Jahre B. Sprengel & Co.: Aufstieg und Niedergang', *Niedersächsisches Jahrbuch für Landesgeschichte* 74 (2002): 305.
219 Hilger, *"Amerikanisierung deutscher Unternehmen"*, 24–5, 244–5.
220 Nina Kleinöder, Stefan Müller and Karsten Uhl, 'Die Humanisierung des Arbeitslebens. Einführung und methodische Überlegungen', in *Humanisierung der Arbeit. Aufbrüche und Konflikte in der rationalisierten Arbeitswelt des 20. Jahrhunderts*, ed. Nina Kleinöder, Stefan Müller and Karsten Uhl (Bielefeld: transcript, 2019), 11, 16.
221 Kleinschmidt, *Der produktive Blick*, 201–2.
222 Susanne Hilger, 'Von der "Amerikanisierung" zur "Gegenamerikanisierung": Technologietransfer und Wettbewerbspolitik in der deutschen Computerindustrie nach dem Zweiten Weltkrieg', *Technikgeschichte* 71, no. 4 (2004): 331; Annette Schuhmann, 'Der Traum vom perfekten Unternehmen. Die Computerisierung der Arbeitswelt in der Bundesrepublik Deutschland (1950er bis 1980er Jahre)', *Zeithistorische Forschungen/Studies in Contemporary History* 9, no. 2 (2012): 240–2.
223 Timo Leimbach, *Die Geschichte der Softwarebranche in Deutschland: Entwicklung und Anwendung von Informations- und Kommunikationstechnologie zwischen den 1950ern und heute* (München: Universitätsbibliothek, 2010), 458; Hilger, 'Von der "Amerikanisierung" zur "Gegenamerikanisierung"', 334–5.
224 Hilger, 'Von der "Amerikanisierung" zur "Gegenamerikanisierung"', 340.
225 Leimbach, *Die Geschichte der Softwarebranche in Deutschland*, 160, 459–60.
226 Simon Doning, 'Appropriating American Technology in the 1960s: Cold War Politics and the GDR Computer Industry', *IEEE Annals of the History of Computing* 32, no.

2 (2010): 32–5, 37, 39; James W. Cortada, 'Information Technologies in the German Democratic Republic (GDR), 1949–1989', *IEEE Annals of the History of Computing* 34, no. 2 (2012): 37–8.

227 Dolores L. Augustine, *Red Prometheus: Engineering and Dictatorship in East Germany, 1945–1990* (Cambridge: MIT Press, 2007), 305, 309; Doning, 'Appropriating American Technology in the 1960s', 39.

228 Hübner, *Arbeiter und Technik in der DDR 1971 bis 1989*, 536.

229 Cortada, 'Information Technologies in the German Democratic Republic (GDR), 1949–1989', 44; Olaf Klenke, *Kampfauftrag Mikrochip: Rationalisierung und sozialer Konflikt in der DDR* (Hamburg: VSA, 2008), 86.

230 Lars Heide, 'Punched Cards for Professional European Offices: Revisiting the Dynamics of Information Technology Diffusion from the United States to Europe, 1889–1918', *History and Technology* 24, no. 4 (2008): 307, 315–16.

231 Rüdiger Hachtmann, 'Rationalisierung, Automatisierung, Digitalisierung. Arbeit im Wandel', in *Geteilte Geschichte: Ost- und Westdeutschland 1970–2000*, ed. Frank Bösch (Göttingen: Vandenhoeck & Ruprecht, 2015), 215; Jürgen Danyel and Annette Schuhmann, 'Wege in die digitale Moderne. Computerisierung als gesellschaftlicher Wandel', in *Geteilte Geschichte: Ost- und Westdeutschland 1970–2000*, ed. Frank Bösch (Göttingen: Vandenhoeck & Ruprecht, 2015), 289–90; Schuhmann, 'Der Traum vom perfekten Unternehmen', 243; Karsten Uhl, 'Challenges of Computerization and Globalization: The Example of the Printing Unions, 1950s to 1980s', in *Since the Boom: Continuity and Change in the Western Industrialized World after 1970*, ed. Sebastian Voigt (Toronto: University of Toronto Press, 2020), 129–52.

232 Thomas Raithel, 'Neue Technologien. Produktionsprozesse und Diskurse', in *Auf dem Weg in eine neue Moderne? Die Bundesrepublik Deutschland in den siebziger und achtziger Jahren*, ed. Thomas Raithel, Andreas Rödder and Andreas Wirsching (Munich: Oldenbourg, 2009), 44.

233 Arnd Sorge, Gert Hartmann, Malcom Warner and Ian Nicholas, *Mikroelektronik und Arbeit in der Industrie: Erfahrungen beim Einsatz von CNC-Maschinen in Großbritannien und Deutschland* (Frankfurt: Campus, 1982), 5, 17, 37, 43, 125, 139, 160; see Uhl, 'Giving Scientific Management a "Human" Face', 511–33.

234 Sorge, Hartmann, Warner and Nicholas, *Mikroelektronik und Arbeit in der Industrie*, 120–1.

235 Wolfgang Coy, *Industrieroboter: Zur Archäologie der zweiten Schöpfung* (Berlin: Rotbuch), 68, 75.

236 Martina Heßler, 'Die Halle 54 bei Volkswagen und die Grenzen der Automatisierung: Überlegungen zum Mensch-Maschine-Verhältnis in der industriellen Produktion der 1980er Jahre', *Zeithistorische Forschungen/Studies in Contemporary History* 11, no. 1 (2014): 72–3.

237 Ulrich Jürgens, Thomas Malsch and Knuth Dohse, *Breaking from Taylorism: Changing Forms of Work in the Automobile Industry* (Cambridge: Cambridge University Press, 1993, repr. 2009), 3, 127, 138.

238 Hübner, *Arbeiter und Technik in der DDR 1971 bis 1989*, 202–5, 443.

239 Hartmut Berghoff, 'Die 1990er Jahre als Epochenschwelle? Der Umbruch der Deutschland AG zwischen Traditionsbruch und Kontinuitätswahrung', *Historische Zeitschrift* 308, no. 2 (2019): 375.

240 Berghahn, *Industriegesellschaft und Kulturtransfer*, 200.

241 Plumpe, 'Das Ende des deutschen Kapitalismus', 7, 21–2.

242 André Steiner, 'Abschied von der Industrie? Wirtschaftlicher Strukturwandel in West- und Ostdeutschland seit den 1960er Jahren', in *Der Mythos von der postindustriellen Welt: Wirtschaftlicher Strukturwandel in Deutschland 1960–1990*, ed. Werner Plumpe and André Steiner (Göttingen: Wallstein, 2016), 32, 52; see Pierenkemper, *Gewerbe und Industrie im 19. und 20. Jahrhundert*, 116.
243 Sorge and Streeck, 'Diversified Quality Production Revisited', 608.
244 Berghoff, 'Die 1990er Jahre als Epochenschwelle?', 367, 398.
245 Sorge and Streeck, 'Diversified Quality Production Revisited', 608.
246 Richter and Streb, 'Catching-Up', 1006–7.
247 Berghoff, 'Die 1990er Jahre als Epochenschwelle', 377, 380, 384.
248 International Federation of Robotics, *World Robotics - Industrial Robots 2018: Statistics, Market Analysis, Forecasts and Case Studies* (Frankfurt: VDM Verlag, 2018), 15, 18–19.
249 Hartmut Hirsch-Kreinsen, *Arbeit 4.0: Pfadabhängigkeit statt Disruption (Soziologisches Arbeitspapier, no. 52)* (Dortmund: TU Dortmund, 2018), 7–8.

Chapter 2

1 Dieter Schott, *Europäische Urbanisierung (1000–2000): Eine umwelthistorische Einführung* (Cologne: Böhlau, 2014), 280–1, 304–5.
2 Brian Ladd, *Urban Planning and Civic Order in Germany, 1860–1914* (Cambridge: Harvard University Press, 1990), 77.
3 Jürgen Reulecke, *Geschichte der Urbanisierung in Deutschland* (Frankfurt: Suhrkamp, 1985), 30–2, 47–8.
4 Ibid., 68; Richard Tilly, 'German Industrialization', in *The Industrial Revolution in National Context: Europe and the USA*, ed. Mikuláš Teich and Roy Porter (Cambridge: Cambridge University Press, 1996), 117.
5 Ladd, *Urban Planning and Civic Order in Germany, 1860–1914*, 1.
6 Mikael Hård and Markus Stippak, 'Progressive Dreams: The German City in Britain and the United States', in *Urban Machinery: Inside Modern European Cities*, ed. Mikael Hård and Thomas J. Misa (Cambridge: MIT Press, 2008), 122.
7 Reulecke, *Geschichte der Urbanisierung in Deutschland*, 44–5.
8 Carsten Jonas, *Die Stadt und ihr Grundriss: Zu Form und Geschichte der deutschen Stadt nach Entfestigung und Eisenbahnanschluss* (Tübingen: Ernst Wasmuth, 2006), 35, 43.
9 Reulecke, *Geschichte der Urbanisierung*, 30.
10 Schott, *Europäische Urbanisierung (1000–2000)*, 303.
11 Dieter Ziegler, *Die Industrielle Revolution*, 55–6; Wolfgang Behringer, *Im Zeichen des Merkur: Reichspost und Kommunikationsrevolution in der Frühen Neuzeit* (Göttingen: Vandenhoeck & Ruprecht, 2003), 512–48.
12 Karl H. Metz, *Ursprünge der Zukunft: Die Geschichte der Technik in der westlichen Zivilisation* (Paderborn: Schöningh, 2006), 249.
13 Kiesewetter, *Industrielle Revolution in Deutschland*, 228.
14 William O. Henderson, *Friedrich List: Economist and Visionary* (London: Frank Cass, 1983)
15 G. Wolfgang Heinze and Heinrich H. Kill, 'The Development of the German Railroad System', in *The Development of Large Technical Systems*, ed. Renate Mayntz and

Thomas P. Hughes (Frankfurt am Main: Campus Verlag, 1988), 115, 117–18; Hård and Jamison, *Hubris and Hybrids,* 177–8; Reulecke, *Geschichte der Urbanisierung*, 31.

16 Dirk van Laak, 'Vom Lebensraum zum Leitungsweg: Die Stadtstraße als soziale Arena', in *Infrastrukturen der Stadt*, ed. Michael Flitner, Julia Lossau and Anna-Lisa Müller (Wiesbaden: Springer, 2017), 146.

17 Wiebke Porombka, *Medialität urbaner Infrastrukturen: Der öffentliche Nahverkehr, 1870–1933* (Bielefeld: transcript, 2013), 157–8.

18 Beate Witzler, *Großstadt und Hygiene: Kommunale Gesundheitspolitik in der Epoche der Urbanisierung* (Stuttgart: Steiner, 1995), 67.

19 Maxwell G. Lay, *Ways of the World: A History of the World's Roads and of the Vehicles that Used Them* (New Brunswick: Rutgers University Press, 1992), 210–11.

20 Clay McShane and Joel A. Tarr, *The Horse in the City: Living Machines in the Nineteenth Century* (Baltimore: Johns Hopkins University Press, 2007), 14.

21 Winfried Wolf, *Berlin – Weltstadt ohne Auto? Verkehrsgeschichte 1848–2015* (Cologne: ISP, 1994), 28.

22 McShane and Tarr, *The Horse in the City,* 58.

23 Barbara Schmucki, 'The Machine in the City: Public Appropriation of the Tramway in Britain and Germany, 1870–1915', *Journal of Urban History* 38, no. 6 (2012): 1061; Ladd, *Urban Planning and Civic Order in Germany, 1860–1914*, 202; Lay, *Ways of the World,* 134.

24 Wolf, *Berlin – Weltstadt ohne Auto*, 34, 36.

25 Schott, *Europäische Urbanisierung (1000–2000)*, 294–5; Barbara Schmucki and Hans-Liudger Dienel, 'Aufstieg und Fall des öffentlichen Personennahverkehrs (ÖPNV) in Deutschland bis heute', in *Mobilität für alle: Geschichte des öffentlichen Personennahverkehrs in der Stadt zwischen technischem Fortschritt und sozialer Pflicht*, ed. Hans-Liudger Dienel and Barbara Schmucki (Stuttgart: Steiner, 1997), 9.

26 McShane and Tarr, *The Horse in the City,* 83; Schmucki, 'The Machine in the City', 1067.

27 McShane and Tarr, *The Horse in the City,* 178.

28 Schmucki, 'The Machine in the City', 1063–4, 1066.

29 Schott, *Europäische Urbanisierung (1000–2000)*, 309–10.

30 Reulecke, *Geschichte der Urbanisierung in Deutschland*, 30.

31 Ladd, *Urban Planning and Civic Order in Germany, 1860–1914*, 152.

32 Andreas Bernard, *Die Geschichte des Fahrstuhls: Über einen beweglichen Ort der Moderne* (Frankfurt: Fischer Taschenbuch, 2006), 91–3.

33 Reulecke, *Geschichte der Urbanisierung*, 54.

34 Brian Ladd, *The Ghosts of Berlin: Confronting German History in the Urban Landscape* (Chicago: University of Chicago Press, 1997), 100.

35 Marion W. Gray, 'Urban Sewage and Green Meadows: Berlin's Expansion to the South 1870–1920', *Central European History* 47, no. 2 (2014): 278.

36 Schott, *Europäische Urbanisierung (1000–2000)*, 253-74.

37 Ladd, *The Ghosts of Berlin*, 101; Ladd, *Urban Planning and Civic Order in Germany, 1860–1914*, 152.

38 Ladd, *The Ghosts of Berlin*, 104.

39 Shahrooz Mohajeri, *100 Jahre Berliner Wasserversorgung und Abwasserentsorgung 1840-1940* (Stuttgart: Franz Steiner, 2005), 147-54; Bernard, *Die Geschichte des Fahrstuhls*, 192-4.

40 Richard J. Evans, *Death in Hamburg: Society and Politics in the Cholera Years, 1830–1910* (Oxford: Clarendon Press, 1987), 177.

41 Peter Münch, *Stadthygiene im 19. und 20. Jahrhundert: Die Wasserversorgung, Abwasser- und Abfallbeseitigung unter besonderer Berücksichtigung Münchens* (Göttingen: Vandenhoeck & Ruprecht, 1993), 32, 341.
42 Reulecke, *Geschichte der Urbanisierung*, 56–7, 61.
43 Wolfgang Schivelbusch, *Disenchanted Night: The Industrialization of Light in the Nineteenth Century* (Berkeley: University of California Press, 1995), 16–17; Stefan Gorißen, 'Fabrik', in *Enzyklopädie der Neuzeit,* vol. 3, ed. Friedrich Jaeger (Stuttgart: Metzler, 2006), 744.
44 Schivelbusch, *Disenchanted Night*, 26–7; see Leslie Tomory, 'London's Water Supply Before 1800 and the Roots of the Networked City', *Technology and Culture* 56, no. 3 (2015): 704-737.
45 Karl Ditt, 'Energiepolitik und Energiekonsum: Gas, Elektrizität und Haushaltstechnik in Großbritannien und Deutschland 1880–1939 und die Beispiele Leeds und York, Dortmund und Münster, *Archiv für Sozialgeschichte* 46 (2006): 114; Ladd, *Urban Planning and Civic Order in Germany, 1860–1914*, 48; Schott, *Europäische Urbanisierung (1000–2000)*, 284.
46 Schivelbusch, *Disenchanted Night*, 32–3.
47 Ditt, 'Energiepolitik und Energiekonsum', 115.
48 Reulecke, *Geschichte der Urbanisierung*, 56–7.
49 Schott, *Europäische Urbanisierung (1000–2000)*, 286–7.
50 Schivelbusch, *Disenchanted Night*, 51.
51 Schott, *Europäische Urbanisierung (1000–2000)*, 285; Schivelbusch, *Disenchanted Night*, 50.
52 Dieter Schott, 'Empowering European Cities: Gas and Electricity in the Urban Environment', in *Urban Machinery: Inside Modern European Cities*, ed. Mikael Hård and Thomas J. Misa (Cambridge: MIT Press, 2008), 166.
53 Münch, *Stadthygiene im 19. und 20. Jahrhundert*, 41.
54 Witzler, *Großstadt und Hygiene*, 70, 75–6, 78; Jürgen Büschenfeld, *Flüsse und Kloaken: Umweltfragen im Zeitalter der Industrialisierung (1870–1918)* (Stuttgart: Klett-Cotta, 1997), 30–1.
55 Shahrooz Mohajeri and Noyan Dinçkal, 'Zentrale Wasserversorgung in Berlin und Istanbul: Einrichtungs-, Diffusions- und Akzeptanzprozesse im Vergleich', *Technikgeschichte* 69, no. 2 (2002): 119.
56 Evans, *Death in Hamburg*, 145.
57 Münch, *Stadthygiene im 19. und 20. Jahrhundert*, 41.
58 Büschenfeld, *Flüsse und Kloaken*, 26–8.
59 Evans, *Death in Hamburg*, 146–60; Anne I. Hardy, *Ärzte, Ingenieure und die städtische Gesundheit: Medizinische Theorien in der Hygienebewegung des 19. Jahrhunderts* (Frankfurt: Campus, 2005), 296; Münch, *Stadthygiene im 19. und 20. Jahrhundert*, 43.
60 Evans, *Death in Hamburg*, 151, 160–1.
61 Mohajeri and Dinçkal, 'Zentrale Wasserversorgung in Berlin und Istanbul', 128–30.
62 Münch, *Stadthygiene im 19. und 20. Jahrhundert*, 27–8; Schott, *Europäische Urbanisierung (1000–2000)*, 226–37.
63 Münch, *Stadthygiene im 19. und 20. Jahrhundert*, 39, 89.
64 Evans, *Death in Hamburg*, 145; Schott, *Europäische Urbanisierung (1000–2000)*, 238–42.
65 Ladd, *Urban Planning and Civic Order in Germany, 1860–1914*, 57; Mohajeri and Dinçkal, 'Zentrale Wasserversorgung in Berlin und Istanbul', 122–3, 125; see Christian Eiden, *Versorgungswirtschaft als regionale Organisation: Die*

Wasserversorgung Berlins und des Ruhrgebiets zwischen 1850 und 1930 (Essen: Klartext, 2006).
66 Hardy, *Ärzte, Ingenieure und die städtische Gesundheit*, 130, 136, 256, 375–6, 380–1.
67 Ladd, *Urban Planning and Civic Order in Germany, 1860–1914*, 60–1.
68 Mohajeri and Dinçkal, 'Zentrale Wasserversorgung in Berlin und Istanbul', 128–32, 135–7.
69 Witzler, *Großstadt und Hygiene*, 69, 79–80.
70 Münch, *Stadthygiene im 19. und 20. Jahrhundert*, 42; Büschenfeld, *Flüsse und Kloaken*, 27–8.
71 Ibid., 138; Witzler, *Großstadt und Hygiene*, 70.
72 Schott, *Europäische Urbanisierung (1000–2000)*, 252.
73 Ute Frevert, '"Fürsorgliche Belagerung": Hygienebewegung und Arbeiterfrauen im 19. und frühen 20. Jahrhundert', *Geschichte und Gesellschaft* 11, no. 4 (1985): 420–446; Mikael Hård, '"The Victorian Eye" and Its Blind Spot: Toward a Cultural Assessment of Technology', *Technology and Culture* 26, no. 2 (2010): 176; Axel C. Hüntelmann, *Hygiene im Namen des Staates: Das Reichsgesundheitsamt 1876–1933* (Göttingen: Wallstein, 2008), 415.
74 Büschenfeld, *Flüsse und Kloaken*, 23, 34, 40–1; Witzler, *Großstadt und Hygiene*, 81–2; Schott, *Europäische Urbanisierung (1000–2000)*, 246–7.
75 Witzler, *Großstadt und Hygiene*, 15, 19–20.
76 Ladd, *Urban Planning and Civic Order in Germany, 1860–1914*, 52; Büschenfeld, *Flüsse und Kloaken*, 36–7.
77 Hardy, *Ärzte, Ingenieure und die städtische Gesundheit*, 132-4, 298.
78 Ladd, *Urban Planning and Civic Order in Germany, 1860–1914*, 55; Witzler, *Großstadt und Hygiene*, 83.
79 Münch, *Stadthygiene im 19. und 20. Jahrhundert*, 39, 89.
80 Büschenfeld, *Flüsse und Kloaken*, 411.
81 Münch, *Stadthygiene im 19. und 20. Jahrhundert*, 40.
82 Ladd, *Urban Planning and Civic Order in Germany, 1860–1914*, 57; Mohajeri, *100 Jahre Berliner Wasserversorgung und Abwasserentsorgung 1840-1940*, 154–67.
83 Eli Rubin, 'From the *Grünen Wiesen* to Urban Space: Berlin, Expansion and the *Longue Durée*: Introduction', *Central European History* 47, no. 2 (2014): 238.
84 Gray, 'Urban Sewage and Green Meadows,' 289.
85 Ladd, *Urban Planning and Civic Order in Germany, 1860–1914*, 57.
86 Witzler, *Großstadt und Hygiene*, 85–8.
87 Büschenfeld, *Flüsse und Kloaken*, 151.
88 Witzler, *Großstadt und Hygiene*, 84, 89.
89 Schott, *Europäische Urbanisierung (1000–2000)*, 251.
90 Büschenfeld, *Flüsse und Kloaken*, 120.
91 Witzler, *Großstadt und Hygiene*, 68–9.
92 Büschenfeld, *Flüsse und Kloaken*, 419.
93 Münch, *Stadthygiene im 19. und 20. Jahrhundert*, 38.
94 Büschenfeld, *Flüsse und Kloaken*, 45.
95 Daniel R. Headrick, *The Tools of Empire: Technology and European Imperialism in the Nineteenth Century* (Oxford: Oxford University Press, 1981).
96 Agnes Kneitz, 'German Water Infrastructure in China: Colonial Quindao 1898–1914', *N.T.M.* 24, no. 4 (2016): 422, 424, 427–8.
97 Ibid., 431–5, 437.

98 Michael Wobring, 'Telekommunikation und Nationsbildung: Die politischen Konzepte früher deutscher Telegrafenplanung vom ausgehenden 18. Jahrhundert bis zur Paulskirche', *Technikgeschichte* 72, no. 3 (2004): 202, 209; Klaus Beyrer, 'Die optische Telegraphie als Beginn der modernen Telekommunikation', in *Vom Flügeltelegraphen zum Internet. Geschichte der modernen Telekommunikation*, ed. Hans-Jürgen Teuteburg and Cornelius Neutsch (Stuttgart: Franz Steiner, 1998), 16, 23.
99 Wobring, 'Telekommunikation und Nationsbildung', 214.
100 Patrice Flichy, *Tele: Geschichte der modernen Kommunikation* (Frankfurt: Campus, 1994), 43.
101 Hård and Jamison, *Hubris and Hybrids*, 212; Misa, *Leonardo to the Internet*, 104.
102 Simone M. Müller, *Wiring the World: The Social and Cultural Creation of Global Telegraph Networks* (New York: Columbia University Press, 2016), 4, 11; see Tom Standage, *The Victorian Internet: The Remarkable Story of the Telegraph and the Nineteenth Century's On-line Pioneers* (New York: Bloomsbury, 2014).
103 Frank Thomas, *Telefonieren in Deutschland: Organisatorische, technische und räumliche Entwicklung eines großtechnischen Systems* (Frankfurt: Campus, 1995), 57.
104 Wolfgang Zängl, *Deutschlands Strom: Die Politik der Elektrifizierung von 1866 bis heute* (Frankfurt: Campus, 1989), 14–5; Schott, *Europäische Urbanisierung (1000–2000)*, 289.
105 Schivelbusch, *Disenchanted Night*, 58.
106 Thomas P. Hughes, *Networks of Power: Electrification in Western Society, 1880–1930* (Baltimore: Johns Hopkins University Press, 1993), 72–3, 77, 183; Schott, *Europäische Urbanisierung (1000–2000)*, 289.
107 Dieter Schott, *Die Vernetzung der Stadt. Kommunale Energiepolitik, öffentlicher Nahverkehr und die 'Produktion' der modernen Stadt, Darmstadt - Mannheim - Mainz 1880-1918* (Darmstadt: Wissenschaftliche Buchgesellschaft, 1999), 709–10.
108 Zängl, *Deutschlands Strom*, 20, 25.
109 Schott, *Europäische Urbanisierung (1000–2000)*, 290.
110 Schott, *Vernetzung*, 729–30.
111 Beate Binder, *Elektrifizierung als Vision: Zur Symbolgeschichte einer Technik im Alltag* (Tübingen: Tübinger Vereinigung für Volkskunde, 1999), 168.
112 Ditt, 'Energiepolitik und Energiekonsum', 125.
113 Schott, *Europäische Urbanisierung (1000–2000)*, 290–1; Binder, *Elektrifizierung als Vision*, 177.
114 Schivelbusch, *Disenchanted Night*, 67; Zängl, *Deutschlands Strom*, 27.
115 Schott, *Europäische Urbanisierung (1000–2000)*, 293.
116 Hughes, *Networks of Power*, 465; Schott, *Vernetzung*, 22.
117 Ibid., 718–19.
118 Ibid., 26–7.
119 Ibid., 731.
120 Zängl, *Deutschlands Strom*, 46, 63; Schott, *Vernetzung*, 732.
121 Ulrich Wengenroth, 'Motoren für den Kleinbetrieb: Soziale Utopien, technische Entwicklung und Absatzstrategien bei der Motorisierung des Kleingewerbes im Kaiserreich', in *Prekäre Selbständigkeit: Zur Standortbestimmung von Handwerk, Hausindustrie und Kleingewerbe im Industrialisierungsprozess*, ed. Ulrich Wengenroth (Stuttgart: Franz Steiner, 1989), 195–8, 204.
122 Frank Dittmann, 'Innovations in the Electric Energy System', *Icon* 13 (2007): 77–8.

123 Vincent Lagendijk, 'Ideas, Individuals and Institutions: Notion and Practices of a European Electricity System', *Contemporary History* 27, no. 2 (2018): 204.
124 Dittmann, 'Innovations in the Electric Energy System', 65; Mats Fridlund and Helmut Maier, 'The Second Battle of the Currents: A Comparative Study of Engineering Nationalism in German and Swedish Electric Power, 1921–1961', *Working Papers from the Department of History of Science and Technology. Royal Institute of Technology, Stockholm* 96, no. 2 (1996): 6.
125 Lagendijk, 'Ideas, Individuals and Institutions', 204–6, 210–12, 214, 219.
126 Hughes, *Networks of Power*, 168; Norbert Gilson, *Konzepte von Elektrizitätsversorgung und Elektrizitätswirtschaft: Die Entstehung eines neuen Fachgebietes der Technikwissenschaften zwischen 1880 und 1945* (Stuttgart: GNT-Verlag, 1994).
127 Hughes, *Networks of Power*, 77, 182–4.
128 Schott, *Vernetzung*, 720.
129 Zängl, *Deutschlands Strom*, 46.
130 Binder, *Elektrifizierung als Vision*, 192–3; Andreas Killen, *Berlin Electropolis: Shock, Nerves, and German Modernity* (Berkeley: University of California Press, 2006), 26.
131 Binder, *Elektrifizierung als Vision*, 208.
132 Schott, *Vernetzung*, 711.
133 Zängl, *Deutschlands Strom*, 34–5; Schott, *Europäische Urbanisierung (1000–2000)*, 297–8; Lay, *Ways of the World*, 134.
134 Schmucki, 'The Machine in the City', 1072–3.
135 Schott, *Europäische Urbanisierung (1000–2000)*, 303.
136 Schmucki and Dienel, 'Aufstieg und Fall des öffentlichen Personennahverkehrs', 10–11.
137 Schott, 'Empowering European Cities', 172.
138 Schmucki, 'The Machine in the City', 1073–4.
139 Ladd, *Urban Planning and Civic Order in Germany, 1860–1914*, 206, 210.
140 Schott, *Vernetzung*, 723.
141 Raphael Emanuel Dorn, *Alle in Bewegung: Räumliche Mobilität in der Bundesrepublik Deutschland 1980–2010* (Göttingen: Vandenhoeck & Ruprecht, 2018), 300–2, 304; Witzler, *Großstadt und Hygiene*, 16.
142 Wolf, *Berlin – Weltstadt ohne Auto*, 32.
143 Schmucki, 'The Machine in the City', 1076; Schott, *Vernetzung*, 735; Schott, *Europäische Urbanisierung (1000–2000)*, 299; Ladd, *Urban Planning and Civic Order in Germany, 1860–1914*, 207.
144 Porombka, *Medialität urbaner Infrastrukturen*, 181–2.
145 Schott, *Europäische Urbanisierung (1000–2000)*, 302.
146 Schott, *Vernetzung*, 734.
147 Killen, *Berlin Electropolis*, 28.
148 Porombka, *Medialität urbaner Infrastrukturen*, 10–12, 21.
149 Wolf, *Berlin – Weltstadt ohne Auto*, 36.
150 Ziegler, *Eisenbahn und Staat im Zeitalter der Industrialisierung*, 381.
151 Porombka, *Medialität urbaner Infrastrukturen*, 163, 172–3, 185.
152 Ibid., 163.
153 Kurt Möser, *Geschichte des Autos* (Frankfurt: Campus, 2002), 115.
154 Anette Schlimm, 'Verkehrseinheit und ruinöser Wettbewerb: Der "Schiene-Straße-Konflikt" in Großbritannien und Deutschland als ein Problem des Social Engineering', *Geschichte und Gesellschaft* 39, no. 3 (2013): 344.

155 van Laak, 'Vom Lebensraum zum Leitungsweg', 147; Killen, *Berlin Electropolis*, 23.
156 Porombka, *Medialität urbaner Infrastrukturen*, 185–6.
157 Barbara Schmucki, 'Automobilisierung: Neuere Forschungen zur Motorisierung', *Archiv für Sozialgeschichte* 35 (1995): 592.
158 Carolyn Birdsall, *Nazi Soundscapes: Sound, Technology and Urban Space in Germany, 1933–1945* (Amsterdam: Amsterdam University Press, 2012), 39.
159 Schmucki and Dienel, 'Aufstieg und Fall des öffentlichen Personennahverkehrs', 12–13.
160 Barbara Schmucki, *Der Traum vom Verkehrsfluss: Städtische Verkehrsplanung seit 1945 im deutsch-deutschen Vergleich* (Frankfurt: Campus, 2001), 103, 402.
161 Schott, *Europäische Urbanisierung (1000–2000)*, 322.
162 Hans Kurella, 'Wohnung und Häuslichkeit', *Neue Deutsche Rundschau* 10 (1899): 819. (In: *German History in Documents and Images, Vol. 5. Wilhelmine Germany and the First World War, 1890–1918*), http://germanhistorydocs.ghi-dc.org/docpage.cfm?docpage_id=996
163 Christoph Bernhardt and Elsa Vonau, 'Zwischen Fordismus und Sozialreform: Rationalisierungsstrategien im deutschen und französischen Wohnungsbau 1900–1933', *Zeithistorische Forschungen/Studies in Contemporary History* 6, no. 2 (2009): 230.
164 Schott, *Vernetzung*, 716.
165 Schott, *Europäische Urbanisierung (1000–2000)*, 316, 320; Hård and Stippak, 'Progressive Dreams', 123–4.
166 Bernhardt and Vonau, 'Zwischen Fordismus und Sozialreform', 232–3.
167 Mark Peach, 'Wohnfords, or German Modern Architecture and the Appeal of Americanism', *Utopian Studies* 8, no. 2 (1997): 53.
168 Bernhardt and Vonau, 'Zwischen Fordismus und Sozialreform', 231–2.
169 Schott, *Europäische Urbanisierung (1000–2000)*, 325–6.
170 David Kuchenbuch, 'Architecture and Urban Planning as Social Engineering: Selective Transfers between Germany and Sweden in the 1930s and 1940s', *Journal of Contemporary History* 51, no. 1 (2016): 26–7.
171 Gerd Kuhn, *Wohnkultur und kommunale Wohnungspolitik in Frankfurt am Main 1880 bis 1930: Auf dem Wege zu einer pluralen Gesellschaft der Individuen* (Bonn: Dietz, 1998), 184.
172 Jeffry M. Diefendorf, *In the Wake of War: The Reconstruction of German Cities after World War II* (New York: Oxford University Press, 1993), 111–12.
173 Dietrich Andernacht, 'Fordistische Aspekte im Wohnungsbau des Neuen Frankfurt', in *Zukunft aus Amerika: Fordismus in der Zwischenkriegszeit; Siedlung, Stadt, Raum*, ed. Stiftung Bauhaus Dessau/Rheinisch-Westfälische Technische Hochschule Aachen (Dessau: Stiftung Bauhaus, 1995), 201.
174 Bernhardt and Vonau, 'Zwischen Fordismus und Sozialreform', 249–51.
175 Andernacht, 'Fordistische Aspekte im Wohnungsbau des Neuen Frankfurt', 195–6.
176 Gerhard Fehl, '"Eine Wohnung, gebaut wie ein Auto": Ford und die "Industrialisierung des Wohnungsbaus" im Nationalsozialismus', in *Zukunft aus Amerika: Fordismus in der Zwischenkriegszeit; Siedlung, Stadt, Raum*, ed. Stiftung Bauhaus Dessau/Rheinisch-Westfälische Technische Hochschule Aachen (Dessau: Stiftung Bauhaus, 1995), 252, 262–3.
177 Diefendorf, *In the Wake of War*, 111–12.
178 Andernacht, 'Fordistische Aspekte im Wohnungsbau des Neuen Frankfurt', 193, 197, 202–3.

179 Bernhardt and Vonau, 'Zwischen Fordismus und Sozialreform', 253; Leif Jerram, *Germany's Other Modernity: Munich and the Making of Metropolis, 1895–1930* (Manchester: Manchester University Press, 2007), 128.
180 Ladd, *The Ghosts of Berlin*, 104.
181 Diefendorf, *In the Wake of War*, 113.
182 Kuchenbuch, 'Architecture and Urban Planning as Social Engineering', 23, 28.
183 Ibid., 28; Fehl, '"Eine Wohnung, gebaut wie ein Auto"', 251.
184 Ibid., 256, 264–5.
185 Mark Hewitson, *Germany and the Modern World, 1880–1914* (Cambridge: Cambridge University Press, 2018), 227.
186 Zängl, *Deutschlands Strom*, 69, 72.
187 Schott, *Vernetzung*, 733.
188 Schott, 'Empowering European Cities', 178.
189 Hewitson, *Germany and the Modern World, 1880–1914*, 90.
190 Schott, 'Empowering European Cities', 171–2; Schott, *Europäische Urbanisierung (1000–2000)*, 286–7.
191 Schivelbusch, *Disenchanted Night*, 48–9, 63–4.
192 Bernard, *Die Geschichte des Fahrstuhls*, 100–1, 195–206, 212–13.
193 Michael Wildt, *Am Beginn der 'Konsumgesellschaft': Mangelerfahrung, Lebenshaltung, Wohlstandshoffnung in Westdeutschland in den fünfziger Jahren* (Hamburg: Ergebnisse, 1994), 143.
194 Sophie Gerber, '"We want to live electrically!" Marketing Strategies of German Power Companies in the 20th Century', in *Past and Present Energy Societies: New Energy Connects Politics, Technologies and Cultures*, ed. Nina Möllers and Karin Zachmann (Bielefeld: transcript, 2012): 85.
195 Wildt, *Am Beginn der 'Konsumgesellschaft'*, 143; Ditt, 'Energiepolitik und Energiekonsum', 144, 149; Schott, 'Empowering European Cities', 184; Hughes, *Networks of Power*, 189.
196 Kuhn, *Wohnkultur und kommunale Wohnungspolitik in Frankfurt am Main 1880 bis 1930*, 168–70; Wilhelm Füßl, *Oskar von Miller 1855–1934: Eine Biographie* (Munich: Beck, 2005), 199–200.
197 Kuhn, *Wohnkultur und kommunale Wohnungspolitik in Frankfurt am Main 1880 bis 1930*, 173, 181.
198 Bernhardt and Vonau, 'Zwischen Fordismus und Sozialreform', 249–51.
199 Kuhn, *Wohnkultur und kommunale Wohnungspolitik in Frankfurt am Main 1880 bis 1930*, 142–3.
200 Wildt, *Am Beginn der 'Konsumgesellschaft'*, 135–6; Kuhn, *Wohnkultur und kommunale Wohnungspolitik in Frankfurt am Main 1880 bis 1930*, 163.
201 Ibid., 150; Schott, 'Empowering European Cities', 184; Andernacht, 'Fordistische Aspekte im Wohnungsbau des Neuen Frankfurt', 199.
202 Kuhn, *Wohnkultur und kommunale Wohnungspolitik in Frankfurt am Main 1880 bis 1930*, 153, 157, 160, 163; Christine Frederick, *The New Housekeeping: Efficiency Studies in Home Management* (Garden City: Doubleday, Page & Co., 1914),
203 Jerram, *Germany's Other Modernity*, 135–6.
204 Wildt, *Am Beginn der 'Konsumgesellschaft'*, 134–5.
205 Jerram, *Germany's Other Modernity*, 128–31, 138.
206 Sophie Gerber, *Küche, Kühlschrank, Kilowatt: Zur Geschichte des privaten Energiekonsums in Deutschland 1945–1990* (Bielefeld: transcript: 2014), 31–2; Wildt, *Am Beginn der 'Konsumgesellschaft'*, 137–8.

207 Ibid., 145–6; Ditt, 'Energiepolitik und Energiekonsum', 148.
208 Martina Heßler, *'Mrs Modern Woman': Zur Sozial- und Kulturgeschichte der Haushalttechnisierung* (Frankfurt am Main: Campus Verlag, 2001), 381.
209 Cf. Herf, *Reactionary Modernism*.
210 Schott, *Europäische Urbanisierung (1000–2000)*, 249
211 Münch, *Stadthygiene im 19. und 20. Jahrhundert*, 41, 43–4.
212 Ibid., 46–50.
213 Kuchenbuch, 'Architecture and Urban Planning as Social Engineering', 39.
214 Schott, *Vernetzung*, 737.
215 Reulecke, *Geschichte der Urbanisierung*, 148, 154.
216 Diefendorf, *In the Wake of War*, 125–6.
217 Mary Fulbrook, *The People's State: East Germany from Hitler to Honecker* (New Haven: Yale University Press, 2005), 51.
218 Emily Pugh, *Architecture, Politics & Identity in Divided Berlin* (Pittsburgh: University of Pittsburgh Press, 2014), 31.
219 Diefendorf, *In the Wake of War*, 139, 141, 143.
220 Pugh, *Architecture, Politics & Identity in Divided Berlin*, 118.
221 Eli Rubin, 'Understanding a Car in the Context of a System: Trabants, Marzahn, and East German Socialism', in *The Socialist Car: Automobility in the Eastern Bloc*, ed. Lewis H. Siegelbaum (Ithaca: Cornell University Press, 2011), 138.
222 Fulbrook, *The People's State*, 62–3.
223 Stefan Wolle, *Die heile Welt der Diktatur: Alltag und Herrschaft in der DDR 1971–1989*, 2nd edn (Bonn: Bundeszentrale für politische Bildung, 1999), 185.
224 Pugh, *Architecture, Politics & Identity in Divided Berlin*, 118, 121–2, 124–5, 136.
225 Annette Kaminsky, *Wohlstand, Schönheit, Glück: Kleine Konsumgeschichte der DDR* (Munich: Beck, 2001), 123.
226 Pugh, *Architecture, Politics & Identity in Divided Berlin*, 294.
227 Rubin, 'Understanding a Car in the Context of a System', 125, 134–5; Eli Rubin, *Amnesiopolis: Modernity, Space, and Memory in East Germany* (Oxford: Oxford University Press, 2016).
228 Pugh, *Architecture, Politics & Identity in Divided Berlin*, 298.
229 Kaminsky, *Wohlstand, Schönheit, Glück*, 125.
230 Pugh, *Architecture, Politics & Identity in Divided Berlin*, 292, 297.
231 Ladd, *The Ghosts of Berlin*, 108.
232 Rubin, 'Understanding a Car in the Context of a System', 135–8.
233 Misa, *Leonardo to the Internet*, 267; Schott, *Europäische Urbanisierung (1000–2000)*, 329.
234 Christoph Maria Merki, *Der holprige Siegeszug des Automobils 1895–1930: Zur Motorisierung des Straßenverkehrs in Frankreich, Deutschland und der Schweiz* (Vienna: Böhlau, 2002), 68.
235 Ibid., 88–93.
236 Ibid., 429.
237 Schott, *Europäische Urbanisierung (1000–2000)*, 330.
238 Bernhard Rieger, *The People's Car: A Global History of the Volkswagen Beetle* (Cambridge: Harvard University Press, 2013).
239 Möser, *Geschichte des Autos*, 208.
240 Frank Steinbeck, *Das Motorrad: Ein deutscher Sonderweg in die automobile Gesellschaft* (Stuttgart: Franz Steiner, 2012), 9.
241 Schmucki, *Der Traum vom Verkehrsfluss*, 402.

242 Christoph Bernhardt, 'Längst beerdigt und doch quicklebendig: Zur widersprüchlichen Geschichte der "autogerechten Stadt"', *Zeithistorische Forschungen/ Studies in Contemporary History* 14, no. 3 (2017): 527–8.
243 Reinhold Bauer, 'Fahrrad, Auto, Stadt: Individualverkehr und städtischer Raum (nicht nur) in Stuttgart', *Technikgeschichte* 85, no. 4 (2018): 261–3; Schmucki, *Der Traum vom Verkehrsfluss*, 99.
244 Barbara Schmucki, 'Fashion and Technological Challenge: Tramways in Germany after 1945', *Journal of Transport History* 31, no. 1 (2010): 6–9; Schmucki, *Der Traum vom Verkehrsfluss*, 101.
245 Schmucki and Dienel, 'Aufstieg und Fall des öffentlichen Personennahverkehrs', 13–14.
246 Bernhardt, 'Längst beerdigt und doch quicklebendig', 530–4; Diefendorf, *In the Wake of War*, 206.
247 Bernhardt, 'Längst beerdigt und doch quicklebendig', 538; Schmucki, *Der Traum vom Verkehrsfluss*, 126, 154, 404, 407.
248 Möser, *Geschichte des Autos*, 99.
249 Schmucki, 'Automobilisierung', 595–6.
250 Schmucki, *Der Traum vom Verkehrsfluss*, 82, 154.
251 Burghard Ciesla, 'Öffentlicher Nahverkehr in einer geteilten Stadt: Grundzüge der Entwicklung in Berlin von 1945 bis 1990', in *Mobilität für alle: Geschichte des öffentlichen Personennahverkehrs in der Stadt zwischen technischem Fortschritt und sozialer Pflicht*, ed. Hans-Liudger Dienel and Barbara Schmucki (Stuttgart: Steiner, 1997), 146.
252 Schmucki, *Der Traum vom Verkehrsfluss*, 60.
253 Ibid., 401; Bernhardt, 'Längst beerdigt und doch quicklebendig', 536.
254 Schmucki, *Der Traum vom Verkehrsfluss*, 134.
255 Schmucki and Dienel, 'Aufstieg und Fall des öffentlichen Personennahverkehrs', 15; Schmucki, 'Fashion and Technological Challenge', 13–16; Schmucki, *Der Traum vom Verkehrsfluss*, 408.
256 Dorn, *Alle in Bewegung*, 69.
257 Heike Weber, 'Towards "Total" Recycling: Women, Waste and Food; Waste Recovery in Germany, 1914-1939', *Contemporary European History* 22, no. 3 (2013): 375.
258 Ruth Oldenziel and Heike Weber, 'Introduction: Reconsidering Recycling', *Contemporary European History* 22, no. 3 (2013): 358–9.
259 Münch, *Stadthygiene im 19. und 20. Jahrhundert*, 52–3.
260 Ibid., 33, 54–6; Stokes, Köster and Sambrook, *The Business of Waste*, 82.
261 Witzler, *Großstadt und Hygiene*, 67.
262 Stokes, Köster and Sambrook, *The Business of Waste,* 23–4, 26; Roman Köster, 'Waste to Assets: How Household Waste Recycling Evolved in West Germany', in *Cycling and Recycling: Histories of Sustainable Practices*, ed. Ruth Oldenziel and Helmuth Trischler (New York: Berghahn, 2016), 169.
263 Münch, *Stadthygiene im 19. und 20. Jahrhundert*, 50–2; Stokes, Köster and Sambrook, *The Business of Waste*, 31–3, 299; Sorge and Streeck, 'Diversified Quality Production Revisited', 599.
264 Stokes, Köster and Sambrook, *The Business of Waste*, 210.
265 Ibid., 80, 85.
266 Roman Köster, 'Abschied von der "verlorenen Verpackung": Das Recycling von Hausmüll in Westdeutschland 1945–1990', *Technikgeschichte* 81, no. 1 (2014): 39–40; Köster, 'Waste to Assets', 170.

267 Andrea Westermann, 'When Consumer Citizens Spoke Up: West Germany's Early Dealings with Plastic Waste', *Contemporary European History* 22, no. 3 (2013): 377–498.
268 Köster, 'Waste to Assets', 169.
269 Christian Möller, 'Der Traum vom ewigen Kreislauf: Abprodukte, Sekundärrohstoffe und Stoffkreisläufe im "Abfall-Regime" der DDR (1945-1990)', *Technikgeschichte* 81, no. 1 (2014): 76–8.
270 Astrid M. Eckert, 'Geteilt, aber nicht unverbunden: Grenzgewässer als deutsch-deutsches Umweltproblem', *Vierteljahrshefte für Zeitgeschichte* 62, no. 1 (2014): 94–6; Stokes, Köster and Sambrook, *The Business of Waste*, 304.
271 Münch, *Stadthygiene im 19. und 20. Jahrhundert*, 52–3, 113.
272 Köster, 'Waste to Assets', 177.
273 Djahane Salehabadi, 'The Scramble for Digital Waste in Berlin', in *Cycling and Recycling: Histories of Sustainable Practices*, ed. Ruth Oldenziel and Helmuth Trischler (New York: Berghahn, 2016), 204, 207.
274 Horst A. Wessel, 'Die Verbreitung des Telephons bis zur Gegenwart', in *Vom Flügeltelegraphen zum Internet: Geschichte der modernen Telekommunikation*, ed. Hans-Jürgen Teuteburg and Cornelius Neutsch (Stuttgart: Franz Steiner, 1998), 68; Frank Thomas, 'The Politics of Growth: The German Telephone System', in *The Development of Large Technical Systems*, ed. Renate Mayntz and Thomas P. Hughes (Frankfurt: Campus, 1988), 181.
275 Ibid., 184–6.
276 Wessel, 'Die Verbreitung des Telephons bis zur Gegenwart', 69.
277 Thomas, *Telefonieren in Deutschland*, 370–1.
278 Ibid., 121.
279 Thomas, 'The Politics of Growth', 186.
280 Wessel, 'Die Verbreitung des Telephons bis zur Gegenwart', 67.
281 Thomas, 'The Politics of Growth', 185.

Chapter 3

1 Helmuth Trischler, '"Big Science" or "Small Science"? Die Luftfahrtforschung im Nationalsozialismus', in *Geschichte der Kaiser-Wilhelm-Gesellschaft im Nationalsozialismus: Bestandsaufnahme und Perspektiven der Forschung, vol. 1*, ed. Doris Kaufmann (Göttingen: Wallstein, 2000), 328–62.
2 Helmuth Trischler, *Luft- und Raumfahrtforschung in Deutschland 1900–1970: Politische Geschichte einer Wissenschaft* (Frankfurt: Campus, 1992), 38.
3 Peter Fritzsche, *A Nation of Fliers: German Aviation and the Popular Imagination* (Cambridge: Harvard University Press, 1992), 7.
4 Trischler, *Luft- und Raumfahrtforschung*, 38–9.
5 Ibid., 43. Thomas Hughes defines the difference between invention and innovation as follows: 'This invention is then brought by research and development to the stage at which it can be introduced to the market . . . The invention is no longer an imaginary device functioning in the inventor's mind. It is important to add that the innovation process is not straightforward; it involves backtracking to identify new subproblems, elicit additional ideas, and make new subinventions.' Thomas P. Hughes, *Networks of Power: Electrification in Western Society, 1880–1930* (Baltimore: Johns Hopkins University Press, 1983), 19–20.

6 Guillaume de Syon, *Zeppelin! Germany and the Airship, 1900–1939* (Baltimore: Johns Hopkins University Press, 2001), 39.
7 Fritzsche, *A Nation of Fliers*, 10–1; de Syon, *Zeppelin*, 40–1, 49.
8 Fritzsche, *A Nation of Fliers*, 2, 21–2.
9 Ibid., 13–14, 16.
10 de Syon, *Zeppelin*, 63.
11 Fritzsche, *A Nation of Fliers*, 25–6.
12 de Syon, *Zeppelin*, 16, 66.
13 Trischler, *Luft- und Raumfahrtforschung*, 44.
14 Fritzsche, *A Nation of Fliers*, 18.
15 de Syon, *Zeppelin*, 60–3.
16 Fritzsche, *A Nation of Fliers*, 2, 43–54; quotation on p. 54; de Syon, *Zeppelin*, 82, 109.
17 Fritzsche, *A Nation of Fliers*, 58.
18 Ibid., 138, 143.
19 Bernhard Rieger, *Technology and the Culture of Modernity in Britain and Germany, 1890–1945* (Cambridge: Cambridge University Press, 2005), 250.
20 Helmuth Trischler and Robert Bud, 'Public technology: Nuclear Energy in Europe', *History and Technology* 34, no. 3–4 (2018): 7.
21 Fritzsche, *A Nation of Fliers*, 2, 21–2.
22 Lutz Budraß, *Flugzeugindustrie und Luftrüstung in Deutschland 1918–1945* (Düsseldorf: Droste, 1998), 21, 27–8.
23 Ibid., 34–7.
24 Helmuth Trischler, 'Historische Wurzeln der Großforschung: Die Luftfahrtforschung vor 1945', in *Großforschung in Deutschland*, ed. Margit Szöllösi-Janze and Helmuth Trischler (Frankfurt: Campus, 1990), 25, 29–30; Budraß, *Flugzeugindustrie*, 28–9.
25 Michael Eckert, 'Strategic Internationalism and the Transfer of Technical Knowledge: The United States, Germany, and Aerodynamics after World War I', *Technology and Culture* 46, no. 1 (2005), 107, 120, 128.
26 Budraß, *Flugzeugindustrie*, 50–2, 55–6.
27 Fritzsche, *A Nation of Fliers*, 106–7.
28 Rieger, *Technology and the Culture of Modernity*, 245–6.
29 Budraß, *Flugzeugindustrie*, 62–5.
30 Fritzsche, *A Nation of Fliers*, 103, 107, 126.
31 Ibid., 134, 146, 177–8.
32 Ibid., 134–5, 146–7, 150, 152.
33 Ibid., 137.
34 Budraß, *Flugzeugindustrie*, 15, 889.
35 Helmuth Trischler, '"Big Science" or "Small Science"? Die Luftfahrtforschung im Nationalsozialismus', in *Geschichte der Kaiser-Wilhelm-Gesellschaft im Nationalsozialismus: Bestandsaufnahme und Perspektiven der Forschung*, ed. Doris Kaufman, vol. 1 (Göttingen: Wallstein, 2000), 340.
36 Trischler, 'Historische Wurzeln der Großforschung', 31–2; Trischler, *Luft- und Raumfahrtforschung*, 262.
37 Trischler, '"Big Science" or "Small Science"', 346–8.
38 Trischler, 'Historische Wurzeln der Großforschung', 33–5.
39 Trischler, '"Big Science" or "Small Science"', 351–2, 361.
40 Ibid., 343–4.
41 Adam Tooze, *The Wages of Destruction: The Making and Breaking of the Nazi Economy* (London: Allen Lane, 2006), 556.

42 Trischler, '"Big Science" or "Small Science"', 345.
43 Johannes Bär, Paul Erker and Geoffrey Giles, 'The Politics of Ambiguity: Reparations, Business Relations, Denazification and the Allied Transfer of Technology', in *Technology Transfer Out of Germany after 1945*, ed. Matthias Judt and Burghard Ciesla (Amsterdam: Harwood, 1996), 137–8.
44 Budraß, *Flugzeugindustrie*, 884–5.
45 Hermione Giffard, *Making Jet Engines in World War II: Britain, Germany, and the United States* (Chicago: University of Chicago Press, 2016), 41, 60.
46 Lutz Budraß, '"The hun is not always ahead of us in secret weapons": Some remarks on a new book on the history of the turbojet', *Technikgeschichte* 85, no. 3 (2018): 183.
47 Giffard, *Making Jet Engines*, 2, 43, 55, 61, 66–7.
48 Budraß, '"The hun is not always ahead of us in secret weapons"', 180.
49 Hermione Giffard, 'Response on Review: What is technical success?', *Technikgeschichte* 85, no. 3 (2018): 190, 193.
50 Giffard, *Making Jet Engines*, 50–1, 57–60.
51 Jens-Christian Wagner, *Produktion des Todes: Das KZ Mittelbau-Dora* (Göttingen: Wallstein, 2001), 227. – Giffard errs: the jet engines were not produced by concentration camp slave labour, cf. Giffard, 'Response on Review', 190.
52 Burghard Ciesla, 'German High Velocity Aerodynamics and Their Significance for the US Air Force 1945–52', in *Technology Transfer Out of Germany after 1945*, ed. Matthias Judt and Burghard Ciesla (Amsterdam: Harwood, 1996), 93–5.
53 Michael J. Neufeld, 'The Nazi Aerospace Exodus: Towards a global, transnational history', *History and Technology* 28, no. 1 (2012): 60.
54 Bär, Erker and Giles, 'The Politics of Ambiguity', 138.
55 Helmuth Trischler, 'Planungseuphorie und Forschungssteuerung in den 1960er Jahren in der Luft- und Raumfahrtforschung', in *Großforschung in Deutschland*, ed. Margit Szöllösi-Janze and Helmuth Trischler (Frankfurt: Campus, 1990), 120.
56 Neufeld, 'The Nazi Aerospace Exodus', 55.
57 Wolfgang Krieger, 'Technologiepolitik der Bundesrepublik Deutschland (1949–1989/90)', in *Technik und Staat: Technik und Kultur, vol. 9*, ed. Armin Hermann and Hans-Peter Sang (Berlin: Springer, 1992), 241–2, 249.
58 Neufeld, 'The Nazi Aerospace Exodus', 58.
59 Jonathan Hagood, 'Arming and Industrializing Perón's "New Argentina": The transfer of German Scientists and Technology after World War II', *Icon* 11 (2005): 63–6.
60 André Steiner, 'The Return of German "Specialists" from the Soviet Union to the German Democratic Republic: Integration and Impact', in *Technology Transfer Out of Germany after 1945*, ed. Matthias Judt and Burghard Ciesla (Amsterdam: Harwood, 1996), 126.
61 Dolores L. Augustine, *Red Prometheus: Engineering and Dictatorship in East Germany, 1945–1990* (Cambridge: MIT Press, 2007), 119.
62 Daniel Brandau, 'Cultivating the Cosmos: Spaceflight thought in imperial Germany', *History and Technology* 28, no. 3 (2012): 225–6, 242, 245–6.
63 Michael J. Neufeld, *The Rocket and the Reich: Peenemünde and the Coming of the Ballistic Missile Era* (Cambridge: Harvard University Press, 1995), 9.
64 Ibid., 5–7.
65 Daniel Brandau, 'Die Plausibilität des Fortschritts: Deutsche Raumfahrtvorstellungen im Jahre 1928', in *Technology Fiction: Technische Visionen und Utopien in der Hochmoderne*, ed. Uwe Fraunholz and Anke Woschech (Bielefeld: transcript, 2012), 66–68, 73–4; Michael B. Petersen, *Missiles for the Fatherland: Peenemünde, National*

Socialism, and the V-2 Missile (Cambridge: Cambridge University Press, 2009), 16–23.
66 Neufeld, *The Rocket and the Reich*, 8; Brandau, 'Cultivating the Cosmos', 234.
67 Petersen, *Missiles for the Fatherland*, 19.
68 Brandau, 'Die Plausibilität des Fortschritts', 75, 80.
69 Petersen, *Missiles for the Fatherland*, 23.
70 Brandau, 'Die Plausibilität des Fortschritts', 69; Petersen, *Missiles for the Fatherland*, 20.
71 Neufeld, *The Rocket and the Reich*, 8, 10–16, 19–24, 38–9.
72 Petersen, *Missiles for the Fatherland*, 1–2.
73 Michael Thad Allen, *The Business of Genocide: The SS, Slave Labor, and the Concentration Camps* (Chapel Hill: University of North Carolina Press, 2002), 208.
74 Allen, *The Business of Genocide*, 214–18; Wagner, *Produktion des Todes*, 186–7.
75 Ibid.
76 Petersen, *Missiles for the Fatherland*, 2.
77 Wagner, *Produktion des Todes*, 288, 367.
78 Allen, *The Business of Genocide*, 228–9.
79 Ibid., 229–30.
80 Ibid., 209.
81 Neufeld, *The Rocket and the Reich*, 167–96.
82 Allen, *The Business of Genocide*, 209–11.
83 Wagner, *Produktion des Todes*, 218–19.
84 Ibid., 110, 203, 225.
85 Bär, Erker and Giles, 'The Politics of Ambiguity', 135–7.
86 Neufeld, 'The Nazi Aerospace Exodus', 57.
87 Petersen, *Missiles for the Fatherland*, 254–5.
88 Tom Lehrer, 'Wernher von Braun', in *That Was The Year That Was* (San Francisco: Reprise Records, 1965).
89 Wagner, *Produktion des Todes*, 671.
90 Asif A. Siddiqi, 'Germans in Russia: Cold war, technology transfer, and national identity', *Osiris* 24 (2009): 123.
91 Wagner, *Produktion des Todes*, 287.
92 Siddiqi, 'Germans in Russia', 126; cf. Matthias Uhl, *Stalins V-2: Der Technologietransfer der deutschen Fernlenkwaffentechnik in die UdSSR und der Aufbau der sowjetischen Raketenindustrie 1945 bis 1959* (Bonn: Bernard und Graefe, 2001).
93 Augustine, *Red Prometheus*, 2.
94 Siddiqi, 'Germans in Russia', 127–8.
95 Augustine, *Red Prometheus*, 2.
96 Siddiqi, 'Germans in Russia', 133, 135; Bär, Erker and Giles, 'The Politics of Ambiguity', 136.
97 Johannes Weyer, 'Verstärkte Rivalitäten statt Rendevous im All? Die wechselhafte Geschichte der deutsch-amerikanischen Zusammenarbeit in der Raumfahrt', in *Technische Visionen – politische Kompromisse: Geschichte und Perspektiven der deutschen Raumfahrt*, ed. Johannes Weyer (Berlin: Edition Sigma, 1993), 93; Neufeld, 'The Nazi Aerospace Exodus', 57.
98 Daniel Brandau, *Raketenträume: Raumfahrt- und Technikenthusiasmus in Deutschland, 1923–1963* (Stuttgart: Schöningh, 2019), 242, 244–5, 282.
99 Augustine, *Red Prometheus*, 208–9.
100 Krieger, 'Technologiepolitik der Bundesrepublik Deutschland', 242–3.

101 Johannes Weyer, 'Erfolgreiches Scheitern und nicht-gewollte Erfolge in der Geschichte der westdeutschen Raumfahrt 1945–1965', in *Technische Visionen – politische Kompromisse: Geschichte und Perspektiven der deutschen Raumfahrt*, ed. Johannes Weyer (Berlin: Edition Sigma, 1993), 11–12.
102 Trischler, 'Die bundesdeutsche Raumfahrt der 60er Jahre', 60–1.
103 Weyer, 'Verstärkte Rivalitäten statt Rendevous im All?', 93–5.
104 Weyer, 'Erfolgreiches Scheitern und nicht-gewollte Erfolge', 12–13, 15–16.
105 Helmuth Trischler, 'Die bundesdeutsche Raumfahrt der 60er Jahre: Forschungs- und technologiepolitische Weichenstellungen', in *Technische Visionen – politische Kompromisse: Geschichte und Perspektiven der deutschen Raumfahrt*, ed. Johannes Weyer (Berlin: Edition Sigma, 1993), 62–3; Trischler, *Luft- und Raumfahrtforschung in Deutschland 1900–1970*, 395, 399, 426; Helmuth Trischler, 'Planungseuphorie und Forschungssteuerung in den 1960er Jahren in der Luft- und Raumfahrtforschung', in *Großforschung in Deutschland* ed. Margit Szöllösi-Janze and Helmuth Trischler (Frankfurt: Campus, 1990), 124.
106 Trischler, 'Die bundesdeutsche Raumfahrt der 60er Jahre', 61–2.
107 Weyer, 'Erfolgreiches Scheitern und nicht-gewollte Erfolge', 17.
108 Trischler, 'Luft- und Raumfahrtforschung in Deutschland 1900–1970', 426–30.
109 Weyer, 'Erfolgreiches Scheitern und nicht-gewollte Erfolge', 19.
110 Trischler, 'Die bundesdeutsche Raumfahrt der 60er Jahre', 67.
111 Weyer, 'Verstärkte Rivalitäten statt Rendevous im All?', 98.
112 Ibid., 89–90, 96; Krieger, 'Technologiepolitik der Bundesrepublik Deutschland', 243–4.
113 Trischler, 'Die bundesdeutsche Raumfahrt der 60er Jahre', 64.
114 Weyer, 'Verstärkte Rivalitäten statt Rendevous im All?', 105–7.
115 Mark Walker, 'A Comparative History of Nuclear Weapons', in *Geschichte der Kaiser-Wilhelm-Gesellschaft im Nationalsozialismus: Bestandsaufnahme und Perspektiven der Forschung*, ed. Doris Kaufman, vol. 1 (Göttingen: Wallstein, 2000), 312.
116 Ibid., 311.
117 Trischler, '"Big Science" or "Small Science"?', 329.
118 Mark Walker, 'A Comparative History of Nuclear Weapons', 324.
119 Walter E. Grunden, Mark Walker and Masakatsu Yamazaki, 'Wartime nuclear weapons research in Germany and Japan', *Osiris* 20 (2005): 108.
120 Anonymous [Johannes Stark], '"White Jews" in Science [July 15, 1937]', in *Physics and National Socialism: An Anthology of Primary Sources*, ed. Klaus Hentschel (Basel: Birkhäuser, 1996), 152–7. This anonymous article was followed by a comment of Stark. It is obvious that he authored both.
121 Grunden, Walker and Yamazaki, 'Wartime Nuclear Weapons Research', 109.
122 Mark Walker, *German National Socialism and the Quest for Nuclear Power, 1939–1949* (Cambridge: Cambridge University Press, 1989), 14–15, 17–19.
123 Ibid., 21, 23–4, 29.
124 Thomas Powers, *Heisenberg's War: The Secret History of the German Bomb* (New York: Alfred A. Knopf, 1993), VII.
125 Walker, *German National Socialism and the Quest for Nuclear Power*, 27, 34, 36, 40–5.
126 Walker, 'A Comparative History of Nuclear Weapons', 317; Grunden, Walker and Yamazaki, 'Wartime nuclear weapons research', 113–14.
127 Paul Lawrence Rose, *Heisenberg and the Nazi Atomic Bomb Project: A Study in German Culture* (Berkeley: University of California Press, 1998), 4.

128 Powers, *Heisenberg's War*, 481–2; Walker rejects both Rose's and Powers' arguments, cf. Walker, 'A Comparative History of Nuclear Weapons', 313.
129 Grunden, Walker and Yamazaki, 'Wartime nuclear weapons research', 115.
130 Walker, 'A comparative history of nuclear weapons', 314, 324.
131 Ibid., 322.
132 Grunden, Walker and Yamazaki, 'Wartime nuclear weapons research', 114.
133 Neufeld, 'The Nazi Aerospace Exodus', 49.
134 Augustine, *Red Prometheus*, 9.
135 Hagood, 'Arming and Industrializing Perón's "New Argentina"', 70–6.
136 Joachim Radkau and Lothar Hahn, *Aufstieg und Fall der deutschen Atomwirtschaft* (Munich: oekom, 2013), 30–6.
137 Cathryn Carson, 'Old Programs, New Politics? Nuclear Reactor Studies after 1945 in the Max-Planck-Instiut für Physik', in *Geschichte der Kaiser-Wilhelm-Gesellschaft im Nationalsozialismus: Bestandsaufnahme und Perspektiven der Forschung*, vol. 2, ed. Doris Kaufmann (Göttingen: Wallstein, 2000), 738.
138 Szöllösi-Janze and Trischler, 'Einleitung', 14–16.
139 Tilmann Hanel and Mikael Hård, 'Inventing traditions: Interests, parables and Nostalgia in the history of nuclear energy', *History and Technology* 31, no. 2 (2015): 91–3; Tilmann Hanel, *Die Bombe als Option: Motive für den Aufbau einer atomtechnischen Infrastruktur in der Bundesrepublik bis 1963* (Essen: Klartext, 2015).
140 Radkau and Hahn, *Aufstieg und Fall der deutschen Atomwirtschaft*, 37.
141 Carson, 'Old programs, new politics?', 735.
142 Hanel and Hård, 'Inventing traditions', 98, 100.
143 Radkau and Hahn, *Aufstieg und Fall der deutschen Atomwirtschaft*, 119–20; Carson, 'Old programs, new politics?', 732.
144 Radkau and Hahn, *Aufstieg und Fall der deutschen Atomwirtschaft*, 13–14, 120–2.
145 Grunden, Walker and Yamazaki, 'Wartime nuclear weapons research', 124.
146 Radkau and Hahn, *Aufstieg und Fall der deutschen Atomwirtschaft*, 13.
147 Ibid., 37.
148 Luciene Fernandes Justo and Gildo Magalhães dos Santos, 'The Otto Hahn Nuclear Ship and the German-Brazilian deals on nuclear energy: A case study in big science', *Icon* 6 (2000): 22, 33, 41, 43.
149 Radkau and Hahn, *Aufstieg und Fall der deutschen Atomwirtschaft*, 49–50.
150 Dolores L. Augustine, *Taking on Technocracy: Nuclear Power in Germany, 1945 to the Present* (New York: Berghahn, 2018), 52.
151 Radkau and Hahn, *Aufstieg und Fall der deutschen Atomwirtschaft*, 115.
152 Augustine, *Taking on Technocracy*, 29–31, 41.
153 Radkau and Hahn, *Aufstieg und Fall der deutschen Atomwirtschaft*, 123.
154 Hanel and Hård, 'Inventing traditions', 94.
155 Augustine, *Taking on Technocracy*, 55.
156 Michael Schüring, 'Advertising the nuclear venture: The Rhetorical and visual public relation strategies of the German Nuclear Industry in the 1970s and 1980s', *History and Technology* 29, no. 4 (2013): 369, 371–2. – The resulting protests against nuclear technology will be discussed in chapter 8.
157 Augustine, *Taking on Technocracy*, 6.
158 Langdon Winner, 'Do artefacts have politics?', in *The Social Shaping of Technology*, ed. Donald A. MacKenzie and Judy Wajcman, 2nd edn (Maidenhaid: Open University Press, 2003), 36.
159 Augustine, *Taking on Technocracy*, 12, 55–6; Augustine, *Red Prometheus*, 119.

160 Augustine, *Taking on Technocracy*, 56–7.
161 Andrew S. Tompkins, *Better Active than Radioactive! Anti-Nuclear Protest in 1970s France and West Germany* (Oxford: Oxford University Press, 2016), 196–7; Augustine, Taking on Technocracy, 3; Radkau and Hahn, *Aufstieg und Fall der deutschen Atomwirtschaft*, 389.
162 Rieger, *Technology and the Culture of Modernity in Britain and Germany*, 246.
163 Joachim Radkau, *Technik in Deutschland. Vom 18. Jahrhundert bis heute* (Frankfurt: Campus, 2008), 340.

Chapter 4

1 Bernhard Rieger, *Technology and the Culture of Modernity in Britain and Germany, 1890–1945* (Cambridge: Cambridge University Press, 2005), 42.
2 Robert Friedel, *A Culture of Improvement: Technology and the Western Millennium* (Cambridge: MIT Press, 2007), 2–4, 10.
3 Max Weber, 'Science as a Vocation', *Daedalus* 87, no. 1 (1958): 116–17.
4 Rieger, *Technology and the Culture of Modernity*, 4, 20, 33.
5 Ibid., 21–5.
6 Alexander Gerschenkron, *Economic Backwardness in Historical Perspective: A Book of Essays* (Cambridge: Belknap Press, 1962), 25.
7 Joyce Appleby, *The Relentless Revolution: A History of Capitalism* (New York/London: Norton, 2010), 171.
8 Eric Dorn Brose, *The Politics of Technological Change in Prussia. Out of the Shadow of Antiquity, 1809–1848* (Princeton: Princeton University Press, 1993), 132.
9 Thomas Großbölting, *'Im Reich der Arbeit'. Die Repräsentation gesellschaftlicher Ordnung in den deutschen Industrie- und Gewerbeausstellungen 1790–1914* (Munich: Oldenbourg, 2008), 123, 138–9, 413.
10 Ibid., 384, 401, 411, 429.
11 Andrew Ure, *The Philosophy of Manufactures: or, an Exposition of the Scientific, Moral, and Commerical Economy of the Factory System in Great Britain*. 2nd ed. (London: Charles Knight, 1835), 18.
12 Karl Marx, *The Capital. A Critique of Political Economy*, vol. 1 (New York: Random House, 1906), 416.
13 Ibid., 483.
14 Ibid., 468.
15 Ibid., 482.
16 Hans-Albert Wulf, *'Maschinenstürmer sind wir keine': Technischer Fortschritt und sozialdemokratische Arbeiterbewegung* (Frankfurt: Campus, 1987), 54.
17 Jeffrey Allan Johnson, *The Kaiser's Chemists: Science and Modernization in Imperial Germany* (Chapel Hill: University of North Carolina Press, 1990), 14.
18 Rudolf Boch, *Handwerker-Sozialisten gegen Fabrikgesellschaft: Lokale Fachvereine, Massengewerkschaft und industrielle Rationalisierung in Solingen 1870 bis 1914* (Göttingen: Vandenhoeck & Ruprecht, 1985), 169, 172–6.
19 Thomas Rohkrämer, *Eine andere Moderne? Zivilisationskritik, Natur und Technik in Deutschland 1880–1933* (Paderborn: Schöningh, 1999), 39, 51–2.
20 Thomas Hänseroth, 'Technischer Fortschritt als Heilsversprechen und seine selbstlosen Bürgen: Zur Konstituierung einer Pathosformel der technischen

Hochmoderne in Deutschland', in *Transzendenz und die Konstitution von Ordnungen*, ed. Hans Vorländer (Berlin: de Gruyter, 2013), 272–3.

21 Dieter Schott, 'Empowering European Cities: Gas and Electricity in the Urban Environment', in *Urban Machinery: Inside Modern European Cities*, ed. Mikael Hård and Thomas J. Misa (Cambridge: MIT Press, 2008), 165.

22 David E. Nye, *The American Technological Sublime* (Cambridge: MIT Press, 1994), 146.

23 Mark Hewitson, *Germany and the Modern World, 1880–1914* (Cambridge: Cambridge University Press, 2018), 89

24 Rohkrämer, *Eine andere Moderne*, 55–6.

25 Hewitson, *Germany and the Modern World*, 84.

26 Heidi J.S. Tworek, 'How not to build a world wireless network: German-British rivalry and visions of global communications in the early twentieth century', *History and Technology* 32, no. 2 (2016): 184.

27 Fernando Esposito, *Mythische Moderne: Aviatik, Faschismus und die Sehnsucht nach Ordnung in Deutschland und Italien* (Munich: Oldenbourg, 2011), 255–6.

28 Tanja Paulitz, *Mann und Maschine: Eine genealogische Wissenssoziologie des Ingenieurs und der modernen Technikwissenschaften, 1850–1930* (Bielefeld: Transcript, 2012), 129.

29 Rohkrämer, *Eine andere Moderne*, 61–3.

30 Hans-Liudger Dienel, 'Zweckoptimismus und -pessimismus der Ingenieure um 1900', in *Der Optimismus der Ingenieure: Triumph der Technik in der Krise der Moderne um 1900*, ed. Hans-Liudger Dienel (Stuttgart: Steiner, 1998), 18.

31 Hänseroth, 'Technischer Fortschritt als Heilsversprechen', 279; Paulitz, *Mann und Maschine*, 257–8.

32 Hänseroth, 'Technischer Fortschritt als Heilsversprechen', 277–8.

33 Rieger, *Technology and the Culture of Modernity*, 277; *Hänseroth*, 'Technischer Fortschritt als Heilsversprechen', 279.

34 Sabine Höhler, *Luftfahrtforschung und Luftfahrtmythos: Wissenschaftliche Ballonfahr in Deutschland 1880–1914* (Frankfurt: Campus, 2001), 115–36.

35 Rohkrämer, *Eine andere Moderne*, 344–52.

36 Edward Ross Dickinson, 'Biopolitics, Fascism, Democracy: Some Reflections on Our Discourse About Modernity', *Central European History* 37, no. 1 (2004): 2, 37, 47.

37 Peter Fritzsche, 'Nazi Modern', *Modernism/modernity* 3, no. 1 (1996): 12.

38 Ibid, 14; Dirk van Laak, *Imperiale Infrastruktur. Deutsche Planungen für eine Erschließung Afrikas 1880 bis 1960* (Paderborn: Schöningh, 2004), 245.

39 Rieger, *Technology and the Culture of Modernity*, 253–4.

40 Nye, *The American Technological Sublime*, 33.

41 Rieger, *Technology and the Culture of Modernity*, 46.

42 van Laak, *Imperiale Infrastruktur*, 231.

43 Jeffrey Herf, *Reactionary Modernism: Technology, Culture, and Politics in Weimar and the Third Reich*, repr. edn (Cambridge: Cambridge University Press, 2003), 152.

44 Dickinson, 'Biopolitics, Fascism, Democracy', 36.

45 Jeffrey Herf, '"Reactionary Modernism" and After: Modernity and Nazi Germany Reconsidered', in *Geschichte der Kaiser-Wilhelm-Gesellschaft im Nationalsozialismus. Bestandsaufnahme und Perspektiven der Forschung*, ed. Doris Kaufman, vol. 1 (Göttingen: Wallstein, 2000), 65.

46 Dickinson, 'Biopolitics, Fascism, Democracy', 5.

47 Ibid., 2.

48 Hanns Günther, *Automaten: Die Befreiung des Menschen durch die Maschine*, 3rd edn (Stuttgart: Dieck & Co., 1930), 5. — The author's real name was Walter de Haas.
49 Mary Nolan, *Visions of Modernity: American Business and the Modernization of Germany* (New York: Oxford University Press, 1994), 10.
50 Herf, *Reactionary Modernism*, 224–5.
51 Fritzsche, 'Nazi Modern', 14.
52 Oswald Spengler, *Man and Technics: A Contribution to a Philosophy of Life* (New York: Alfred A. Knopf, 1963), 87–8.
53 Ibid., 12.
54 Fritzsche, 'Nazi Modern', 16.
55 Spengler, *Man and Technics*, 13–4, 73.
56 Ibid., 90.
57 Ibid., 100, 103.
58 Ibid., 99–102; see van Laak, *Imperiale Infrastruktur*, 232.
59 Nye, *The American Technological Sublime*, 158.
60 Fritzsche, 'Nazi Modern', 16.
61 Rieger, *Technology and the Culture of Modernity*, 48–9.
62 John C. Guse, 'Nazi Technological Thought Revisited', *History and Technology* 26, no. 1 (2010): 6.
63 Peter Fritzsche, 'Nazi Modern', 2.
64 Rohkrämer, *Eine andere Moderne*, 349.
65 Andrew Denning, '"Life Is Movement, Movement Is Life!" Mobility Politics and the Circulatory State in Nazi Germany', *American Historical Review* 123, no. 5 (2018): 1480, 1498–9.
66 Helmut Stein, 'Menschenleere Fabriken', in *Völkischer Beobachter*, 27 February 1944.
67 Achim Eberspächer, *Das Projekt Futurologie: Über Zukunft und Fortschritt in der Bundesrepublik Deutschland 1952–1982* (Stuttgart: Ferdinand Schöningh, 2019), 62–70.
68 Rieger, *Technology and the Culture of Modernity*, 285.
69 Rüdiger Graf, 'Totgesagt und nicht gestorben. Die Persistenz des Fortschritts im 20. und 21. Jahrhundert', *Traverse* 23, no. 3 (2016): 93.
70 Dolores L. Augustine, *Taking on Technocracy: Nuclear Power in Germany, 1945 to the Present* (New York: Berghahn, 2018), 1; see Joachim Radkau, *Technik in Deutschland: Vom 18. Jahrhundert bis heute* (Frankfurt: Campus, 2008), 362.
71 Augustine, *Taking on Technocracy*, 29.
72 Franz-Josef Brüggemeier, *Grubengold: Das Zeitalter der Kohle von 1750 bis heute* (Munich: Beck, 2018), 359.
73 Julia Kurig, *Bildung für die technische Moderne: Pädagogische Technikdiskurse zwischen den 1920er und den 1950er Jahren in Deutschland* (Würzburg: Königshausen & Neumann, 2015), 366–7, 439–40.
74 Jürgen Habermas, 'Praktische Folgen des wissenschaftlich-technischen Fortschritts', in *Theorie und Praxis: Sozialphilosophische Studien* (Frankfurt: Suhrkamp, 1971), 339–45, 348–9, 357. Quote on p. 348.
75 Alexander Gall, 'Wunder der Technik, Wunder der Natur: Zur Vermittlungsleistung eines medialen Topos', in *Wunder. Poetik und Politik des Staunens im 20. Jahrhundert*, ed. Alexander C.T. Geppert and Till Kössler (Frankfurt: Suhrkamp, 2011), 283.
76 Jürgen Danyel and Annette Schuhmann: 'Wege in die digitale Moderne: Computerisierung als gesellschaftlicher Wandel', in *Geteilte Geschichte: Ost- und*

Westdeutschland 1970–2000, ed. Frank Bösch (Göttingen: Vandenhoeck & Ruprecht, 2015), 290.

77 Raymond G. Stokes, *Constructing Socialism: Technology and Change in East Germany 1945–1990* (Baltimore: Johns Hopkins University Press, 2000), 48.

78 Ibid., 195.

79 Ibid., 128.

80 Dolores L. Augustine, *Red Prometheus. Engineering and Dictatorship in East Germany, 1945–1990* (Cambridge: MIT Press, 2007), 201, 204.

81 Martin Schwarz, '"Werkzeuge der Geschichte": Automatisierungsdiskurse der 1950er und 1960er Jahre im deutsch-deutschen Vergleich', *Technikgeschichte* 82, no. 2 (2015): 139, 142. The end of the major automation euphoria was marked in West Germany by the 1973 oil crisis and in East Germany by the transition of power from Ulbricht to Honecker in 1971, see ibid., 143.

82 Martin Schwarz, 'Fabriken ohne Arbeiter: Automatisierungsvisionen von Ingenieuren im Spiegel der Zeitschrift "automatic" 1956–1972', in *Ingenieure in der technokratischen Hochmoderne*, ed. Uwe Fraunholz and Sylvia Wölfel (Münster: Waxmann, 2012), 171.

83 Johannes Platz, '"Revolution der Roboter" oder "Keine Angst vor Robotern"? Die Verwissenschaftlichung des Automatisierungsdiskurses und die industriellen Beziehungen von den 50ern bis 1968', in *Entreprises et crises économiques au XXe siècle*, ed. Laurent Commaille (Metz: Universite de Lorraine, 2009), 39, 54.

84 See Karl Böhm/Rolf Dörge: *Unsere Welt von morgen* (Berlin: Neues Leben, 1960), 54, 62.

85 See ibid., 56.

86 Detlef Hensche, "Technische Revolution und Arbeitnehmerinteresse. Zu Verlauf und Ergebnissen des Arbeitskampfes in der Druckindustrie 1978," *Blätter für deutsche und internationale Politik*, no. 4 (1978): 415.

87 Interview with a rotary printer, twenty-nine years old, 1977–1978, in Margareta Steinrücke, *Generationen im Betrieb: Fallstudien zur generationenspezifischen Verarbeitung betrieblicher Konflikte* (Frankfurt: Campus, 1986), 196–7.

88 Claudia Weber, *Rationalisierungskonflikte in Betrieben der Druckindustrie* (Frankfurt: Campus, 1982), S. 143.

89 Interview with a bookbinder, forty-six years old, 1977–1978, in Steinrücke, *Generationen*, 151–2.

90 'Die menschenleere Fabrik ist in zehn Jahren Wirklichkeit', *Frankfurter Rundschau*, 7 June 1982.

91 Danyel and Schuhmann, 'Wege in die digitale Moderne', 288; Andreas Wirsching, 'Durchbruch des Fortschritts? Die Diskussion über die Computerisierung in der Bundesrepublik', *ZeitRäume. Potsdamer Almanach des Zentrums für Zeithistorische Forschung* 5 (2009): 209–10.

92 Ibid., 207–8, 210–11.

93 Gall, 'Wunder der Technik', 284.

94 Elke Seefried, 'Bruch im Fortschrittsverständnis? Zukunftsforschung zwischen Steuerungseuphorie und Wachstumskritik', in *Vorgeschichte der Gegenwart: Dimensionen des Strukturbruchs nach dem Boom* (Göttingen: Vandenhoeck & Ruprecht, 2016), 425–6, 333–6, 441.

95 Hartmut Hirsch-Kreinsen, Peter Ittermann and Jonathan Niehaus, ed., *Digitaliseirung industrieller Arbeit. Die Vision Industrie 4.0 und ihre sozialen Herausforderungen* (Baden-Baden: Nomos, 2015).

Chapter 5

1 Anson Rabinbach, 'Ermüdung, Energie und der menschliche Motor', in *Physiologie und industrielle Gesellschaft: Studien zur Verwissenschaftlichung des Körpers im 19. und 20. Jahrhundert*, ed. Philipp Sarasin and Jakob Tanner (Frankfurt: Suhrkamp, 1998), 295; Anson Rabinbach, *The Human Motor: Energy, Fatigue, and the Origins of Modernity* (Berkeley: University of California Press, 1992), 52.
2 Oliver J. T. Harris, John Robb and Sarah Tarlow, 'The Body in the Age of Knowledge', in *The Body in History: Europe form the Palaeolithic to the Future*, ed. John Robb and Oliver J. T. Harris (Cambridge: Cambridge University Press, 2013), 186.
3 See Kurt Möser, *Fahren und Fliegen in Frieden und Krieg: Kulturen individueller Mobilitätsmaschinen 1880–1930* (Heidelberg: Verlag Regionalkultur, 2009).
4 Adelheid Voskuhl, '"Bewegung" und "Rührung": Musik spielende Androiden und ihre kulturelle Bedeutung im späten 18. Jahrhundert', in *Artifizielle Körper – lebendige Technik: Technische Modellierungen des Körpers in historischer Perspektive* (Zurich: Chronos, 2005), 89–94. See Adelheid Voskuhl, *Androids in the Enlightenment: Mechanics, Artisans, and Cultures of the Self* (Chicago: University of Chicago Press, 2013).
5 Johann Wolfgang von Goethe, *Berliner Ausgabe, Vol. 16. Poetische Werke: Autobiographische Schriften IV* (Berlin: Aufbau-Verlag, 1964), 147–8; Johann Wolfgang von Goethe, *The Autobiography of Goethe: Truth and Poetry – from my own Life. Together with his Annals; or Day and Year Papers* (London: Georg Bell and Sons, 1882), 320; see Jessica Riskin, 'The Defecating Duck, or, the Ambiguous Origins of Artificial Life', *Critical Inquiry* 29, no. 4 (2003): 619.
6 Bianca Westermann, *Anthropomorphe Maschinen: Grenzgänge zwischen Biologie und Technik seit dem 18. Jahrhundert* (Paderborn: Wilhelm Fink, 2012), 87, 147.
7 Westermann, *Anthropomorphe Maschinen*, 98–101, 148.
8 Jessica Riskin, *The Restless Clock: A History of the Centuries-Long Argument over what makes living things tick* (Chicago: University of Chicago Press, 2016), 301–2.
9 Fritz Kummer, 'Die Maschinenmenschen. Als Verkehrspolizisten, Rechenmeister und Warenverkäufer', *Metallarbeiter-Zeitung: Wochenblatt des Deutschen Metallarbeiter-Verbandes* 47, no. 20 (1929): 155.
10 Westermann, *Anthropomorphe Maschinen*, 97, 150.
11 Petra Frerichs, Martina Morschhäuser and Margareta Steinrücke, *Traueninteressen im Betrieb: Arbeitssituation und Interessenvertretung von Arbeiterinnen und weiblichen Angestellten im Zeichen neuer Technologien* (Opladen: Westdeutscher Verlag, 1989), 49–50, 99.
12 Julie Wosk, *Women and the Machine: Representations from the Spinning Wheel to the Electronic Age* (Baltimore: Johns Hopkins University Press, 2001), xiv.
13 Andrew Ure, *The Philosophy of Manufactures: or, an Exposition of the Scientific, Moral, and Commerical Economy of the Factory System in Great Britain*. 2nd ed. (London: Charles Knight, 1835), 15.
14 Karl Marx, *The Capital: A Critique of Political Economy*, vol. 1 (New York: Random House, 1906), 460–3.
15 Ibid.
16 Ibid., 372.
17 Harris, Robb and Tarlow, 'The Body in the Age of Knowledge', 186.

18 Ulrich Wengenroth, *Technik der Moderne: Ein Vorschlag zu ihrem Verständnis. Version 1.0* (Munich, 6.11.2015), 197. https://www.edu.tum.de/fileadmin/tuedz01/fggt/Wengenroth-offen/TdM-gesamt-1.0.pdf
19 Wolfhard Weber, *Arbeitssicherheit: Historische Beispiele, aktuelle Analysen* (Reinbek: Rowohlt, 1988), 88–90; Arne Anderson, 'Arbeiterschutz in Deutschland im 19. und frühen 20. Jahrhundert', *Archiv für Sozialgeschichte* 31 (1991): 63–4, 74–5; Sabine Schmitt, *Der Arbeiterinnenschutz im deutschen Kaiserreich: Zur Konstruktion der schutzbedürftigen Arbeiterin* (Stuttgart: Metzler, 1995); quotation from Kathleen Canning, *Languages of Labor and Gender: Female Factory Work in Germany, 1850–1914* (Ithaca: Cornell University Press, 1996), 284.
20 Gertjan de Groot and Marlou Schrover, 'General Introduction', in *Women Workers and Technological Change in Europe in the Nineteenth and Twentieth Centuries*, ed. Gertjan de Groot and Marlou Schrover (London: Taylor & Francis, 1995), 10.
21 Wosk, *Women and the Machine*, 39.
22 Barbara Orland, 'Männer in der Wäscherei: Technik und geschlechterhierarchische Arbeitsteilung im Waschgewerbe des 19. Jahrhunderts', in *Geschlechterhierarchie und Arbeitsteilung: Zur Geschichte ungleicher Erwerbschancen von Männern und Frauen*, ed. Karin Hausen (Göttingen: Vandenhoeck & Ruprecht, 1993), 113–14.
23 Brigitte Kassel, *Frauen in einer Männerwelt: Frauenerwerbsarbeit in der Metallindustrie und ihre Interessenvertretung durch den Deutschen Metallarbeiter-Verbund 1891–1933* (Cologne: Bund-Verlag, 1997), 625–6; Andrea Neugebauer, '"Frauen, welche ein Hauswesen zu versorgen haben, werden nicht angenommen": Frauenarbeit in den Opelwerken von 1880 bis 1945', *Zeitschrift für Unternehmensgeschichte* 44, no. 2 (1999): 174–5; Gertraude Krell, *Das Bild der Frau in der Arbeitswissenschaft* (Frankfurt: Campus, 1984), 109.
24 Emil Hänsgen, 'Fraueneinsatz im Maschinenbau', *Werkstattstechnik und Werksleiter* 35, no. 23/24 (1941): 407. – Some US factories employed gramophone music already before 1918 to stimulate workers, Hugo Münsterberg, *Psychologie und Wirtschaftsleben: Ein Beitrag zur angewandten Experimental-Psychologie* (Leipzig: Barth, 1919), 139.
25 Rüdiger Hachtmann, *Industriearbeit im 'Dritten Reich': Untersuchungen zu Lohn- und Arbeitsbedingungen in Deutschland 1933–1945* (Göttingen: Vandenhoeck & Ruprecht, 1989), 85.
26 Rüdiger Hachtmann, 'Gewerkschaften und Rationalisierung: Die 1970er-Jahre – ein Wendepunkt?', in *'Nach dem Strukturbruch'? Kontinuität und Wandel von Arbeitsbeziehungen seit den 1970er-Jahren*, ed. Knud Andresen, Ursula Bitzegeio and Jürgen Mittag (Bonn: Dietz, 2011), 187.
27 Eike-Christian Heine, 'Die technisierten Körper der Erdarbeiter um 1900', *Body Politics* 6, no. 9 (2018): 229–58; Eike-Christian Heine, 'Connect and Divide: On the history of the Kiel Canal', *Journal of Transport History* 35, no. 2 (2014): 200–19; Eike-Christian Heine, *Vom großen Graben: Die Geschichte des Nord-Ostsee-Kanals* (Berlin: Kadmos, 2015).
28 Lars Bluma, 'The hygienic movement and German mining, 1890–1914', in *A History of the Workplace: Environment and Health at Stake* (London: Routledge, 2015), 20.
29 Rabinbach, *The Human Motor*, 1–2, 190.
30 Ibid., 2, 243.
31 Jennifer Karns Alexander, *The Mantra of Efficiency: From Waterwheel to Social Control* (Baltimore: Johns Hopkins University Press, 2008), 101.
32 Ibid., 122–3.

33 Karsten Uhl, 'The Ideal of Lebensraum and the Spatial Order of Power at German Factories, 1900–45', *European Review of History/ Revue europeenne d'histoire* 20, no. 2 (2013): 287–307; Karsten Uhl, *Humane Rationalisierung? Die Raumordnung der Fabrik im fordistischen Jahrhundert* (Bielefeld: transcript, 2014).

34 Andreas Killen, *Berlin Electropolis: Shock, Nerves, and German Modernity* (Berkeley: University of California Press, 2006), 198; Noyan Dinçkal, '"Sport ist die körperliche und seelische Selbsthygiene des arbeitenden Volkes": Arbeit, Leibesübungen und Rationalisierungskultur in der Weimarer Republik, in *Body Politics* 1, no. 1 (2013): 71–97.

35 Cf. Astrid Kusser, 'Reversible Relationen: Körper- und Medienbewegungen in der "Welt als Ausstellung"', in *Um/Ordnungen: Fotografische Menschenbilder zwischen Konstruktion und Destruktion*, ed. Klaus Krüger, Leena Crasemann and Matthias Weiß (Munich: Fink, 2010), 61–3.

36 Alf Lüdtke, 'Organizational Order or Eigensinn? Workers' Privacy and Workers' Politics in Imperial Germany', in *Rites of Power: Symbolism, Ritual, and Politics since the Middle Ages*, ed. Sean Wilentz (Philadelphia: University of Pennsylvania Press, 1985), 310–11.

37 Sabine Kienitz, *Beschädigte Helden: Kriegsinvaliden und Körperbilder 1914–1923* (Stuttgart: Schöningh, 2008), 183.

38 Harris, Robb and Tarlow, 'The Body in the Age of Knowledge', 178–9; Kienitz, *Beschädigte Helden*, 189; Elsbeth Bösl, '"An unbroken man despite losing an arm": Corporeal reconstruction and embodied difference – prosthetics in Western Germany after the Second World War (c. 1945–1960)', in *War and the Body: Militarisation, practice and experience*, ed. Kevin McSorley (London: Routledge, 2013), 170.

39 Kienitz, *Beschädigte Helden*, 186.

40 Simon Bihr, '"Entkrüppelung der Krüppel": Der Siemens-Schuckert-Arbeitsarm und die Kriegsinvalidenfürsorge in Deutschland während des Ersten Weltkrieges', *NTM. Zeitschrift für Geschichte der Wissenschaften, Technik und Medizin* 21, no. 2 (2013): 107–141.

41 Kienitz, *Beschädigte Helden*, 190–1, 350.

42 Eva Horn, 'Der Krüppel. Maßnahmen und Medien zur Wiederherstellung des versehrten Leibes in der Weimarer Republik', in *KörperTopoi: Sagbarkeit – Sichtbarkeit – Wissen*, ed. Dietmar Schmidt (Weimar: Verlag und Datenbank für Geisteswissenschaft, 2002), 122.

43 Heather R. Perry, *Recycling the Disabled: Army, Medicine and Modernity in WWI Germany* (Manchester: Manchester University Press, 2014), 56, 71.

44 Kienitz, *Beschädigte Helden*, 184–7.

45 Ibid., 188.

46 Bösl, '"An unbroken man despite losing an arm"', 170–1, 176.

47 Oliver J. T. Harris, Maryon McDonald and John Robb, 'The Body in the Age of Technology', in *The Body in History: Europe form the Palaeolithic to the Future*, ed. John Robb and Oliver J. T. Harris (Cambridge: Cambridge University Press, 2013), 198.

48 Dierk Spreen, *Upgradekultur: Der Körper in der Enhancement-Gesellschaft* (Bielefeld: transcript, 2015): 49–50.

49 Rabinbach, *The Human Motor*, 11.

50 Michael Polanyi, *The Tacit Dimension* (London: Routledge, 1966).

51 Fritz Böhle and Hartmut Schulze, 'Gefühl bei der Arbeit mit CNC-Maschinen', in *Arbeit als Subjektivierendes Handeln: Handlungsfähigkeit bei Unwägbarkeiten*

und Ungewissheit, ed. Fritz Böhle (Wiesbaden: Springer, 2017), 146; Ursula Carus and Hartmut Schulze, 'Subjektivierendes Arbeitshandeln bei der Arbeit mit CNC-Maschinen', in *Arbeit als Subjektivierendes Handeln: Handlungsfähigkeit bei Unwägbarkeiten und Ungewissheit*, ed. Fritz Böhle (Wiesbaden: Springer, 2017), 97, 102.

52 Interview with a rotary printer, twenty-nine years old, 1977–1978, in Margareta Steinrücke, *Generationen im Betrieb: Fallstudien zur generationenspezifischen Verarbeitung betrieblicher Konflikte* (Frankfurt: Campus, 1986), 196–7.

53 Karsten Uhl, 'Challenges of Computerization and Globalization: The Example of the Printing Unions, 1950s to 1980s', in *Since the Boom: Continuity and Change in the Western Industrialized World after 1970*, ed. Sebastian Voigt (Toronto: University of Toronto Press 2020), S. 129–52.

54 Wolfgang Hindrichs et al., *Der lange Abschied vom Malocher: Sozialer Umbruch in der Stahlindustrie und die Rolle der Betriebsräte von 1960 bis in die neunziger Jahre* (Essen: Klartext, 2000), 7, 16, 30–1.

55 Weber, *Arbeitssicherheit*, 203; Joachim Radkau, *Technik in Deutschland: Vom 18. Jahrhundert bis heute* (Frankfurt: Campus, 2008), 374.

56 Jennifer Karns Alexander overrates the efficiency debates of the 1920s and thus mistakes those expert debates for social reality, see Alexander, *The Mantra of Efficiency*, 124.

57 Harris, Robb and Tarlow, 'The Body in the Age of Knowledge', 180.

58 Joachim Radkau, *Das Zeitalter der Nervosität. Deutschland zwischen Bismarck und Hitler* (Munich: Hanser, 1998), 193.

59 Ruth Oldenziel and Mikael Hård, *Consumers, Tinkerers, Rebels: The People who Shaped Europe* (Basingstoke: Palgrave Macmillan, 2013), 102–4, 107; Stefan Poser, *Glücksmaschinen und Maschinenglück: Grundlagen einer Technik- und Kulturgeschichte des technisierten Spiels* (Bielefeld: transcript, 2016), 81.

60 Wolfgang Schivelbusch, *The Railway Journey: The Industrialization of Time and Space in the 19th Century* (Leamington Spa: Berg, 1986), 72–3, 78.

61 Oldenziel and Hård, *Consumers, Tinkerers, Rebels*, 109–16.

62 Möser, *Fahren und Fliegen in Frieden und Krieg*, 356; Anne-Katrin Ebert, 'Liberating Technologies? Of Bicycles, Balance and the "New Woman" in the 1890s', *ICON* 16 (2010): 33; Ebert, *Radelnde Nationen*, 75.

63 Möser, *Fahren und Fliegen in Frieden und Krieg*, 163, 167, 183, 189.

64 Anne-Katrin Ebert, *Radelnde Nationen: Die Geschichte des Fahrrads in Deutschland und den Niederlanden bis 1940* (Frankfurt: Campus, 2010): 56.

65 Ebert, 'Liberating Technologies?', 33–4, 36, 42, 44; Ebert, *Radelnde Nationen*, 77.

66 Ibid., 82.

67 Ebert, 'Liberating Technologies?', 26.

68 Anne-Katrin Ebert, 'Zwischen "Radreiten" und "Kraftmaschine": Der bürgerliche Radsport am Ende des 19. Jahrhunderts', *Werkstatt Geschichte* 44 (2006): 42.

69 Stefan Poser, 'Playful Celebrations of Technology: Technology at Amusement Parks', *ICON* 19 (2013): 253; Joachim Radkau, 'Die wilhelminische Ära als nervöses Zeitalter, oder: Die Nerven als Netz zwischen Tempo- und Körpergeschichte', *Geschichte und Gesellschaft* 20, no. 2 (1994): 233.

70 Möser, *Fahren und Fliegen in Frieden und Krieg*, 155.

71 Wiebke Porombka, *Medialität urbaner Infrastrukturen. Der öffentliche Nahverkehr, 1870–1933* (Bielefeld: transcript, 2013), 195; Radkau, Das Zeitalter der Nervosität, 207.

72 Harris, McDonald and Robb, 'The Body in the Age of Technology', 202; Möser, *Fahren und Fliegen in Frieden und Krieg*, 221, 393–7.
73 Kurt Möser, '"Der Kampf des Automobilisten mit seiner Maschine": Eine Skizze der Vermittlung der Autotechnik und des Fahrenlernens im 20. Jahrhundert', in *Technikvermittlung und Technikpopularisierung: Historische und didaktische Perspektiven*, ed. Lars Bluma, Karl Pichol and Wolfhard Weber (Münster: Waxmann, 2004), 89, 93–4; Kurt Möser, *Geschichte des Autos* (Frankfurt and New York: Campus, 2002), 307; Dietmar Fack, 'Das deutsche Kraftfahrschulwesen und die technisch-rechtliche Konstitution der Fahrausbildung 1899–1943', *Technikgeschichte* 67, no. 2 (2000): 111–38.
74 Sasha Disko, *The Devil's Wheels: Men and Motorcycling in the Weimar Republic* (New York: Berghahn, 2016), 281–2.
75 Ibid., 283.
76 Ibid., 284.
77 Ibid., 158–9, 170–1.
78 Poser, *Glücksmaschinen und Maschinenglück*, 52.
79 Christian Kehrt, *Moderne Krieger: Die Technikerfahrungen deutscher Militärpiloten 1910–1945* (Paderborn: Schöningh, 2010), 38, 319, 454.
80 Ibid., 331.
81 Möser, *Fahren und Fliegen in Frieden und Krieg*, 219.
82 Klaus Theweleit, *Männerphantasien, Vol. 2: Männerkörper – zur Psychologie des weißen Terrors* (Munich: Piper, 2000), 200.
83 Bernhard Rieger, *Technology and the Culture of Modernity in Britain and Germany, 1890–1945* (Cambridge: Cambridge University Press, 2005), 279.
84 Fernando Esposito, *Mythische Moderne: Aviatik, Faschismus und die Sehnsucht nach Ordnung in Deutschland und Italien* (Munich: Oldenbourg, 2011), 246, 270, 275–6; Viktor Otto, 'An der Herz-Lungen-Maschine von Avantgarde und Aviatik: Mensch-Maschine-Hybride in Literatur und Technologie 1909-1945', *Zeitschrift für Germanstik* 14, no. 2 (2004): 352–3.
85 Esposito, *Mythische Moderne*, 290.
86 Evelyn Zegenhagen, '"Im übrigen sind die Erfahrungen, die wir mit jungen Frauen gemacht haben, die allerbesten": Geschlechterverhältnisse und -konkurrenz in der deutschen Luftfahrt 1918 bis 1945', *Technikgeschichte* 73, no. 3/4 (2006): 255–6, 263; Evelyn Zegenhagen, *"Schneidige deutsche Mädel". Fliegerinnen zwischen 1918 und 1945* (Göttingen: Wallstein, 2007).
87 Zegenhagen, '"Im übrigen sind die Erfahrungen"', 260–1.
88 Möser, '"Der Kampf des Automobilisten"', 96, 98, 100.
89 Poser, *Glücksmaschinen und Maschinenglück*, 185–6.
90 Wengenroth, *Technik der Moderne*, 195.
91 Andrew Pickering, 'The History of Economics and the History of Agency', in *The State of the History of Economics: Proceedings of the History of Economics Society*, ed. James P. Henderson (London: Routledge, 1997), 9–10.
92 Joan Campbell, *Joy in Work, German Work: The National Debate, 1800–1945* (Princeton: Princeton University Press, 1989); Alf Lüdtke, '"Deutsche Qualitätsarbeit", "Spielereien" am Arbeitsplatz und "Fliehen" aus der Fabrik: Industrielle Arbeitsprozesse und Arbeiterverhalten in den 1920er Jahren – Aspekte eines offenen Forschungsfeldes', in *Arbeiterkulturen zwischen Alltag und Politik: Beiträge zum europäischen Vergleich in der Zwischenkriegszeit*, ed. Friedhelm Boll (Vienna: Europa-Verlag, 1986), 155–97.

Chapter 6

1 Joachim Radkau, *Technik in Deutschland: Vom 18. Jahrhundert bis heute* (Frankfurt: Campus, 2008), 334.
2 Raphael Emanuel Dorn, *Alle in Bewegung: Räumliche Mobilität in der Bundesrepublik Deutschland 1980–2010* (Göttingen: Vandenhoeck & Ruprecht, 2018), 276–8.
3 Christian Zumbrägel, 'Urban Energy Consumption, Mobility and Environmental Legacies', in *Concepts of Urban-Environmental History*, ed. Sebastian Haumann, Martin Knoll and Detlev Mares (Bielefeld: transcript, 2020), 174.
4 David Blackbourn, *The Conquest of Nature: Water, Landscape, and the Making of Modern Germany* (New York: Norton, 2006), 158–61, 281–91, 191, 198.
5 David Blackbourn, *Landschaften der deutschen Geschichte: Aufsätze zum 19. und 20. Jahrhundert* (Bielefeld: Vandenhoeck & Ruprecht, 2016), 223.
6 Zumbrägel, 'Urban Energy Consumption', 174–5.
7 Blackbourn, *The Conquest of Nature*, 199–200, 207.
8 Christian Zumbrägel, *"Viele Wenige machen ein Viel": Eine Technik- und Umweltgeschichte der Kleinwasserkraft (1880–1930)* (Stuttgart: Schöningh, 2018), 282–6, 294.
9 Blackbourn, *The Conquest of Nature*, 192–3, 220.
10 Blackbourn, *Landschaften der deutschen Geschichte*, 212; Blackbourn, *The Conquest of Nature*, 201.
11 Vincent Lagendijk, 'Europe's Rhine Power: Connections, Borders, and Flows', *Water History* 8, no. 1 (2016): 24, 35.
12 Mark Cioc, *The Rhine: An Eco-Biography, 1815–2000* (Seattle: University of Washington Press, 2002), 10, 204.
13 Theodore R. Schatzki, 'Nature and Technology in History', *History and Theory*, theme issue 42 (2003): 82–93.
14 Frank Uekötter, *Umweltgeschichte im 19. und 20. Jahrhundert* (Munich: Oldenbourg, 2007), 8–9; Jeffrey K. Wilson, *The German Forest: Nature, Identity, and the Contestation of a National Symbol, 1871–1914* (Toronto: University of Toronto Press, 2012), 55.
15 Ibid., 54–5, 240.
16 Ibid., 55–6.
17 Uekötter, *Umweltgeschichte im 19. und 20. Jahrhundert*, 24.
18 Radkau, *Technik in Deutschland*, 335.
19 Jens Ivo Engels, *Naturpolitik in der Bundesrepublik: Ideenwelt und politische Verhaltensstile in Naturschutz und Umweltbewegung 1950–1980* (Paderborn: Schöningh, 2006), 31, 72.
20 Jan de Vries, 'The Industrial Revolution and the Industrious Revolution', *The Journal of Economic History* 54, no. 2 (1994): 251.
21 Verena Lehmbrock, *Der denkende Landwirt: Agrarwissen und Aufklärung in Deutschland 1750–1820* (Vienna: Böhlau, 2020), 157, 257.
22 Reiner Prass, *Grundzüge der Agrargeschichte, Volume 1: Vom Dreißigjährigen Krieg bis zum Beginn der Moderne (1650–1880)* (Cologne: Böhlau, 2016), 91–3, 103–4, 113, 121–4, 142–3, 156; see de Vries, 'The Industrial Revolution and the Industrious Revolution'.
23 Rita Gudermann, 'Der Take-off der Landwirtschaft im 19. Jahrhundert und seine Konsequenzen für Umwelt und Gesellschaft', in *Agrarmodernisierung und ökologische*

Folgen. Westfalen vom 18. bis zum 20. Jahrhundert, ed. Karl Ditt, Rita Gudermann and Norwich Rüße (Paderborn et al.: Schöningh, 2001), 65.

24 Prass, *Grundzüge der Agrargeschichte*, 121.

25 Fernand Braudel, *Civilizations and Capitalism, Volume 1: The Structures of Everyday Life* (Berkeley: University of California Press, 1992), 74; Hans-Jürgen Teuteberg, *Die Rolle des Fleischextrakts für die Ernährungswissenschaften und den Aufstieg der Suppenindustrie: Kleine Geschichte der Fleischbrühe* (Stuttgart: Franz Steiner, 1990), 23.

26 Dieter Ziegler, *Die industrielle Revolution*, 3rd ed. (Darmstadt: WBG, 2012), 16; Prass, *Grundzüge der Agrargeschichte*, 155.

27 Hans-Jürgen Teuteberg, 'Britische Frühindustrialisierung und kurhannoversches Reformbewußtsein im späten 18. Jahrhundert', *Vierteljahrschrift für Sozial- und Wirtschaftsgeschichte* 86, no. 2 (1999): 178–9; Lehmbrock, *Der denkende Landwirt*, 113.

28 Gudermann, 'Der Take-off der Landwirtschaft', 68, 71–2; Kiesewetter, *Industrielle Revolution in Deutschland*, 146–8; Lehmbrock, *Der denkende Landwirt*, 37, 245, 262, 268.

29 Prass, *Grundzüge der Agrargeschichte*, 165–6, 174–5; Gudermann, 'Der Take-off der Landwirtschaft', 58, 72; Ziegler, *Die Industrielle Revolution*, 25.

30 Michael Kopsidis and Heinrich Hockmann, 'Technical change in Westphalian peasant agriculture and the rise of the Ruhr, circa 1830–1880', *European Review of Economic History* 14, no. 2 (2010): 225–6; Prass, *Grundzüge der Agrargeschichte*, 149, 168, 192.

31 Gudermann, 'Der Take-off der Landwirtschaft', 71–3; Lehmbrock, *Der denkende Landwirt*, 164.

32 Frank Uekötter, *Die Wahrheit ist auf dem Feld: Eine Wissensgeschichte der deutschen Landwirtschaft* (Göttingen: Vandenhoeck & Ruprecht, 2010), 290; Prass, *Grundzüge der Agrargeschichte*, 182, 193.

33 Kiesewetter, *Industrielle Revolution in Deutschland*, 158; Gunter Mahlerwein, *Grundzüge der Agrargeschichte, Volume 3: Die Moderne (1880–2010)* (Cologne: Böhlau, 2016), 70–2; Uekötter, *Die Wahrheit ist auf dem Feld*, 304; Kopsidis and Hockmann, 'Technical Change': 228.

34 Thomas Rohkrämer, *Eine andere Moderne? Zivilisationskritik, Natur und Technik in Deutschland 1880–1933* (Paderborn: Schöningh, 1999), 41.

35 Kopsidis and Hockmann, 'Technical change', 212.

36 Gudermann, 'Der Take-off der Landwirtschaft', 129.

37 Ziegler, *Die Industrielle Revolution*, 106; Mahlerwein, *Grundzüge der Agrargeschichte*, 73; Ulrich Kluge, *Agrarwirtschaft und ländliche Gesellschaft im 20. Jahrhundert* (Munich: Oldenbourg, 2005), 6.

38 Kiesewetter, *Industrielle Revolution in Deutschland*, 151; Prass, *Grundzüge der Agrargeschichte*, 170; Mahlerwein, *Grundzüge der Agrargeschichte*, 73.

39 Peter Exner, '"Die Technik läßt sie nicht mehr los, ob sie wollen oder nicht wollen": Die Verwissenschaftlichung der Agrarproduktion in den Landwirtschaftsschulen (1920er–1970er Jahre), in *Agrarmodernisierung und ökologische Folgen. Westfalen vom 18. bis zum 20. Jahrhundert*, ed. Karl Ditt, Rita Gudermann and Norwich Rüße (Paderborn: Schöningh, 2001), 169; Uekötter, *Die Wahrheit ist auf dem Feld*, 290.

40 Kiesewetter, *Industrielle Revolution in Deutschland*, 149; Uekötter, *Die Wahrheit ist auf dem Feld*, 170.

41 Frank Uekötter, 'Why Panaceas Work: Recasting Science, Knowledge, and Fertilizer Interests in German Agriculture', *Agricultural History* 88, no. 1 (2014): 81.

42 Uekötter, *Die Wahrheit ist auf dem Feld*, 160.
43 Dieter Ziegler, 'Beyond the Leading Regions: Agricultural Modernization and Rural Industrialization in North-Western Germany', in *Regions, Industries, and Heritage: Perspectives on Economy, Society, and Culture in Modern Western Europe*, ed. Juliane Czierpka, Kathrin Oerters and Nora Thorade (Basingstoke: Palgrave Macmillan, 2015), 161–2; Prass, *Grundzüge der Agrargeschichte*, 167.
44 Mahlerwein, *Grundzüge der Agrargeschichte*, 101–2.
45 Alice Weinreb, *Modern Hungers: Food and Power in Twentieth-Century Germany* (New York: Oxford University Press, 2017), 17–18.
46 Ursula Heinzelmann, *Beyond Bratwurst: A History of Food in Germany* (London: Reaktion Books, 2014), 244.
47 Weinreb, *Modern Hungers*, 39–40.
48 Uekötter, *Die Wahrheit ist auf dem Feld*, 199–201, 204; Uekötter, 'Why Panaceas Work', 81.
49 Mahlerwein, *Grundzüge der Agrargeschichte*, 106.
50 Uekötter, *Die Wahrheit ist auf dem Feld*, 164.
51 Barbara Orland, 'Turbo-Cows: Producing a Competitive Animal in the Nineteenth and Early Twentieth Centuries', in *Industrializing Organisms: Introducing Evolutionary History*, ed. Susan R. Schrepfer and Philip Scranton (New York: Routledge, 2003): 169–72; Prass, *Grundzüge der Agrargeschichte*, 167.
52 Orland, 'Turbo-Cows', 178, 180–3; Mahlerwein, *Grundzüge der Agrargeschichte*, 90–1.
53 Burkhard Theine, 'Technisierung in der Stall- und Feldwirtschaft: Entlastung der Landwirte – Belastung der Umwelt?' in *Agrarmodernisierung und ökologische Folgen. Westfalen vom 18. bis zum 20. Jahrhundert*, ed. Karl Ditt, Rita Gudermann and Norwich Rüße (Paderborn: Schöningh, 2001), 199.
54 Frank Dittmann, 'Vom "Strippen," Saugen und Drücken: Zur Geschichte des maschinellen Melkens', *Technikgeschichte* 66, no. 4 (1999): 259–75; for the US history of milking, see Kendra Smith-Howard, *Pure and Modern Milk: An Environmental History Since 1900* (New York: Oxford University Press, 2013).
55 Mahlerwein, *Grundzüge der Agrargeschichte*, 74.
56 Exner, '"Die Technik"', 183.
57 Wolfgang Zängl, *Deutschlands Strom: Die Politik der Elektrifizierung von 1866 bis heute* (Frankfurt: Campus, 1989), 73; Edmund N. Todd, 'Electric Ploughs in Wilhelmine Germany: Failure of an Agricultural System', *Social Studies in Science* 22, no. 2 (1992): 272; David Nye, *Electrifying America: Social Means of a New Technology* (Cambridge: MIT Press, 1990), 287, 299.
58 Klaus Herrmann, 'Technisierung der Landwirtschaft: Landmaschinen-Industrie im Ruhrgebiet', in *Technikgeschichte im Ruhrgebiet, Technikgeschichte für das Ruhrgebiet*, ed. Manfred Rasch and Dietmar Bleidick (Essen: Klartext, 2004), 836.
59 Gudermann, 'Der Take-off der Landwirtschaft', 78; Theine, 'Technisierung in der Stall- und Feldwirtschaft', 197; Herrmann, 'Technisierung der Landwirtschaft', 827–8; Mark Hewitson, *Germany and the Modern World, 1880–1914* (Cambridge: Cambridge University Press, 2018), 84.
60 Uekötter, *Die Wahrheit ist auf dem Feld*, 307–9.
61 Todd, 'Electric Ploughs in Wilhelmine Germany', 265.
62 Herrmann, 'Technisierung der Landwirtschaft', 833.
63 Uekötter, *Die Wahrheit ist auf dem Feld*, 307–9, 320; Kurt Möser, *Geschichte des Autos* (Frankfurt: Campus, 2002), 117.

64 Mahlerwein, *Grundzüge der Agrargeschichte*, 75–6; Theine, 'Technisierung in der Stall- und Feldwirtschaft', 198, 200, 205.
65 Uekötter, *Die Wahrheit ist auf dem Feld*, 278–80, 282–4, 286–7, 291, 321; Mahlerwein, *Grundzüge der Agrargeschichte*, 74–5; Kluge, *Agrarwirtschaft und ländliche Gesellschaft im 20. Jahrhundert*, 21.
66 Möser, *Geschichte des Autos*, 118–21; Uekötter, *Die Wahrheit ist auf dem Feld*, 297.
67 Ibid., 280–1, 294–5.
68 Rohkrämer, *Eine andere Moderne*, 351; Mahlerwein, *Grundzüge der Agrargeschichte*, 78.
69 Möser, *Geschichte des Autos*, 120.
70 Tiago Saraiva, *Fascist Pigs: Technoscientifc Organisms and the History of Fascism* (Cambridge: MIT Press, 2016), 107–10, 123–4, 135.
71 Veronika Settele, 'Mensch, Tier und Technik: "Doing Technology" in deutschen Schweineställen und die Veränderung des Verhältnisses zwischen Mensch und Tier seit 1945', *Technikgeschichte* 87, no. 2 (2020): 136, 152.
72 Mahlerwein, *Grundzüge der Agrargeschichte*, 94; Uekötter, *Die Wahrheit ist auf dem Feld*, 345.
73 Settele, 'Mensch, Tier und Technik', 161–2.
74 Mahlerwein, *Grundzüge der Agrargeschichte*, 94–5.
75 Settele, 'Mensch, Tier und Technik', 162.
76 Uekötter, *Die Wahrheit ist auf dem Feld*, 179; Clemens Zimmermann, 'Ländliche Gesellschaft und Agrarwirtschaft im 19. und 20. Jahrhundert: Transformationsprozesse als Thema der Agrargeschichte', in *Agrargeschichte: Positionen und Perspektiven*, ed. Werner Troßbach and Clemens Zimmermann (Stuttgart: Lucius & Lucius, 1998), 157–9; Exner, '"Die Technik"', 182; Kluge, *Agrarwirtschaft und ländliche Gesellschaft im 20. Jahrhundert*, 42.
77 Uekötter, *Die Wahrheit ist auf dem Feld*, 330.
78 Ibid., 298, 302, 440.
79 Mahlerwein, *Grundzüge der Agrargeschichte*, 79.
80 Uekötter, *Die Wahrheit ist auf dem Feld*, 306.
81 Gerda Krohn's report upon growing up on a farm in northern Germany (2020), 7–8. Private collection of Karsten Uhl.
82 Uekötter, *Die Wahrheit ist auf dem Feld*, 173–5, 289.
83 Mahlerwein, *Grundzüge der Agrargeschichte*, 95; Todd, 'Electric Ploughs in Wilhelmine Germany', 274.
84 Exner, '"Die Technik"', 185, 195.
85 Uekötter, *Die Wahrheit ist auf dem Feld*, 324–5; Elsbeth Bösl, '"Landfrau und Kamerad Maschine": Agrarexpertinnen der frühen Bundesrepublik über Technik und Geschlecht', *Ariadne* 63 (2013): 52–63.
86 Uekötter, *Die Wahrheit ist auf dem Feld*, 299–300.
87 Mahlerwein, *Grundzüge der Agrargeschichte*, 80–2; Kluge, *Agrarwirtschaft und ländliche Gesellschaft im 20. Jahrhundert*, 42–3.
88 Mahlerwein, *Grundzüge der Agrargeschichte*, 112.
89 Bösl, '"Landfrau und Kamerad Maschine"', 54.
90 Veronika Settele, 'Mensch, Kuh, Maschine: Kapitalismus im westdeutschen Kuhstall, 1950–1980', *Mittelweg 36. Zeitschrift des Hamburger Insituts für Sozialforschung* 26, no. 1 (2017): 50.
91 Exner, '"Die Technik"', 183.
92 Bösl, '"Landfrau und Kamerad Maschine"', 57.

93 Ibid.; Settele, 'Mensch, Kuh, Maschine': 57; Mahlerwein, *Grundzüge der Agrargeschichte*, 80.
94 Ibid., 95.
95 Settele, 'Mensch, Kuh, Maschine', 47, 56, 59, 62–3.
96 Theine, 'Technisierung in der Stall- und Feldwirtschaft', 218.
97 Mahlerwein, *Grundzüge der Agrargeschichte*, 151.
98 Ibid., 80.
99 Theine, 'Technisierung in der Stall- und Feldwirtschaft', 198; Mahlerwein, *Grundzüge der Agrargeschichte*, 74, 80; Uekötter, *Die Wahrheit ist auf dem Feld*, 304.
100 Mahlerwein, *Grundzüge der Agrargeschichte*, 80; Uekötter, *Die Wahrheit ist auf dem Feld*, 326–7, 339.
101 Ibid., 328–9, 356.
102 Ibid., 332; Pfister, 'Rationalisierungsprozesse', 131.
103 Uekötter, *Die Wahrheit ist auf dem Feld*, 341, 346, 360–1; Mahlerwein, *Grundzüge der Agrargeschichte*, 96.
104 Uekötter, *Die Wahrheit ist auf dem Feld*, 354.
105 Ibid., 349–53, 386.
106 Kluge, *Agrarwirtschaft und ländliche Gesellschaft im 20. Jahrhundert*, 42.
107 Karin Zachmann and Per Østby, 'Food, Technology, and Trust: An Introduction', *History and Technology* 27, no. 1 (2011): 5.
108 Ziegler, 'Beyond the Leading Regions', 149, 163–4.
109 Ziegler, *Die Industrielle Revolution*, 114–5, 133–5; Ziegler, 'Beyond the Leading Regions', 153, 157–9.
110 Otto Keck, 'The National System for Technical Innovation in Germany', in *National Innovation Systems: A Comparative Analysis*, ed. Richard R. Nelson (New York: Oxford University Press, 1993), 125.
111 Dirk Schaal, *Rübenzuckerindustrie und regionale Industrialisierung: Der Industrialisierungsprozess im mitteldeutschen Raum* (Münster: LIT, 2005), 235–7; Prass, *Grundzüge der Agrargeschichte*, 193.
112 Kiesewetter, *Industrielle Revolution in Deutschland*, 155–7; Prass, *Grundzüge der Agrargeschichte*, 159–60.
113 Teuteberg, *Die Rolle des Fleischextrakts*, 84.
114 Statistics of the Federal Ministry of Food and Agriculture, https://www.bmel-statistik.de/ernaehrung-fischerei/versorgungsbilanzen/fleisch/
115 Teuteberg, *Die Rolle des Fleischextrakts*, 51; Uwe Spiekermann, 'Redefining Food: The Standardization of Products and Production in Europe and the United States, 1880–1914', *History and Technology* 27, no. 1 (2011): 14.
116 Uwe Spiekermann, 'Die gescheiterte Neugestaltung der Alltagskost Nähr- und Eiweißpräparate im späten Kaiserreich', *Technikgeschichte* 78, no. 3 (2011): 205–6.
117 Michael Wildt, *Am Beginn der 'Konsumgesellschaft': Mangelerfahrung, Lebenshaltung, Wohlstandshoffnung in Westdeutschland in den fünfziger Jahren* (Hamburg: Ergebnisse, 1994), 169.
118 Teuteberg, *Die Rolle des Fleischextrakts*, 1, 4, 10–11, 95, 117.
119 Ibid., 12–14, 16, 18.
120 Weinreb, *Modern Hungers*, 21, 64.
121 Jürgen Bönig, *Die Einführung von Fließbandarbeit in Deutschland bis 1933: Zur Geschichte einer Sozialinnovation*, Volume 2 (Münster: LIT, 1993), 624.
122 Exner, '"Die Technik"', 184.

123 Michael Mende, 'Verschwundene Stellmacher, gewandelte Schmiede: Die Handwerker des ländlichen Fahrzeugbaus zwischen expandierender Landwirtschaft und Motorisierung 1850–1960', in *Prekäre Selbständigkeit: Zur Standortbestimmung von Handwerk, Hausindustrie und Kleingewerbe im Industrialisierungsprozess*, ed. Ulrich Wengenroth (Stuttgart: Franz Steiner, 1989), 118–20.

124 Möser, *Geschichte des Autos*, 98.

Chapter 7

1 Joachim Radkau, *Technik in Deutschland: Vom 18. Jahrhundert bis heute* (Frankfurt: Campus, 2008), 14.

2 Mikael Hård, '"The Victorian Eye" and Its Blind Spot: Toward a Cultural Assessment of Technology', *Technology and Culture* 26, no. 2 (2010): 176.

3 Nelly Oudshoorn and Trevor Pinch, 'Introduction: How Users and Non-Users Matter', in *How Users Matter: The Co-Construction of Users and Technologies*, ed. Nelly Oudshoorn and Trevor Pinch (Cambridge: MIT Press, 2003), 16.

4 Karin Hausen, 'Große Wäsche: Technischer Fortschritt und sozialer Wandel in Deutschland vom 18. bis ins 20. Jahrhundert', *Geschichte und Gesellschaft* 13, no. 3 (1987): 278–9, 282, 284.

5 Edwin Redslob, 'Telephones and Electric Light (c. 1890)', in *German History in Documents and Images, vol. 4: Forging an Empire - Bismarckian Germany, 1866–1890*, ed. James Retallack (www.germanhistorydocs.ghi-dc.org).

6 Wolfgang Schivelbusch, *Disenchanted Night: The Industrialization of Light in the Nineteenth Century* (Berkeley: University of California Press, 1995), 69, 159, 186–7.

7 Stefan Poser, 'Playful Celebrations of Technology: Technology at Amusement Parks', *ICON* 19 (2013): 244–63; Stefan Poser, *Glücksmaschinen und Maschinenglück: Grundlagen einer Technik- und Kulturgeschichte des technisierten Spiels* (Bielefeld: transcript, 2016).

8 Redslob, 'Telephones and Electric Light (c. 1890)'.

9 Martina Heßler, *'Mrs. Modern Woman': Zur Sozial- und Kulturgeschichte der Haushaltstechnisierung* (Frankfurt: Campus, 2001), 56–7.

10 David Kuchenbuch, *Geordnete Gemeinschaft*: Architekten als Sozialingenieure - Deutschland und Schweden im 20. Jahrhundert (Bielefeld: transcript, 2010), 89; Gerd Kuhn, *Wohnkultur und kommunale Wohnungspolitik in Frankfurt am Main 1880 bis 1930: Auf dem Wege zu einer pluralen Gesellschaft der Individuen* (Bonn: Dietz, 1998), 177; for a critical discussion of the concept of script, see Oudshoorn and Pinch, Introduction, 9–11.

11 Christoph Bernhardt and Elsa Vonau, 'Zwischen Fordismus und Sozialreform: Rationalisierungsstrategien im deutschen und französischen Wohnungsbau 1900–1933', *Zeithistorische Forschungen/Studies in Contemporary History* 6, no. 2 (2009): 238.

12 Martina Heßler, 'The Frankfurt Kitchen: The Model of Modernity and the "Madness" of Traditional Users, 1926 to 1933', in *Cold War Kitchen: Americanization, Technology and European Users*, ed. Ruth Oldenziel and Karin Zachmann (Cambridge: MIT Press, 2009), 176.

13 Kuhn, *Wohnkultur und kommunale Wohnungspolitik*, 180.

14 Heßler, 'The Frankfurt Kitchen', 175.

15 Kuhn, *Wohnkultur und kommunale Wohnungspolitik*, 174.
16 Martina Heßler, 'Educating Men how to Develop Technology: The Role of Professional Housewives in the Diffusion of Electrical Domestic Appliances in the Interwar-Period in Germany', *ICON* 7(2001): 97.
17 Michael Wildt, *Am Beginn der 'Konsumgesellschaft': Mangelerfahrung, Lebenshaltung, Wohlstandshoffnung in Westdeutschland in den fünfziger Jahren* (Hamburg: Ergebnisse, 1994), 137.
18 Heßler, 'The Frankfurt Kitchen', 178–9.
19 Bernd Stöver, *Der Kalte Krieg 1947–1991: Geschichte eines radikalen Zeitalters* (Munich: C.H. Beck, 2007), 178.
20 Karin Zachmann, 'Technikgeschichte des Kalten Krieges: Eine einführende Skizze', *Technikgeschichte* 80, no. 3 (2013): 202.
21 Emily Pugh, *Architecture, Politics & Identity in Divided Berlin* (Pittsburgh: University of Pittsburgh Press, 2014), 39.
22 Sophie Gerber, '"We want to live electrically!" Marketing Strategies of German Power Companies in the 20th Century', in *Past and Present Energy Societies: New Energy Connects Politics, Technologies and Cultures*, ed. Nina Möllers and Karin Zachmann (Bielefeld: transcript, 2012), 79, 93.
23 Hausen, 'Große Wäsche', 273, 302.
24 Heßler, '*Mrs. Modern Woman*', 55; Wildt, *Am Beginn der 'Konsumgesellschaft'*, 131; Sophie Gerber, *Küche, Kühlschrank, Kilowatt: Zur Geschichte des privaten Energiekonsums in Deutschland 1945–1990* (Bielefeld: Transcript, 2014), 28; Ruth Schwartz Cowan, *More Work for Mother: The Ironies of Household Technology from the Open Hearth to the Microwave* (New York: Basic Books, 1983).
25 Alice Weinreb, *Modern Hungers: Food and Power in Twentieth-Century Germany* (Oxford: Oxford University Press, 2017), 165; Ruth Oldenziel and Karin Zachmann, ed., *Cold War Kitchen: Americanization, Technology and European Users* (Cambridge: MIT Press, 2009).
26 Weinreb, *Modern Hungers*, 169–70.
27 Heßler, '*Mrs. Modern Woman*', 383–4, 395.
28 Weinreb, *Modern Hungers*, 181; Kaminsky, *Wohlstand, Schönheit, Glück*, 87.
29 Weinreb, *Modern Hungers*, 166, 185, 191.
30 Ulrike Thoms, 'Essen in der Arbeitswelt: Das betriebliche Kantinenwesen seit seiner Entstehung um 1850', in *Die Revolution am Esstisch*, ed. Hans-Jürgen Teuteberg (Stuttgart: Franz Steiner, 2004), 214–15; Peter Hübner, 'Betriebe als Träger der Sozialpolitik: Betriebliche Sozialpolitik', in *Geschichte der Sozialpolitik in Deutschland seit 1945, volume 9: Deutsche Demokratische Republik 1961–1971*, ed. Christoph Kleßmann (Baden-Baden: Nomos, 2006), 753.
31 Weinreb, *Modern Hungers*, 170.
32 Lydia Langer, *Revolution im Einzelhandel: Die Einführung der Selbstbedienung in Lebensmittelgeschäften der Bundesrepublik Deutschland (1949–1973)* (Cologne: Böhlau, 2013), 272.
33 Sibylle Meyer and Eva Schulze, *Von Liebe sprach damals keiner: Familienalltag in der Nachkriegszeit* (Munich: C.H. Beck, 1985), 174.
34 Weinreb, *Modern Hungers*, 170.
35 Dr. Oetker, *Schul-Kochbuch für den Elektroherd* (Bielefeld: Ceres-Verlag, 1956), 5.
36 AEG, ed., *AEG-Elektroherde: Gebrauchsanleitung und Rezepte* (Berlin: AEG, 1959), 5, 14.
37 Weinreb, *Modern Hungers*, 173.
38 Ibid., 170–1.

39 Wildt, *Am Beginn der 'Konsumgesellschaft'*, 155.
40 Ruth Oldenziel and Mikael Hård, *Consumers, Tinkerers, Rebels: The People who Shaped Europe* (Basingstoke: Palgrave Macmillan, 2013), 171–3.
41 Wildt, *Am Beginn der 'Konsumgesellschaft'*, 155–6.
42 Weinreb, *Modern Hungers*, 171.
43 Wildt, *Am Beginn der 'Konsumgesellschaft'*, 152, 158, 172.
44 Heike Weber, 'Stecken, Drehen, Drücken: Interfaces von Alltagstechniken und ihre Bediengesten', *Technikgeschichte* 76, no. 3 (2009): 247–8.
45 Arne Krüger, *Kälte macht das Kochen leicht* (Mainz-Kostheim: Duofrost, 1995), 7–8, 12.
46 Arne Krüger, *Gaumenfreuden aus dem Linde Microtherm* (Mainz-Kostheim: Duofrost, 1987), 7–8.
47 Gerber, *Küche, Kühlschrank, Kilowatt*, 30.
48 Weinreb, *Modern Hungers*, 183–4; Zängl, *Deutschlands Strom*, 348.
49 Langer, *Revolution im Einzelhandel*, 310, 387.
50 Wildt, *Am Beginn der 'Konsumgesellschaft'*, 176.
51 Ibid., 179, 188.
52 Langer, *Revolution im Einzelhandel*, 129, 272.
53 Annette Kaminsky, *Wohlstand, Schönheit, Glück: Kleine Konsumgeschichte der DDR* (München: C.H. Beck, 2001), 51.
54 Weinreb, *Modern Hungers*, 195.
55 Detlev Siegfried, *Time Is on My Side: Konsum und Politik in der westdeutschen Jugendkultur der 60er Jahre* (Göttingen: Wallstein, 2006), 100, 102.
56 Siegfried, *Time Is on My Side*, 77, 80, 97, 100; Heike Weber, '"Mobile electronic media": Mobility history at the intersection of transport and media history', *Transfers* 1, no. 1 (2011): 35.
57 Siegfried, *Time Is on My Side*, 93–4, 98, 104–6.
58 Monika Röter, 'Alltägliche Objekte als aussagekräftige Zeugen der Vergangenheit: Musikschrank und Stereoanlage erzählen von den 1960er Jahren', *Geschichte in Wissenschaft und Unterricht* 64, no. 5/6 (2013): 320, 326–7, 331; Andreas Wirsching: 'Konsum statt Arbeit? Zum Wandel von Individualität in der modernen Massengesellschaft', *Vierteljahrshefte für Zeitgeschichte*, 57, no. 2 (2009): 192.
59 Siegfried, *Time Is on My Side*, 105.
60 Kurt Möser, '"Der Kampf des Automobilisten mit seiner Maschine": Eine Skizze der Vermittlung der Autotechnik und des Fahrenlernens im 20. Jahrhundert', in *Technikvermittlung und Technikpopularisierung: Historische und didaktische Perspektiven*, ed. Lars Bluma, Karl Pichol and Wolfhard Weber (Münster: Waxmann, 2004), 91, 97.
61 Kurt Möser, 'Autobasteln: Modifying, Maintaining and Repairing Private Cars in the GDR, 1970–1990', in *The Socialist Car: Automobility in the Eastern Bloc*, ed. Lewis H. Siegelbaum (Ithaca: Cornell University Press, 2011), 159–60, 162–4.
62 Eli Rubin, 'Understanding a Car in the Context of a System: Trabants, Marzahn, and East German Socialism', in *The Socialist Car: Automobility in the Eastern Bloc*, ed. Lewis H. Siegelbaum (Ithaca: Cornell University Press, 2011), 129.
63 Möser, 'Autobasteln', 159.
64 Eli Rubin, 'The Trabant. Consumption, Eigen-Sinn, and Movement', *History Workshop Journal*, no. 68 (2009): 37.
65 Möser, 'Autobasteln', 168.
66 Reinhild Kreis, 'Do it yourself mit Pioniergeist: Selbermachen in deutsch-amerikanischer Perspektive', in *Do it yourself: Mach's doch selbst* (Oberschönenfeld:

Schwäbisches Volkskundemuseum Oberschönenfeld, 2016), 23, 26, 28; Reinhild Kreis, 'Why Not Buy? Making Things Oneself in an Age of Consumption', *Bulletin of the GHI* 56, no. 1 (2015): 87; Jonathan Voges, 'Maintaining, Repairing, Refurbishing: The Western German Do-it-Yourselfers and their Homes', *European History Yearbook* 18 (2017): 111–12.

67 Ibid., 116.

68 Ibid., 111–12.

69 Jonathan Voges, *'Selbst ist der Mann': Do-it-yourself und Heimwerken in der Bundesrepublik Deutschland* (Göttingen: Vandenhoeck & Ruprecht, 2017), 81–3.

70 Cf. Joan Campbell, *Joy in Work, German Work: The National Debate, 1800–1945* (Princeton: Princeton University Press, 1989).

71 Voges, *'Selbst ist der Mann'*, 81–3.

72 Ibid., 226–7, 231–6, 242–6.

73 Kreis, 'Why not Buy?', 93; Voges, *'Selbst ist der Mann'*, 220.

74 Voges, 'Maintaining, Repairing, Refurbishing', 119–20.

75 Voges, *'Selbst ist der Mann'*, 200–4, 256–61, 316, 319, 324.

76 Ibid., 208; Reinhild Kreis, 'A "Call to Tools": DIY between State Building and Consumption Practices in the GDR', *International Journal for History, Culture and Modernity* 6, no. 1 (2018): 64–5.

77 Voges, *'Selbst ist der Mann'*, 210–11, 217.

78 Julia Fleischhack, *Eine Welt im Datenrausch: Computeranlagen als gesellschaftliche Herausforderung in der Bundesrepublik Deutschland (1965–1975)* (Zurich: Chronos, 2016), 32.

79 Frank Bösch, 'Wege in die digitale Gesellschaft: Computer als Gegenstand der Zeitgeschichtsforschung', in *Wege in die digitale Gesellschaft: Computernutzung in der Bundesrepublik 1955–1990*, ed. Frank Bösch (Göttingen: Wallstein, 2018), 13.

80 Werner Faulstich, 'Die Anfänge einer neuen Kulturperiode: Der Computer und die digitalen Medien', in *Die Kultur der achtziger Jahre*, ed. Werner Faulstich (Munich: Wilhem Fink, 2005), 238.

81 Bösch, 'Wege in die digitale Gesellschaft', 18.

82 Matthias Röhr, 'Gebremste Vernetzung: Digitale Kommunikation in der Bundesrepublik der 1970er/80er Jahre' in *Wege in die digitale Gesellschaft: Computernutzung in der Bundesrepublik 1955–1990*, ed. Frank Bösch (Göttingen: Wallstein, 2018), 262–3, 267; Julia Gül Erdogan, 'Technologie, die verbindet: Die Enstehung und Vereinigung von Hackerkulturen in Deutschland', in *Wege in die digitale Gesellschaft: Computernutzung in der Bundesrepublik 1955–1990*, ed. Frank Bösch (Göttingen: Wallstein, 2018), 229.

83 Röhr, 'Gebremste Vernetzung', 260.

84 Gleb J. Albert, 'Subkultur, Piraterie und neue Märkte: Die transnationale Zirkulation von Heimcomputersoftware, 1986–1995', in *Wege in die digitale Gesellschaft: Computernutzung in der Bundesrepublik 1955–1990*, ed. Frank Bösch (Göttingen: Wallstein, 2018), 272.

85 Julia Gül Erdogan, '"Computer Wizards" und Haecksen: Geschlechtsspezifische Rollenzuschreibungen in der privaten und subkulturellen Computernutzung in den UA und der Bundesrepublik', *Technikgeschichte* 87, no. 2 (2020): 112.

86 Erdogan, 'Technologie, die verbindet', 228, 235; Erdogan, '"Computer Wizards" und Haecksen', 108–9.

87 Martin Schmitt, Julia Erdogan, Thomas Kasper and Janine Funke, 'Digitalgeschichte Deutschlands', *Technikgeschichte* 83, no. 1 (2016): 46.

88 Jürgen Danyel, 'Zeitgeschichte der Informationsgesellschaft', *Zeithistorische Forschungen/Studies in Contemporary History* 9, no. 2 (2012):198–9.
89 Bösch, 'Wege in die digitale Gesellschaft', 35; Faulstich, 'Die Anfänge einer neuen Kulturperiode', 232–4; Poser, *Glücksmaschinen und Maschinenglück*, 313.
90 Gleb J. Albert, '"Mikro-Clochards" im Kaufhaus: Die Entdeckung der Computerkids in der Bundesrepublik', *Nach Feierabend: Zürcher Jahrbuch für Wissensgeschichte* 12 (2016): 64–5, 68.
91 Ibid., 72.
92 Gleb J. Albert, 'Computerkids als mimetische Unternehmer: Die Cracker-Szene zwischen Subkultur und Ökonomie (1985–1995)', *WerkstattGeschichte*, no. 74 (2016): 51, 61, 65–6.
93 Albert, 'Subkultur, Piraterie und neue Märkte', 276.
94 Albert, 'Computerkids als mimetische Unternehmer', 51, 61, 65–6; quotation on p. 62.
95 Ibid., 53.
96 Albert, 'Subkultur, Piraterie und neue Märkte', 275.
97 Albert, 'Computerkids als mimetische Unternehmer', 55–7.
98 Danyel, 'Zeitgeschichte der Informationsgesellschaft', 206–7.
99 Erdogan, 'Technologie, die verbindet', 227, 229, 232.

Chapter 8

1 Joachim Radkau, 'Mythos German Angst: Zum neuesten Aufguss einer alten Denunziation der Umweltbewegung', *Blätter für deutsche und internationale Politik*, no. 5 (2011): 73, 78.
2 David Edgerton, *The Shock of the Old: Technology and Global History since 1900* (London: Profile Books, 2006).
3 Joachim Radkau, *Das Zeitalter der Nervosität: Deutschland zwischen Bismarck und Hitler* (Munich: Hanser, 1998), 200.
4 Eric Dorn Brose, *The Politics of Technological Change in Prussia. Out of the Shadow of Antiquity, 1809–1848* (Princeton: Princeton University Press, 1993), 262.
5 Eric J. Hobsbawm, 'The Machine Breakers', *Past & Present*, no. 1 (1952): 62.
6 Michael Spehr, *Maschinensturm: Protest und Widerstand gegen technische Neuerunge am Anfang der Industrialisierung* (Münster: Westfälisches Dampfboot, 2000), 61–4, 104, 106, 167; Thomas Rohkrämer, *Eine andere Moderne? Zivilisationskritik, Natur und Technik in Deutschland 1880–1933* (Paderborn: Schöningh, 1999), 17; Rolf Peter Sieferle, *Fortschrittsfeinde? Opposition gegen Technik und Industrie von der Romantik bis zur Gegenwart* (Munich: C.H. Beck, 1984), 66–8.
7 Spehr, *Maschinensturm*, 149–51, 163, 166.
8 Sieferle, *Fortschrittsfeinde*, 78.
9 Spehr, *Maschinensturm*, 166; Hans-Albert Wulf, *'Maschinenstürmer sind wir keine': Technischer Fortschritt und sozialdemokratische Arbeiterbewegung* (Frankfurt: Campus, 1987), 27.
10 Ulrike Gilhaus, *'Schmerzenskinder der Industrie': Umweltverschmutzung, Umweltpolitik und sozialer Protest im Industriezeitalter in Westfalen 1845–1914* (Paderborn: Schöningh, 1995), 451, 454.
11 Rohkrämer, *Eine andere Moderne?*, 25.
12 Bernhard Rieger, *Technology and the Culture of Modernity in Britain and Germany, 1890–1945* (Cambridge: Cambridge University Press, 2005), 16.

13 Rohkrämer, *Eine andere Moderne?*, 71.
14 Mary Nolan, *Visions of Modernity: American Business and the Modernization of Germany* (New York: Oxford University Press, 1994), 172–3.
15 Ibid., 177; Jürgen Bönig, *Die Einführung von Fließbandarbeit in Deutschland bis 1933: Zur Geschichte einer Sozialinnovation*, Volume 2 (Münster: LIT, 1993), 695.
16 CD [no full name given], 'Mensch und Maschine', *Kölnische Zeitung*, 12. February 1939: 3.
17 See Johannes Platz, '"Revolution der Roboter" oder "Keine Angst vor Robotern"? Die Verwissenschaftlichung des Automatisierungsdiskurses und die industriellen Beziehungen von den 50ern bis 1968', in *Entreprises et crises économiques au XXe siècle*, ed. Laurent Commaille (Metz, 2009), 39, 54; Martina Heßler, 'Zur Persistenz der Argumente im Automatisierugsdiskurs', *Aus Politik und Zeitgeschichte* 66, no. 18/19 (2016): 19.
18 IG Druck und Papier, *Notzeitung*, 6 March 1978. Archive in House of the History of the Ruhr [Archiv im Haus der Geschichte des Ruhrgebiets], Bochum, file MüJe 140.
19 Jürgen Alberts, Susanne Herzog, Karl Unger and Joachim Wilmersdorf, *Zeitungsstreik* (Hamburg: VSA-Verlag, 1978), 71.
20 Claudia Weber, *Rationalisierungskonflikte in Betrieben der Druckindustrie* (Frankfurt: Campus, 1982), 130, 132–3.
21 Andie Rothenhäusler, 'Die Debatte um die Technikfeindlichkeit in der BRD in den 1980er Jahren', *Technikgeschichte* 80, no. 4 (2013): 283–4.
22 David F. Noble, *Maschinenstürmer oder die komplizierten Beziehungen der Menschen zu ihren Maschinen* (Berlin: Wechselwirkung, 1986), 34, 67.
23 Ibid., 69.
24 Ulrich Briefs, *Informationstechnologien und Zukunft der Arbeit: Ein politisches Handbuch zu Mikroelektronik und Computertechnik* (Cologne: Pahl-Rugenstein, 1984), 17.
25 Rothenhäusler, 'Die Debatte um die Technikfeindlichkeit', 280, 282.
26 Frank Bösch, 'Wege in die digitale Gesellschaft: Computer als Gegenstand der Zeitgeschichtsforschung', in *Wege in die digitale Gesellschaft: Computernutzung in der Bundesrepublik 1955–1990*, ed. Frank Bösch (Göttingen: Wallstein, 2018), 32; Julia Fleischhack, *Eine Welt im Datenrausch: Computeranlagen als gesellschaftliche Herausforderung in der Bundesrepublik Deutschland (1965–1975)* (Zurich: Chronos, 2016), 158; Marcel Berlinghoff, '"Totalerfassung" im "Computerstaat": Computer und Privatheit in den 1970er und 1980er Jahren', in *Im Sog des Internets: Öffentlichkeit und Privatheit im digitalen Zeitalter*, ed. Ulrike Ackermann (Frankfurt: Humanities Online, 2013), 97–8; Eva Oberloskamp, 'Auf dem Weg in den Überwachungsstaat? Elektronische Datenverarbeitung und die Anfänge des bundesdeutschen Datenschutzes in den 1970er Jahren', in *Ausnahmezustände: Entgrenzungen und Regulierungen in Europa während des Kalten Krieges*, ed. Cornelia Rauh and Dirk Schumann (Göttingen: Wallstein, 2015), 171.
27 Berlinghoff, '"Totalerfassung" im "Computerstaat", 99, 104–5; Oberloskamp, 'Auf dem Weg in den Überwachungsstaat', 168–70, 173–4.
28 Ibid., 175; Berlinghoff, '"Totalerfassung" im "Computerstaat", 107–9; Wirsching, 'Durchbruch des Fortschritts', 209–12.
29 Sieferle, *Fortschrittsfeinde*, 99–103.
30 Wolfgang Schivelbusch, *The Railway Journey: The Industrialization of Time and Space in the 19th Century* (Leamington Spa: Berg, 1986), 130.

31 Uwe Fraunholz, *Motorphobia: Anti-automobiler Protest in Kaiserreich und Weimarer Republik* (Göttingen: Vandenhoeck & Ruprecht, 2002), 268–71.
32 Oswald Spengler, *Man and Technics: A Contribution to a Philosophy of Life* (New York: Alfred A. Knopf, 1963), 95; cf. Kurt Möser, *Geschichte des Autos* (Frankfurt and New York: Campus, 2002), 200.
33 Barbara Schmucki, *Der Traum vom Verkehrsfluss: Städtische Verkehrsplanung seit 1945 im deutsch-deutschen Vergleich* (Frankfurt: Campus, 2001), 119–20; Möser, *Geschichte des Autos*, 198–200.
34 Christopher Neumaier, 'Eco-Friendly versus Cancer-Causing: Perceptions of Diesel Cars in West Germany and the United States, 1970–1990', *Technology and Culture* 55, no. 2 (2014): 430, 453.
35 Peter Itzen, 'Aus Verkehrsunfällen lernen? Der Tod auf deutschen Straßen und die vergangenen Träume des 20. Jahrhunderts', *Zeithistorische Forschungen/Studies in Contemporary History* 14, no. 3 (2017): 521–2.
36 Möser, *Geschichte des Autos*, 201; Sina Fabian, *Boom in der Krise: Konsum, Tourismus, Autofahren in Westdeutschland und Großbritannien 1970*–1990 (Göttingen: Wallstein, 2016), 399–407; Heike Bergmann, 'Angeschnallt und los: Die Gurtdebatte der 1970er und 1980er Jahre in der BRD', *Technikgeschichte* 76, no. 2: 105–30.
37 Spehr, *Maschinensturm*, 165.
38 Radkau, 'Mythos German Angst', 79.
39 Jürgen Büschenfeld, *Flüsse und Kloaken: Umweltfragen im Zeitalter der Industrialisierung (1870–1918)* (Stuttgart: Klett-Cotta, 1997), 34, 37, 41.
40 Frank Uekötter, *The Age of Smoke: Environmental Policy in Germany and the United States, 1880–1970* (Pittsburgh: University of Pittsburgh Press, 2009), 2.
41 Franz-Josef Brüggemeier, *Das unendliche Meer der Lüfte: Luftverschmutzung, Industrialisierung und Risikodebatten im 19. Jahrhundert* (Essen: Klartext, 1996), 72–3.
42 Büschenfeld, *Flüsse und Kloaken*, 97.
43 Brüggemeier, *Das unendliche Meer der Lüfte*, 51, 74.
44 Büschenfeld, *Flüsse und Kloaken*, 156, 401.
45 Gilhaus, '*Schmerzenskinder der Industrie*', 352.
46 Brüggemeier, *Das unendliche Meer der Lüfte*, 216, 244, 251.
47 Gilhaus, '*Schmerzenskinder der Industrie*', 353–4, 357.
48 Brüggemeier, *Das unendliche Meer der Lüfte*, 304–5.
49 Büschenfeld, *Flüsse und Kloaken*, 45.
50 Franz-Josef Brüggemeier, *Grubengold: Das Zeitalter der Kohle von 1750 bis heute* (Munich: Beck, 2018), 117.
51 Gilhaus, '*Schmerzenskinder der Industrie*', 468.
52 Uekötter, *The Age of Smoke*, 43.
53 Brüggemeier, *Das unendliche Meer der Lüfte*, 307.
54 Mark Cioc, *The Rhine: An Eco-Biography, 1815–2000* (Seattle: University of Washington Press, 2002), 123.
55 Ibid., 88, 91, 96, 124, 129, 132.
56 Andrea Westermann, 'When Consumer Citizens Spoke Up: West Germany's Early Dealing with Plastic Waste', *Contemporary European History* 22, no. 3 (2013): 484.
57 Spengler, *Man and Technics*, 94.
58 Frank Uekötter, 'Polycentrism in Full Swing: Air Pollution Control in Nazi Germany', in *How Green Were the Nazis? Nature, Environment, and Nation in the Third* Reich, ed. Franz-Josef Brüggemeier, Marc Cioc and Thomas Zeller (Athens: Ohio University Press, 2005), 102.

59 Engels, *Naturpolitik in der Bundesrepublik*, 154, 159–60, 207.
60 Brüggemeier, *Grubengold*, 371–3; Uekötter, *The Age of Smoke*, 16–7, 262.
61 Uekötter, *The Age of Smoke*, 264–5.
62 Sophie Gerber, '"We want to live electrically!" Marketing Strategies of German Power Companies in the 20th Century', in *Past and Present Energy Societies: New Energy Connects Politics, Technologies and Cultures*, ed. Nina Möllers and Karin Zachmann (Bielefeld: transcript, 2012), 99.
63 Christian Pfister, 'Das "1950er Syndrom": Die umweltgeschichtliche Epochenschwelle zwischen Industriegesellschaft und Konsumgesellschaft', in *Das 1950er Syndrom: Der Weg in die Konsumgesellschaft*, ed. Christian Pfister (Bern: Haupt, 1995), 94.
64 Ruth Oldenziel and Mikael Hård, *Consumers, Tinkerers, Rebels: The People who Shaped Europe* (Basingstoke: Palgrave Macmillan, 2013), 244, 260.
65 Büschenfeld, *Flüsse und Kloaken*, 407.
66 Frank Biess, *German Angst: Fear and Democracy in the Federal Republic of Germany* (Oxford: Oxford University Press, 2020), 293.
67 Frank Biess, '"Everybody has a Chance": Nuclear Angst, Civil Defence, and the History of Emotions in Postwar West Germany', *German History* 27, no. 2 (2009): 223.
68 Helmuth Trischler and Robert Bud, 'Public technology: Nuclear Energy in Europe', *History and Technology* 34, no. 3–4 (2018): 12; Biess, '"Everybody has a Chance"', 225.
69 Karsten Uhl, 'Deckgeschichten: "Von der Hölle zu den Sternen" – das KZ Mittelbau-Dora in Nachkriegsnarrativen', *Technikgeschichte* 72, no. 3 (2005): 251–2.
70 Dolores L. Augustine, *Taking on Technocracy: Nuclear Power in Germany, 1945 to the Present* (New York: Berghahn, 2018), 42.
71 Radkau, 'Mythos German Angst', 74; Biess, *German Angst*, 298–9, 305.
72 Trischler and Bud, 'Public Technology', 12.
73 Rohkrämer, *Eine andere Moderne?*, 14.
74 Biess, *German Angst*, 299.
75 Augustine, *Taking on Technocracy*, 74, 87–8; Melanie Arndt, *Tschernobyl: Auswirkungen des Reaktorunfalls auf die Bundesrepublik Deutschland und die DDR* (Erfurt: Landeszentrale für politische Bildung Thüringen, 2011), 102
76 Biess, '"Everybody has a Chance"', 242.
77 Biess, *German Angst*, 299–300.
78 Augustine, *Taking on Technocracy*, 93.
79 Jens Ivo Engels, 'Geschichte und Heimat: Der Widerstand gegen das Kernkraftwerk Wyhl', in *Wahrnehmung, Bewusstsein, Identifikation: Umweltprobleme und Umweltschutz als Triebfedern regionaler Entwicklung*, ed. Kerstin Kretschmer (Freiberg: Technische Universität Bergakademie, 2003), 129; Andrew S. Tompkins, *Better Active than Radioactive! Anti-Nuclear Protest in 1970s France and West Germany* (Oxford: Oxford University Press, 2016), 238; Biess, *German Angst*, 300.
80 Radkau, 'Mythos German Angst', 80.
81 Augustine, *Taking on Technocracy*, 8.
82 Biess, *German Angst*, 301.
83 Augustine, *Taking on Technocracy*, 6, 126, 150, 152.
84 Joachim Radkau and Lothar Hahn, *Aufstieg und Fall der deutschen Atomwirtschaft* (Munich: oekom, 2013), 14; Augustine, *Taking on Technocracy*, 242; Trischler and Bud, 'Public Technology', 14.
85 Tompkins, *Better Active than Radioactive!*, 2, 7.
86 Oldenziel and Hård, *Consumer, Tinkerers, Rebels*, 268.
87 Arndt, *Tschernobyl*, 78–9.

88 Nicole Hesse, 'Windwerkerei: Praktiken der Windenergienutzung in der frühen deutschen Umweltbewegung', *Technikgeschichte* 83, no. 2 (2016): 127–8, 131, 140.
89 Augustine, *Taking on Technocracy*, 242.
90 Hesse, 'Windwerkerei', 133; Biess, *German Angst*, 302.
91 Augustine, *Taking on Technocracy*, 62.
92 Biess, *German Angst*, 292, 304; Tompkins, *Better Active than Radioactive!*, 6–7, 32; Augustine, *Taking on Technocracy*, 162; Radkau and Hahn, *Aufstieg und Fall der deutschen Atomwirtschaft*, 392.
93 Arndt, *Tschernobyl*, 111.
94 Radkau and Hahn, *Aufstieg und Fall der deutschen Atomwirtschaft*, 389; Augustine, *Taking on Technocracy*, 3, 212–13, 231; Tompkins, *Better Active than Radioactive!*, 235.
95 Rieger, *Technology and the Culture of Modernity in Britain and Germany*, 285.

Concluding remarks

1 Edward Ross Dickinson, 'Biopolitics, fascism, democracy: Some reflections on our discourse about modernity', *Central European History* 37, no. 1 (2004): 5.

Annotated Bibliography

1 Friedel, *A Culture of Improvement*, 3–4.
2 Misa, *Leonardo to the Internet*, 19, 26.
3 Ibid., 265.
4 Rieger, *Technology and the Culture of Modernity*, 47.
5 Herf, *Reactionary Modernism*, 155.
6 Saraiva, *Fascist Pigs*, 6.
7 Zeller, *Driving Germany*, 2–3, 237, 239.
8 Stokes, *Constructing Socialism*, 2.
9 Augustine, *Red Prometheus*, 201.
10 Magnusson, *The Contest for Control*, 139.
11 Schlombs, *Productivity Machines*, 8.
12 Schivelbusch, *The Railway Journey*, 60.
13 Fritzsche, *A Nation of Fliers*, 137.

Annotated Bibliography

This bibliography introduces some fifty monographs and edited volumes in English that are essential works in the field of this book. While the main focus is on recent research, it is necessary also to present some classical works published in the 1980s or 1990s. Although the history of technology is an established field in German history, most studies deal with specific themes and much shorter time periods than this volume. In the fields of global and Western history, some recent excellent overviews have enriched the scholarship of the history of technology, in particular Thomas Misa's *Leonardo to the Internet* (2011), Mikael Hård's and Andrew Jamison's *Hubris and Hybrids* (2005) and Robert Friedel's *A Culture of Improvement* (2007). All three books integrate the efforts of 'ordinary, anonymous workers and tinkerers' that were long neglected in the history of technology. Friedel gives particular emphasis to 'small, gradual improvements' of technologies and thus covers a very long period of Western history – from the Middle Ages to our present time.[1] The other two books begin with the early modern period. According to Misa, the 'composite invention' of printing was a watershed for the history of technology. Afterwards, technology became 'cumulative and irreversible, permanent and for all time' whereas before, inventions were often forgotten after some time, and technology transfer frequently failed.[2]

Although technology users are also important to Misa's story, he is somewhat sceptical about an overemphasis of users' agency that tends to neglect the 'societal effects of large-scale, mature technological systems'.[3] Hård and Jamison also point to the fact that technology is not neutral. While Misa draws upon German history in some of his chapters, Hård's and Jamison's book is even more relevant to the student of German history because it covers lesser-known aspects of special German developments. These three books give some attention to global aspects but are foremost accounts of Western history. For the global history of technology, David Edgerton's *Shock of the Old* (2006) is still the best choice. Most of all, it is a challenge to simplified views on innovation. Instead, Edgerton points out the manifold ways in which old technologies persisted.

The six volumes of the *Making Europe: Technology and Transformations, 1850–2000* series have both a focus on globalization and on the pivotal role of Germany within the European history of technology. The respective volumes are concerned with experts, users and organizations. They cover issues such as colonialism, infrastructure and communication. One of the series' volumes, Ruth Oldenziel's and Mikael Hård's *Consumers, Tinkerers, Rebels* (2013), is one of the most impressive examples of a new cultural history of technology. Their book focuses on technology users and their pivotal role in shaping Europe. All of these studies broaden the scope of the history of technology to everyday technologies and the appropriation of technology. For instance, the sewing machine, bicycles, waste and recycling became central topics of investigation. For example, the history of technology is no more only interested in locomotive engines and railway-system building, but also in train compartments and in the passengers as technology users.

In German, two compelling studies discuss technology in German history, but both have a strong focus on the forces of production and rather neglect other topics. Joachim Radkau's classic book *Technik in Deutschland* (2008) gives a brilliant overview of technological development in Germany since the eighteenth century, but mostly reflects the historiographical discussion by the time of its first appearance in 1989. Christian Kleinschmidt's *Technik und Wirtschaft in Deutschland* (2006) explicitly and conceptually focuses on productive technology; it provides a state-of-the-art overview of economic history, dedicating some attention to technology.

Some monographs explore the history of technology in Germany during a specific period. While most research work on the nineteenth century focuses on industrialization, Bernhard Rieger's *Technology and the Culture of Modernity in Britain and Germany* is a fascinating comparative account of assessments of technology. Rieger shows that around 1900, many contemporaries conceived of technological artefacts as 'modern wonders'. As a result of this understanding, the 'new totalizing category' of technology that combined very different artefacts evolved.[4] In this context, Rieger points out to what extent modern nationalism was built upon certain assertions of technological progress. His study shows that there were very different proponents of modernity. Thus, Rieger disagrees with Jeffrey Herf's classical book on *Reactionary Modernism* (first published in 1984). Herf argues that a strand of reactionary modernists, most notably the Nazis, integrated modern technology into their vision of Germanness without 'succumbing to Enlightenment rationality'.[5] While Herf's book is still a worthwhile read that helps with understanding the history of ideas about technology and German nationalism in the first half of the twentieth century, recent studies point to the fact that the assumption of a 'normal' path to modernity is somewhat misleading. Instead, there were obviously manifold alternative modernities. Recently, for example, Tiago Saraiva criticized the concept of 'reactionary modernism' for suggesting 'an unsolved problematic contradiction at the heart of Nazi ideology between romanticism and technical rationality'. According to Saraiva, technologies were at the core of fascism.[6] His comparative study of fascism in Germany, Italy and Portugal is a fascinating account of the technoscientific means by which fascist regimes manipulated organisms, such as pigs or potatoes, to fit into their ideological needs.

By exploring the history of the autobahn, Thomas Zeller's *Driving Germany* (2007) shows how a 'central icon of the Nazi state' was transformed after 1945 to fit under the new conditions. This case study of the specific artefact of Germany's highway system makes clear that there were very different 'versions of modernity' in the Third Reich and post-war West Germany, respectively. Although there was much continuity after 1945, 'the panoramic autobahn of the 1930s was obsolete' by 1970. While Nazi modernists envisioned a 'landscape-sensitive autobahn', the West German concept of modernity focussed on rapid and safe transportation.[7] To a certain degree, Frank Biess also deals with a specific post-war heritage of Nazi Germany in his book *German Angst* (2020): trauma and fear. While Biess is not a historian of technology, three of his chapters are very valuable contributions to the history of emotions and technology in post-war Germany. Fears of nuclear war, the consequences of automation and nuclear power shaped West German politics from the 1950s onwards. Biess demonstrates that these fears were embedded into rational debates about the risks of technologies. The integration of emotional aspects into discourses of modern technology both supported the development of German democracy and eased the acceptance of technology in general.

In the other German Cold War state, the socialist GDR, technology was even more crucial for the development of the state. Two excellent overview works – by Dolores Augustine and Raymond Stokes – explore the relationship between politics, society and technology. Stokes' *Constructing Socialism* (2000) points to the technological origins of the GDR's social and economic issues. Additionally, 'islands of technological excellence' supported the state's stability.[8] Augustine's *Red Prometheus* (2007) not only deals with engineers and technocrats in the GDR but also offers a comprehensive account of East Germany's history of technology. Very convincingly, she shows to what extent technology was an essential part of East German culture: 'Technology played a growing role in the conception of a socialist modernity, as well as in the national identity of East Germans.'[9]

The different stages of industrialization still inspire many historians of technology. Thus, the literature is manifold, from early industry to the age of computers. The economic historian Gary Herrigel's *Industrial Construction* (1996) is still – twenty-five years after its first appearance – a notable work. Herrigel demonstrates the importance of regional differences to industrial history. His work has helped to overcome the sole focus on urban large-scale industry and broadened the scope to rural industries. From the historian of technology's point of view, his research is a valuable contribution to keep the balance between innovation and the persistence of old technologies. Also, Eric Dorn Brose's contribution to early industrialization, *The Politics of Technological Change in Prussia* (1993), investigates the role of the Prussian state in industrial promotion. While Brose points out that state bureaucrats were decisive for technology transfer to Prussia, he reminds us of the fact that the actors' intentions were often very different to the actual outcome of their actions. In this case, most bureaucrats envisioned a less radical change and hoped for the establishment of rather small-scale rural industries.

Although industrialization transformed Germany in the late nineteenth century, the transition to the factory system was 'gradual, piecemeal and uneven', as Lars Magnusson demonstrates in his comparative study on the English, German and Swedish metal industries.[10] His book proves that cultural appropriations of technology need to be taken into account to understand social and technological change fully. Magnusson's case study of the German city of Solingen exemplifies the importance of contemporary beliefs regarding the necessity to preserve small-scale industry. Likewise, Ulrich Wengenroth's *Enterprise and Technology* argues for a comprehensive picture of innovation. In the late nineteenth century, the German success in steel industries was not based upon technological supremacy. Instead, structural and organizational issues of German businesses and economy made the difference in comparison to the British competitors.

While the steel industry is a good example of the German speciality of science-based industry, chemistry is probably the best one. Jeffrey Allan Johnson's classic study *The*

Kaiser's Chemists demonstrates to what extent the state bureaucracy created institutions that supported this new industry and technological innovation, but at the same time integrated the old elites into the novel system. More recently, Werner Abelshauser edited a volume on the history of the most successful chemical corporation BASF. His *German Industry and Global Enterprise* (2004) explores 150 years of chemistry research and production. While good research has always been the key to success, during the course of the twentieth century, progress in process technology became more and more important as a competitive force rather than chemical innovation alone.

The 1920s obsession with efficiency has inspired many studies in the history of technology. Mary Nolan's *Visions of Modernity* (1994) points out that a comprehensive understanding of Weimar Germany needs to account for Americanization and the German fascination with Fordism. Because of the remarkable obsession of 1920s-era Germans with efficiency, Jennifer Karns Alexander devotes one chapter of her transnational history *The Mantra of Efficiency* (2008) to a German exhibition about efficient worker seating. The Weimar German concern with the optimal relationship between the machine and the working body was a decisive move towards expanding the concept of efficiency into the work routines at the factories. Although Alexander tends to overemphasize the effects of intended social control, this broad view of the history of industrial efficiency and its specific German aspects is very inspiring. Corinna Schlombs' *Productivity Machines* (2019) tells the history of ongoing industrial Americanization in West Germany after 1945. Her history of early computerization makes clear once again that US technology and industrial inventions were 'appropriated only with modifications that suited local conditions in Germany'.[11] Furthermore, specific features of German industry, most notably the strong machine-tool industry with a focus on a general-purpose approach, helped West Germany avoid the problematic inflexibility of Detroit automation.

With regard to urban development, nineteenth-century Germans firstly followed the British model, but German cities soon established comprehensive approaches to city planning. Brian Ladd's pioneer study *Urban Planning and Civic Order in Germany, 1860–1914* (1990) demonstrates to what extent different urban technologies became interwoven in municipal programmes from the late nineteenth century onwards. The issues of hygiene, energy and water supply as well as transport transformed urban life in Germany rapidly. Another classic work, Richard Evans' *Death in Hamburg* (1987), is a case study of the social outcomes and the political aspects of urban sanitary politics in the nineteenth century. Technologies of hygiene, established in an incomplete manner, even resulted in the Hamburg cholera pandemic.

More recent studies of urban technologies particularly focus on issues of housing. For a long time, historical research was mainly interested in the famous reform projects of 1920s modernist architecture. In this respect, Leif Jerram's study on *Germany's Other Modernity* (2007), a case study of early twentieth-century Munich, is notable. It demonstrates that less spectacular versions of modern housing reform, often pragmatic mixtures of modernity and traditionalism, were more widespread than radical approaches in interwar Germany. During the Cold War, urban planning became even more openly political. Both German states were in competition to prove that their approach to modern urbanity was supreme. Emily Pugh's study on *Architecture, Politics*

& Identity in Divided Berlin (2014) is a good example of comparative West and East German politics of urban housing and transport. Elie Rubin's *Amnesiopolis* (2016) is a case study of the large prefabricated-housing estate Marzahn in East Berlin that explores everyday life in socialist modernity.

Most research on the history of electricity has a similar focus on the technology users and the social and technological framing as well as the outcome of technological innovation. Yet, still to mention as one of the greatest studies is Thomas Hughes's *Networks of Power* (1993) that explores national cultures of constructing large technological systems. To fully grasp the national differences in the history of electrification, Hughes' book is essential, and his chapter on Berlin explains many particularities of the development of electricity in Germany. Wolfgang Schivelbusch's *Disenchanted Night* (1995) shows how gas and electric light changed life in the nineteenth century. He considers cultural assumptions of technology users and investigates the transfer of technology from Britain to Germany. More recently, Andreas Killen demonstrates the everyday outcomes of urban electrification around 1900. His *Berlin Electropolis* (2006) explores the manifold effects of electrification on household technology, transport and entertainment (namely cinema) and connects it to contemporary complaints about the modern illness of nervousness.

The volume *Past and Present Energy Societies* (2012), edited by Nina Möllers and Karin Zachmann, is interested in European household energy consumption with a focus on twentieth-century German history. Its contributions explore the whole range of cultural history of technology, from representations of technology to consumption practices. Of course, the kitchen was the first place in the household that was electrified. And even seemingly unspectacular kitchen technologies became contested during the Cold War. Oldenziel's and Zachmann's edited volume on the *Cold War Kitchen* (2009) demonstrates the different outcome of Americanization in various European countries. With regards to the kitchen, Alice Weinreb's monograph *Modern Hungers* (2017) on the political history of food in twentieth-century Germany investigates both the everyday outcomes of politics of household technification and the persistence of traditional patterns, not least the gendered division of labour. The technified consumer household leaves a lot of waste. Recently, historians of technology have begun to pay more attention to this matter. Raymond Stokes, Roman Köster and Stephen Sambrook's *The Business of Waste* (2013) is a notable comparative study of Great Britain and Germany. It explores the establishment of a large technological system after 1945 and also accounts for the changing habits of common people.

Arguably, the transport revolution beginning in the nineteenth century was among the most radical changes to everyday life engendered by new technologies. Schivelbusch's classic monograph on *The Railway Journey* (1986) reflects upon the changing perceptions of passengers. Time and space received different meanings, while the railroad has 'created a new landscape'.[12] As both a cultural and a social historian, Schivelbusch points out how the new transport technology contributed to the structures of the class society. Motorized individual transport began to be a mass phenomenon in Germany with the motorcycle in the 1920s. Sasha Disko's *The Devil's Wheels* (2016) explores the gender history of this new artefact. Her research on manliness and motorcycling demonstrates how the history of technology can be combined with the history of the body. Contrary to the motorcycle, the automobile became a mass phenomenon in Germany only from the 1950s onwards. Nevertheless, its most successful product, the Volkswagen Beetle, soon became a global success. Bernhard Rieger's *The People's Car* (2013) introduces the global history of this German technological artefact that was mostly appreciated for its liability and low costs.

In the early twentieth century, the German contributions to high tech, namely aviation and aerospace, were deeply intertwined with the development of German nationalism. Peter Fritzsche's *A Nation of Fliers* (1992) demonstrates how 'machine dreams mangled with national dreams'.[13] While Fritzsche investigates both airplanes and airships, Guillaume de Syon's *Zeppelin* (2001) solely focuses on the airship and its important history in Germany. In a comparative manner, Hermione Giffard explores the history of armaments production in her study *Making Jet Engines in World War II* (2016). Giffard revises the traditional view of the quantitative success of the German wartime jet engine production. For Giffard, successful mass production was less the outcome of ingenious engineering and more the result of a lack of resources and skilled labour. Near the end of the war, it was a sheer necessity for the Nazis to design a pragmatic product according to the necessities of simplified manufacturing done by forced labourers. Consequently, the jets did not become wonder weapons to change the course of the war. The same is true for the often-mysterious German rocket programme. Michael Neufeld's and Michael Petersen's books shed light on the prehistory of rocket tinkering during the 1920s and the later large-scale approach at the Peenemünde Army Research Centre. While the Peenemünde research rested upon forced and slave labour, from late 1943 onwards, rocket production was transferred to the air raid secure tunnels of the Mittelbau-Dora concentration camp. Michael Thad Allen's *The Business of Genocide* (2002) explores the devastating combination of efficiency, high tech and slave labour. Immediately after the end of the war, German high-tech artefacts and experts were of utmost value to the former war opponents. The volume *Technology Transfer Out of Germany After 1945* (1996), edited by Matthias Judt and Burghard Ciesla, investigates this sought after Nazi expertise.

Although Germany had eminent physicists (despite the fact that the Nazis expelled many Jewish experts), the Nazi programme for nuclear power was a total failure. Mark Walker's *German National Socialism and the Quest for Nuclear Power* (1996) makes clear that the Nazis did not even come close to nuclear success because they failed to expand the project to industrial scale. After the war, the civil use of nuclear power was mostly a product of Americanization. Yet, the recent books of Dolores Augustine (*Taking on Technocracy*, 2018) and Andrew Tompkins (*Better Active than Radioactive!*, 2016) demonstrate that the public response towards nuclear power was partly a specifically German development. People in the US and France also protested against nuclear power plants, and there was transnational cooperation between the protesters. Yet, the German anti-nuclear movement was particularly strong and – surprisingly – successful after the Fukushima disaster.

The strength of the German environmental movement is explained to some degree by the harshness of 250 years of technological transformation of landscapes that took place in the country. David Blackbourn's *The Conquest of Nature* (2006) is a fascinating study in the very productive entanglement of environmental history and the history of technology. Marc Cioc's case study on *The Rhine* (2002) demonstrates that the intense relationship between nature and technology makes any search for untouched nature in modern history senseless. Furthermore, Cioc explores how environmentalists were attacked as being anti-modern foes of technology already in the nineteenth century. Frank Uekotter's research, most notably *The Age of Smoke* (2009) and *The Greenest Nation?* (2014), investigates the long history of German environmentalism. His comparative and transnational approach prevents him from affirming a mere success story of German environmentalism. Also very valuable is the volume *How Green Were the Nazis?* (2005), edited by Franz-Josef Brüggemeier, Marc Cioc and Thomas Zeller. It shows that, as with modernity in general, environmentalism has a variety of manifestations and can be used for manifold political goals.

Abelshauser, Werner, ed. *German Industry and Global Enterprise: BASF – The History of a Company*. Cambridge: Cambridge University Press, 2004.

Alexander, Jennifer Karns. *The Mantra of Efficiency: From Waterwheel to Social Control.* Baltimore: Johns Hopkins University Press, 2008.

Allen, Michael Thad. *The Business of Genocide: The SS, Slave Labour, and the Concentration Camps*. Chapel Hill: University of North Carolina Press, 2002.

Augustine, Dolores L. *Red Prometheus: Engineering and Dictatorship in East Germany, 1945–1990*. Cambridge: MIT Press, 2007.

Augustine, Dolores L. *Taking on Technocracy: Nuclear Power in Germany, 1945 to the Present*. New York: Berghahn, 2018.

Biess, Frank. *German Angst: Fear and Democracy in the Federal Republic of Germany.* Oxford: Oxford University Press, 2020.

Blackbourn, David. *The Conquest of Nature: Water, Landscape, and the Making of Modern Germany*. New York: Norton, 2006.

Brose, Eric Dorn. *The Politics of Technological Change in Prussia: Out of the Shadow of Antiquity, 1809–1848*. Princeton: Princeton University Press, 1993.

Brüggemeier, Franz-Josef, Marc Cioc and Thomas Zeller, ed. *How Green Were the Nazis? Nature, Environment, and Nation in the Third Reich*. Athens: Ohio University Press, 2005.

Cioc, Marc. *The Rhine: An Eco-Biography, 1815–2000*. Seattle: University of Washington Press, 2002.

de Syon, Guillaume. *Zeppelin! Germany and the Airship, 1900–1939*. Baltimore: Johns Hopkins University Press, 2001.

Disko, Sasha. *The Devil's Wheels: Men and Motorcycling in the Weimar Republic*. New York: Berghahn, 2016.

Edgerton, David. *The Shock of the Old: Technology and Global History since 1900*. London: Profile Books, 2006.

Evans, Richard J. *Death in Hamburg: Society and Politics in the Cholera Years, 1830–1910*. Oxford: Clarendon Press, 1987.

Friedel, Robert. *A Culture of Improvement: Technology and the Western Millennium*. Cambridge: MIT Press, 2007.

Fritzsche, Peter. *A Nation of Fliers: German Aviation and the Popular Imagination*. Cambridge: Harvard University Press, 1992.

Giffard, Hermione. *Making Jet Engines in World War II: Britain, Germany, and the United States*. Chicago: University of Chicago Press, 2016.

Hård, Mikael, and Andrew Jamison. *Hubris and Hybrids: A Cultural History of Technology and Science*. New York and London: Routledge, 2005.
Herf, Jeffrey. *Reactionary Modernism: Technology, Culture, and Politics in Weimar and the Third Reich*, repr. edn. Cambridge: Cambridge University Press, 2003.
Herrigel, Gary. *Industrial Constructions: The Sources of German Industrial Power*. Cambridge: Cambridge University Press, 1996.
Hughes, Thomas P. *Networks of Power: Electrification in Western Society, 1880–1930*. Baltimore: Johns Hopkins University Press, 1993.
Jerram, Leif. *Germany's Other Modernity: Munich and the Making of Metropolis, 1895–1930*. Manchester: Manchester University Press, 2007.
Johnson, Jeffrey Allan. *The Kaiser's Chemists: Science and Modernization in Imperial Germany*. Chapel Hill: University of North Carolina Press, 1990.
Judt, Matthias and Burghard Ciesla. *Technology Transfer Out of Germany after 1945*. Amsterdam: Harwood, 1996.
Killen, Andreas. *Berlin Electropolis: Shock, Nerves, and German Modernity*. Berkeley: University of California Press, 2006.
Kleinschmidt, Christian. *Technik und Wirtschaft im 19. und 20. Jahrhundert*. Munich: Oldenbourg, 2006.
Ladd, Brian. *Urban Planning and Civic Order in Germany, 1860–1914*. Cambridge: Harvard University Press, 1990.
Magnusson, Lars. *The Contest for Control: Metal Industries in Sheffield, Solingen, Remscheid and Eskilstuna during Industrialization*. Oxford: Berg, 1994.
Misa, Thomas. *Leonardo to the Internet: Technology and Culture from the Renaissance to the Present*, 2nd edn. Baltimore: Johns Hopkins University Press, 2011.
Möllers, Nina and Karin Zachmann, ed. *Past and Present Energy Societies: New Energy Connects Politics, Technologies and Cultures*. Bielefeld: transcript, 2012.
Neufeld, Michael J. *The Rocket and the Reich: Peenemünde and the Coming of the Ballistic Missile Era*. Cambridge: Harvard University Press, 1995.
Nolan, Mary. *Visions of Modernity: American Business and the Modernization of Germany*. New York: Oxford University Press, 1994.
Oldenziel, Ruth and Mikael Hård. *Consumers, Tinkerers, Rebels: The People who Shaped Europe*. Basingstoke: Palgrave Macmillan, 2013.
Oldenziel, Ruth and Karin Zachmann, ed., *Cold War Kitchen: Americanization, Technology and European Users*. Cambridge: MIT Press, 2009.
Petersen, Michael B. *Missiles for the Fatherland: Peenemünde, National Socialism, and the V-2 Missile*. Cambridge: Cambridge University Press, 2009.
Pugh, Emily. *Architecture, Politics & Identity in Divided Berlin*. Pittsburgh: University of Pittsburgh Press, 2014.
Radkau, Joachim. *Technik in Deutschland: Vom 18. Jahrhundert bis heute*. Frankfurt: Campus, 2008.
Rieger, Bernhard. *Technology and the Culture of Modernity in Britain and Germany*. Cambridge: Cambridge University Press, 2009.
Rieger, Bernhard. *The People's Car: A Global History of the Volkswagen Beetle*. Cambridge: Harvard University Press, 2013.
Rubin, Eli. *Amnesiopolis: Modernity, Space, and Memory in East Germany*. Oxford: Oxford University Press, 2016.
Saraiva, Tiago. *Fascist Pigs: Technoscientifc Organisms and the History of Fascism*. Cambridge: MIT Press, 2016.

Schivelbusch, Wolfgang. *Disenchanted Night: The Industrialization of Light in the Nineteenth Century*. Berkeley: University of California Press, 1995.

Schivelbusch, Wolfgang. *The Railway Journey: The Industrialization of Time and Space in the 19th Century*. Leamington Spa: Berg, 1986.

Schlombs, Corinna. *Productivity Machines: German Appropriations of American Technology from Mass Production to Computer Automation*. Cambridge: MIT Press, 2019.

Schot, Johan and Philip Scranton, ed. *Making Europe: Technology and Transformations, 1850–2000*. 6 vols. Basingstoke: Palgrave Macmillan, 2013–19.

Stokes, Raymond G. *Constructing Socialism: Technology and Change in East Germany 1945–1990*. Baltimore: Johns Hopkins University Press, 2000.

Stokes, Raymond G., Roman Köster and Stephen C. Sambrook. *The Business of Waste: Great Britain and Germany, 1945 to the Present*. New York: Cambridge University Press, 2013.

Tompkins, Andrew S. *Better Active than Radioactive! Anti-Nuclear Protest in 1970s France and West Germany*. Oxford: Oxford University Press, 2016.

Uekoetter, Frank. *The Age of Smoke: Environmental Policy in Germany and the United States, 1880–1970*. Pittsburgh: University of Pittsburgh Press, 2009.

Uekoetter, Frank. *The Greenest Nation? A New History of German Environmentalism*. Cambridge: MIT Press, 2014.

Walker, Mark. *German National Socialism and the Quest for Nuclear Power, 1939–1949*. Cambridge: Cambridge University Press, 1989.

Weinreb, Alice. *Modern Hungers: Food and Power in Twentieth-Century Germany*. Oxford: Oxford University Press, 2017.

Wengenroth, Ulrich. *Enterprise and Technology: The German and British Steel Industries, 1865–1895*. Cambridge: Cambridge University Press, 1994.

Zeller, Thomas. *Driving Germany: The Landscape of the German Autobahn, 1930–1970*. Oxford: Berghahn Books, 2007.

Index

Printed in the USA
CPSIA information can be obtained
at www.ICGtesting.com
LVHW021632221223
767237LV00003B/106

9 781350 289949